Advances in Mexican Limnology:
Basic and Applied Aspects

Developments in Hydrobiology 163

Series editor
H. J. Dumont

Advances in Mexican Limnology: Basic and Applied Aspects

Edited by

Javier Alcocer[1] & S.S.S. Sarma[2]

[1] *Limnology Laboratory, CyMA, UIICSE;* [2] *Aquatic Zoology Laboratory, UMF;*
FES Iztacala, UNAM, Mexico City, Mexico

Reprinted from Hydrobiologia, volume 467 (2002)

Springer Science+Business Media, B.V.

Library of Congress Cataloging-in-Publication Data

A C.I.P. Catalogue record for this book is available from the Library of Congress.

ISBN 978-1-4020-0621-0   ISBN 978-94-010-0415-2 (eBook)
DOI 10.1007/978-94-010-0415-2

TABLE OF CONTENTS

vi

Hydrobiologia **467**: vii, 2002.
J. Alcocer & S.S.S. Sarma (eds), Advances in Mexican Limnology: Basic and Applied Aspects.

Preface

Mexican limnological studies, with a long and honorable history of about 100 years, have in recent times gained importance in teaching at undergraduate and postgraduate levels and research. Fast industrialization and rapid urbanization has resulted in pollution in many inland waterbodies and some have even been intentionally made into terrestrial zones. In order to halt further deterioration in natural epicontinental water bodies in Mexico, the National Association of Limnology was formed in 1997 with a modest membership of 200 workers from all over the Mexican Republic. Under the auspices of this august association, the first National Limnological Conference was organized during 1999. Some of the works presented during the conference have been included in the present proceedings. Some additional works not presented during the conference were also considered for inclusion in order to give a broad vision of limnological research in Mexico.

The present volume comprises aspects of both basic and applied limnology. They include works on physical, chemical and biological limnology as well as experimental approaches in selected areas. No contributions dealing specifically with aquatic conservation and biodiversity were received; therefore these aspects are considered within various other included works. Most manuscripts deal with lentic aquatic resources. This is not surprising since Mexican limnology follows the general study trend of that from temperate limnology. Despite this, we must emphasize that lotic resources in Mexico are quite important both locally and regionally. This does not mean that rivers are not under limnological research in Mexico, just that their study has only recently begun.

Although the present proceedings do not contain much information on lotic ecosystems, we did receive some manuscripts that could not be included due to the failure of some authors to meet the deadline.

We hope that our attempt will stimulate a larger section of limnologists to further research in this field. We also hope that policy-framing governmental authorities in Mexico will benefit from it, and consider some of the aspects described so that further damage to the epicontinental water bodies can be halted, and remedial measures can be considered in the future.

I (JA) express my gratitude to my tutor, Professor William D. Williams[†], for inducing my passion for limnology. The editors thank the following people who widely support and encourage the progress of Limnology in the FES Iatacala, UNAM, and the creation of the present volume:

Dr Felipe Tirado, *Director*
Dr Ignacio Peñalosa, *Secretario General Académico*
Dr Patricia Dávila, *Jefa de la División de Investigación y Posgrado*

JAVIER ALCOCER
S.S.S. SARMA
Guest Editors

[†] Deceased.

Hydrobiologia **467**: ix–x, 2002.
J. Alcocer & S.S.S. Sarma (eds), Advances in Mexican Limnology: Basic and Applied Aspects.

Referees

Refereeing is a poorly recognized labor; however, it is essential if scientific journals are to maintain high standards. In this regard, we would like to acknowledge all the reviewers for their excellent work that made this special volume possible. For the selection of articles we relied mainly on comments we received from both national and international experts for each manuscript.

Belk, Denton
USA

Biggs, Jeremy
Pond Action, Oxford Brookes University,
United Kingdom

Chacón Torres, Arturo
INIRENA, Universidad Michoacana de
San Nicolas de Hidalgo
Mexico

Chiappa Carrara, Xavier
FES Zaragoza, UNAM
Mexico

Comín, Francisco A.
Departamento de Ecología, Facultad de Biología,
Universidad de Barcelona
Spain

DeAngelis, Don
Department of Biology, University of Miami
USA

Dodds, Walter K.
Division of Biology, Kansas State University
USA

Erséus, Christer
Department of Invertebrate Zoology, Swedish
Museum of Natural History
Sweden

Esquivel Herrera, Alfonso
Universidad Autónoma Metropolitana Xochimilco
Mexico

Ferrara, María de Jesús
Station Marine d'Endoume, Lab. Claude
Charpy-Roubaud
France

Filonov, Anatoliy
Universidad de Guadalajara
Mexico

Fritz, Sheri
Department of Geosciences, University of Nebraska
USA

Fugate, Michael
WM Keck Science Center
USA

Gons, Herman
Netherlands Institute of Ecology, Centre for
Limnology
The Netherlands

Green, Jim
United Kingdom

Gulati, Ramesh D.
Netherlands Institute of Ecology, Centre of Limnology
The Netherlands

Herbst, David B.
Sierra Nevada Aquatic Research Laboratory,
University of California
USA

Herrera-Silveira, Jorge
CINVESTAV Mérida, Instituto Politécnico Nacional
Mexico

Hull, Andrew
Liverpool John Moores University, The Pond Life
Project
United Kingdom

Kerekes, Joseph
Canada

López Blanco, Jorge
Instituto de Geografía, UNAM
Mexico

Lugo Vázquez, Alfonso
Laboratorio de Limnología, CyMA, UIICSE, FES
Iztacala, UNAM
Mexico

Marchese, Mercedes
INALI, José Macia 1933, (3016) Santo Tome
(Santa Fe),
Argentina

Marín, Luis E.
Instituto de Geofísica, UNAM
Mexico

Merino Ibarra, Martín
Instituto de Ciencias del Mar y Limnología, UNAM
Mexico

Monreal Gómez, Adela
Instituto de Ciencias del Mar y Limnología, UNAM
Mexico

Mooers, Christopher N. K.
Ocean Pollution Research Center, Rosenstiel School
of Marine and Atmospheric Science, Division of
Applied Marine Physics, University of Miami
USA

Pacheco Ávila, Julia
Universidad de Yucatan
Mexico

Parenti, Lynne R.
National Museum of Natural History Smithsonian
Institution
USA

Perry, Eugene
Department of Geology, Northern Illinois University
USA

Rodrigues Coelho, Vania
Columbia University, Biosphere 2 Center
USA

Romero, David
Centro de Investigación sobre Fijación del Nitrógeno,
UNAM
Mexico

Salazar-Vallejo, Sergio
CIQRO
Mexico

Sarma, Nandini
Laboratorio de Limnología, CyMA, UIICSE, FES
Iztacala, UNAM
Mexico

Scheffer, Marten
Department of Environmental Sciences, Wageningen
Agricultural University
The Netherlands

Schmitter-Soto, Juan Jacobo
El Colegio de la Frontera Sur (ECOSUR)
Mexico

Smirnov, N.N.
A. N. Severstov Institute for Problems of Ecology &
Evolution, Russian Academy of Sciences
Russia

Suárez Morales, Eduardo
El Colegio de la Frontera Sur (ECOSUR)
Mexico

Timms, Brian V.
Australia

Vilaclara Fatjó, Gloria
Laboratorio de Limnología, CyMA, UIICSE, FES
Iztacala, UNAM
Mexico

Villalobos Hiriart, José Luis
Departamento de Zoología, Instituto de Biología,
UNAM
Mexico

Wetzer, Regina
Research and Collections, Crustacea, Natural History
Museum of Los Angeles County
USA

Wuest, Alfred Johny
APEC / EAWAG
Switzerland

Hydrobiologia **467**: 1–44, 2002.
J. Alcocer & S.S.S. Sarma (eds), Advances in Mexican Limnology: Basic and Applied Aspects.
© 2002 *Kluwer Academic Publishers.*

1

A re-evaluation of the *Macrothrix rosea-triserialis* group, with the description of two new species (Crustacea Anomopoda: Macrothricidae)

Henri J. Dumont[1], Marcelo Silva Briano[2] & K. K. Subhash Babu[3]
[1]*Institute of Animal Ecology, Ghent University, K. L. Ledeganckstraat 35, B-9000 Gent, Belgium*
E-mail: henri.dumont@rug.ac.be
[2]*Universidad Autónoma de Aguascalientes, Centro de Ciencias Básicas, Depto. de Biología, Av. Universidad 940,*
Fracc. Primo Verdad, C. P. 20100, Aguascalientes, Mexico
E-mail: msilva@cbasico.basico.uaa.mx
[3]*Christian College, Dept. of Zoology, Irinjalakuda 680 125, Kerala, India*

Key words: Cladocera, Macrothricidae, taxonomy, males, trunk limbs, species groups and subgroups, world distribution

Abstract

We redescribe *Macrothrix rosea* (females and males) based on material collected in Belgium. We also compare seven populations of *Macrothrix 'triserialis'* from different parts of the world, including a topotypical population of *M. triserialis* s. str. from Sri Lanka, and males from South India (here first described), relying heavily on the structure of the trunk limbs, beside classical features of morphology. *M. rosea* and *M. triserialis* are extremely closely related: males are easily separated, but the identification of females requires micro-characters such as the relative length of the apical segment of the setae natatoriae and the adornment of the first antenna and of the longest swimming seta of the antenna. *M. rosea* and *triserialis* together constitute a sub-group of the *rosea*-group.

Macrothrix triserialis-like animals occur in the tropical–subtropical belts of four continents. We compare populations from Asia, South America and Africa, and find differences in microcharacters of the trunk limbs, but cannot decide whether these represent random variation or sound taxonomical differences.

One of the basic characters of the *Macrothrix rosea-triserialis* subgroup is that the setae natatoriae of the postabdomen are implanted on a prominence, the 'heel'. Other characters include the fact that the Fryer' forks are adorned with one or two big teeth only, and that the scrapers of trunk limb two form a row of eight without any doublings. Possibly, scraper five, and scraper four to a lesser degree has an enlarged subapical tooth. The exopodite of trunk limb three has four plumose setae, the back and front row of the endopodite six setae and/or receptors, the exopodite of trunk limb four has two setae, and the back row of the endopodite six setae, plus one on the gnathobase. The pre-epipodite of trunk limb five consists of three lobes, the 'endopodite' is small, and the 'exopodite' is reduced to a single seta. The male postabdomen has a tubular ending, without true end-claws, although a rudiment of an end-claw is seen in *M. triserialis*.

Two new species are described: *M. tabrizensis* and *M. agsensis*. A comparison, including the males of *Macrothrix triserialis*, *M. rosea*, *M. smirnovi* and *M. tabrizensis* confirms the relationship of all these taxa, but also reveals a morphological series in the shape of the postabdomen, from a complete absence of end-claws, over rudiments of a pair of end-claws, to complete endclaws. Absence of end-claws is here considered to represent an evolved character state. *Macrothrix smirnovi* Ciros & Elías (1987) is less closely related to the *rosea-triserialis* group, and is considered to form a sub-group in its own right. It shows a short 'heel' on the postabdomen, but carries a supplementary seta behind scraper 4 of the endopodite of trunk limb two, and has a male with a postabdomen that closely resembles that of the female. These are primitive characters, which are also found in *Wlassicsia*, *Bunops* and *Onchobunops* and provide a possible phylogenetic link between *Macrothrix* and these three genera, although the genetic distance between them is considered to be quite large.

Introduction

During the past few years, taxonomic studies on macrothricid anomopods have revealed the existence of complexes of species. One example was suggested by Frey (1988). He noticed that *Guernella raphaelis* from different parts of the world (Africa: Congo River, Asia: Sri Lanka) and *Macrothrix paulensis* (South America: Brazil, and South of North America: Florida, U.S.A.), showed differences in several micro-characters which had not been given previous consideration. Although he did not arrive at a clear conclusion, he proposed the existence of related taxa, different at the species level. As in many other anomopods, incomplete and superficial morphological descriptions lie at the heart of much taxonomic confusion. Yet, the morphology of the Macrothricidae had been studied in detail. Fryer (1974) made an important contribution to the functional morphology of the trunk limbs in this group. Smirnov (1976) attempted the first world fauna of Macrothricidae. A recent identification guide for that family (Smirnov, 1992) provided more progress, but a coherent and standard recipe for describing a *Macrothrix* was still lacking. Recent descriptions and redescriptions (Ciros-Pérez & Elías-Gutiérrez, 1997; Silva-Briano et al., 1999; Kotov, 1999) have begun to pave the way for a better understanding of the nature of a number of species and species-groups.

In the present investigation, we attempt to shed new light on a group of *Macrothrix* species that occurs almost worldwide, and that, for priority reasons, should be called the *rosea*-group, although the tropical–subtropical *M. triserialis* is potentially more widespread and, therefore, more often cited. Eight populations of *Macrothrix triserialis* from different parts of the world, including *M. triserialis* s. s. are studied here, with an aim at clarifying their taxonomic status. It turns out that this group is especially speciose, and the taxa described hereinafter are possibly only a sample of the true diversity that is extant.

The gross taxonomic characters of this group, according to Smirnov (1992), are: female with dorsoposterior corner of caparace pointed, its dorsal margin with fine serration. Antennule with small indentations. Antennal setae 0-0-1-3/1-1-3, spines 0-1-0-1/0-1-1. Largest seta with two largest spines in the area of the joint between its proximal and distal segments, distally followed by many small spines. Labrum with few serrations. Thoracic limb one with three setae of different size on the Inner Distal Lobe (IDL). Inner spines

(Fryer's forks) on endites 1 and 2 present. Seta natatoria with a short distal segment. Postabdomen with teeth which shorten distally on the preanal margin, and groups of spines along the anal margin. Ocellus small and situated closer to base of antennule than to eye. Length c. 0.5 mm. Descriptions and figures of the male of *rosea* can be found in Flössner (1972) of '*triserialis*' in Sars (1901) (from South America) (!), and in Korinek (1984) (from Africa) and summarise to: antennule bent, postabdomen with a narrowed distal part and **no** endclaws. Thoracic limb one with a hook.

Clearly, this diagnosis is excessively vague and partly irrelevant, and needs extensive emendation.

Materials and methods

Samples, preserved in the plankton collection of the University of Ghent (see the individual species descriptions for data on their origin) were obtained from different parts of the world. All are fixed in formalin 5%. Permanent glass slides of isolated specimens were produced after transfer of specimens to glycerin and mounting with DePex Gurr mounting medium. SEM photographs were made after critical-point drying and gold-coating of other specimens, using a JEOL JSM 840 Scanning Electron Microscope.

Specimens were hand-dissected, and drawing made using a Leitz microscope equipped with a camera lucida. All drawings were made under oil immersion, occasionally using phase contrast. For the identification of parts of the trunk limb, and the numbering system of setae, spines, and other elements of the trunk limb, we used the system of Dumont & Silva-Briano (1997), in which such appendages are invariably numbered from outside to inside. An exception is trunk limb one, in which we numbered the nine setae from endite 1 towards endite 3 (i.e. from the inside to the outside). A list of abbreviations used is given hereunder:

Cl – clasper on first trunk limb of male; E1–E3 – endites 1–3 of trunk limb one; EE – external endite of trunk limb three; Ej. H. – ejector hooks at base of trunk limb one; Ep – epipodite; Ex – exopodite; FC – filtratory comb of plumose setae; GT – gnathobase; IDL – inner distal lobe, or endite four of trunk limb one; IE – internal endite of trunk limb three; Lab. – labrum; ODL – outer distal lobe or exopodite of trunk limb one; PEP – pre-epipodite of trunk limb five; Ro. – rostrum; R – receptor; Sc – scrapers of endopodite

Table 1. Overview of diagnostic characters used in the *Macrothrix rosea*-group

		rosea	*triserialis* Sri-Lanka	*triserialis* Kerala	*cf triserialis* Kenya	*elegans* Brazil	*cf triserialis* Mexico	*tabricensis*	*agsensis*	*smirnovi*
First antenna, female	1	With spines not distinctly grouped	With spines not distinctly grouped	With spines not distinctly grouped	With spines not distinctly grouped	With spines not distinctly grouped	With spines not distinctly grouped	About nine rows of spines	About four rows of long, thin spines	About eight rows of rather long spines
	2	No bigger spines near apex	Few bigger spines near apex	Few bigger spines near apex	Few bigger spines near apex	Few bigger spines near apex	Few bigger spines near apex	Apical crown of small spines		With halfway seta, and complex field of spinules
First antenna, male	3	With one (two?) basal setae and rows of lateral spinules	?	With one (two?) basal setae and rows of lateral spinules	?	?	?	With strong additional seta on papilla about halfway the antenna and rows of spinules	?	With halfway seta, and complex field of spinules
Second antenna, longest seta	4	With out 20 medium-sized spines along apical two-thirds	With three strong spines in middle section	--	With sparse medium-sized spines along apical two-thirds	With two strong spines in middle section	With 2–4 strong spines in middle section	With numerous spines, medium-sized all along	With numerous small spines all along	With five strong spines decreasing in size apical, in middle section
Trunk limb 1										
Seta 8	5	With very long and fine setules	With normal setules	With normal setules	With short setules	With normal setules	With normal setules	With short setules	With apical tuft of setules	With short setules
Setae 3–4	6	Setulated	Setulated	Setulated	Setulated	Setulated	Setulated	Setulated	With five thick spines	Setulated
Seta 5 epipodite	7	As long as seta 4	As long as seta 4	As long as seta 4	As long as seta 4	As long as seta 4	As long as seta 4	**Shortened**	As long as seta 4	As long as seta 4
	8	Sinuous, elongated	Sinuous, elongated	Sinuous, elongated	Sinuous, elongated	Sinuous, elongated	**Short**	**Short and globular**	Elongated	Elongated
Trunk limb 2										
Seta of exopodite Scraper 5	9	Curved over scrapers 1–2 With enlarged subapical tooth	Curved over scrapers 1–2 With enlarged subapical tooth	Curved over scrapers 1–2 With enlarged subapical tooth	Curved over scrapers 1–2 With **normal** subapical tooth	Curved over scrapers 1–2 With enlarged subapical tooth	Curved over scrapers 1–2 With **normal** subapical tooth	Pointing outwards With strong but not enlarged subapical tooth	Curved over scrapers 1–2 With normal subapical tooth	Curved over scrapers 1–2 With strong but not enlarged teeth
Seta 4 of filter comb	10	Normal	Enlarged	Normal	Enlarged	Enlarged	Enlarged	Enlarged	Enlarged	
Scraper doublings?	11	No	No	No	No	No	No	No	One seta behind scraper 4	
Trunk limb 3										
Seta 3 of exopodite	12	Longer than seta 2	Much longer than seta 2 (which is glabrous)	Longer than seta 2	Twice as long as seta 2	Longer than seta 2	Twice as long as seta 2 (which is plumose)	Longer than seta 2	**Shorter** than seta 2	**Shorter** than seta 2
Back row of setae on endopodite	3	Two long, four short	Two long, four short	Two long, four short	Two long, four short	Two long, four short	Two long, four short	All six long	All six long	All six long
Trunk limb 4										
Setae of exopodite	14	Of equal length	Of equal length	Of equal length	One long, one short	One long, one short	Of equal length	Both short	Of subequal length	Of subequal length
External scraper of endopodite	15	Spinulated	Spinulated	Spinulated	With long setules	Spinulated	Spinulated	Spinulated	**Glabrous**	**Glabrous**
Posterior row of plumose setae on endopodite	16	Composed of 5+1 setae	5+1 setae	5+1 setae	5+1 setae	5+1 setae	5+1 setae	**3 setae**	**5 setae**	**5 setae**

Continued on p. 4

Table 1. Continued

		rosea	triserialis Sri-Lanka	triserialis Kerala	cf triserialis Kenya	elegans Brazil	cf triserialis Mexico	tabrizensis	agsensis	smirnovi
Trunk limb 5										
Pre-epipodite	17	Triple	Triple	Triple	Triple	Triple	Triple	**Double**	Triple	**Double**
epipodite	18	Elongate	Elongate	Elongate	Elongate	**Very elongate**	Globular	Globular	Elongate	Elongate
Exopodite seta	19	Plumose, pointed	Glabrous	?	Plumose, rounded	Plumose, short	Short	Two lobes Widened (bulging) at base	Plumose	Rounded
long gnathobasic seta	20	Normal	Normal	Normal	Normal	Normal	Normal		Normal	Normal
Postabdomen female										
Seta natatoria apical segment	21	Well developed	Vestigial	Vestigial	Vestigial	Vestigial	Vestigial	Well developed	Well developed	Well developed
Rows of marginal spines between anus and seta natatoria	22	Spines increase in size towards seta	Spines increase in size towards seta	Spines increase in size towards seta	Spines increase in size towards seta	Spines increase in size towards seta	Spines increase in size towards seta	Spines only slightly increasing in size towards seta	Spines of same size throughout	Spines of same size throughout
Postabdomen male	23	No endclaw, apically tapering	?	No endclaw but small dorsal 'leaflets' present; apically rounded	Korinek (1984) gives a figure which suggests the same shape as in true triserialis	Sars (1901) gives a figure which is similar to that of true triserialis	?	End-claws well developed, but postabdomen hook-shaped, different from that of female	?	Postabdomen shaped as in female

of trunk limb two; Sca – 'scales', pointed or rounded hillocks behind scrapers 2–4 of trunk limb two.

We provide descriptions of all taxa, and discuss their degree of relatedness, partly in tabular form. Ample illustration, both as line drawings and as SEM micrographs are provided in the form of an atlas at the end of the paper. Diagnostic characters are highlighted in two different ways: by arrows pointing at them, by rendering them in a gray-tone, or by a combination of both.

Results

1. Macrothrix rosea (Jurine, 1820)

Material examined: Achel, Molenvijver, Belgium, 9 Sep 1933, leg. W. Luyten (see Luyten, 1934, for details). Number of specimens analyzed: three females, two males.

Smirnov (1992) lists Liévin (1848) as the first describer, because Jurine's description is not usable beyond the genus level. However, this applies to a great number of pioneer descriptions, and cannot be accepted. Liévin (1848) is to be considered the first revisor, but the oldest name available is clearly *Monoculus roseus* Jurine, 1820. *M. rosea* has a holarctic range, but is generally rare.

Description of the parthenogenetic female

Shape: (Fig. 1) oval, length 0.63 mm, height 0.38 mm, dorsal margin curved from tip of rostrum to posterior dorsal corner; the latter produced into a spine (arrow on figure), with fine serration along the valve. Ventral margin serrated. Intestine without convolutions.

Head: (Fig. 1) rostral region with lateral depression, arched in lateral view, head-shield not reticulate, dorsal margin not serrated. Head-pore window-like, rounded. Ventral margin of head smooth. Compound eye located close to margin. Ocellus smaller than compound eye, located near tip of rostrum. Labrum bi-segmented, with transversal lines along the edge, distal part a lobe with some hair-setae. This structure was found to apply to all species studied, and will not be repeated further on.

First antenna: (Figs 1 and 4) rod-like. Ventro-dorsal margin with short spines. One sensory seta located ventro-laterally of proximal end of antennular length. Nine terminal aesthetascs in two groups (two and seven) of different length. A distal external row of spines.

Second antenna: (Fig. 2). Basipod with an inner ap-

ical seta and a stout apical spine on outer surface. Swimming setae 0-0-1-3/1-1-3; spines 0-1-0-1/0-0-1. Largest seta bi-segmented, unilaterally setulated, with fine spines at its basis and about 20 stronger spines yet shorter than the diameter of the seta, from median region to apex (Fig. 11).

Trunk limb 1: (Figs 12 and 13) epipodite (EP) long and elongated, sinuous. ODL with long apical, bi-segmented seta, its distal segment unilaterally setulated, and with short lateral seta. IDL with three curved spine-setae of different length; the longest bi-segmented, with serrated distal segment. The middle one bi-segmented, serrated distally. The smallest bi-segmented, serrated distally. Endite 3 with four setae. Seta 9 long, curved back towards setae 6–8. Seta 8 short and thin 1, with long and fine setules all along (arrow). Setae 6–7 strong bi-segmented spines, with a transverse row of spinules. Seta 7 shorter than seta 6. Endite 2 with three setae. Setae 3–5 similar in length, bi-segmented, with dense setulation at their tip. Endite 1 with two setae. Setae 8–9 long of similar length, bi-segmented, serrate distally. Fryer's forks on endites 1 and 2 with one and two large teeth, respectively. Ejector hooks are present but only shown in Figure 34. The basic structure of limb one is similar in all species dealt with here, and therefore, only deviations from this basic pattern will be recorded in subsequent descriptions.

Trunk limb 2: (Fig. 14) Exopod oval, bearing a plumose apical seta. This seta sharply curves back over the scrapers, reaching scraper 3. Endopod with eight scrapers. Scrapers 1–2 long and seta-like, serrate. Scraper 1 implanted on a segment. Scraper 3, long, bi-segmented, with a comb of teeth of similar size along its distal segment. Scrapers 4–5 similar in length, bi-segmented, with a comb of strong tooth, of which the subapical one is larger than the others. Scrapers 6 and 8 short, denticulated. Scraper 7 longer than 6 and 8, bi-segmented, with fine teeth. No sensillum or hairy seta behind scraper 8 or scraper 4. Rounded hillocks behind scrapers 2–4. Gnathobase conical, with three apical and one short internal seta. Gnathobasic seta 1 naked. Brushing seta 2 crooked, its apical half bent outwardly and adorned with long setules. Seta 3 apically hooked. A filter comb of four long unsegmented, plumose setae. The innermost seta not conspicuously thicker than the other three.

Trunk limb 3: (Fig. 15) exopod with **four** hyaline setae, of which 1–3 plumose and 4 serrate. Seta 1 shorter than Seta 2. Seta 3 longest, about 1/3 longer than seta 2. Seta IV slightly shorter than seta 2. En-

6

dopod differentiated into an external (EE) and internal endite (IE). Endopodite with two rows of appendages. Anterior row composed of two strong scrapers (1–2), an inwardly curved plumose seta (3) on the EE, and two bottle-shaped receptors and three plumose setae on the IE, i.e. six setal elements and two scrapers in all. The posterior row begins with a fat, spinulated seta 1′; seta 2′ longer than 1′, with very long internal setules. Setae 3′–6′ short and plumose, forming one line, although there is a small gap between 3′ and the three internal setae. Gnathobase blunt, with rows of thick setules, and outwardly pointing, bottle-shaped receptor. Two papillae at its posterior end. This basic structure is repeated in most species, and only deviations from it will be discussed.

Trunk limb 4: (Fig. 16) exopod small, with **two** plumose setae (I–II) of similar length. Endopod a single bloc, with two rows of appendages. Anterior row with an external scraper and three flaming-torch setae (broad-based setae with complex, contorted setules) and a small, conical receptor. A posterior row of five similarly shaped filter-setae. Gnathobase with a large, outwardly turned setose seta (the 'horsetail' seta). Horsetail seta reinforced by a distal, naked support seta. This overall structure too is generalised.

Trunk limb 5: (Fig. 17) pre-epipodite composed of three apically rounded, hyaline lobes with long setules at their tip. Epipodite large, elongate, tapering at its apex. 'Exopod' reduced to a single plumose seta. 'Endopod' a small, rounded, setulated hillock. Gnathobase with three plumose setae of which Seta 1 the longest, and Seta 3 (more like a short rounded lobe than a true seta) the smallest.

Postabdomen: (Fig. 3) elongated in lateral view. Ventral margin smooth. Dorsal margin not clearly bilobed, without flange. Preanal zone with numerous rows of spines, increasing in size towards the seta natatoria. The latter implanted onto a distinctive prominence, the 'heel'. Terminal claws small, evenly curved, with lateral teeth. Setae natatoriae with long proximal segment and a distal segment which amounts to about 1/5–1/6 of the total length of the seta, and carries few setules.

Description of the male

Shape: (Fig. 6) oval-acute, length 0.22 mm, height 0.37 mm, dorsal margin curved from tip of rostrum, and straight in posterior dorsal corner, not produced into a spine. Fine serration on posterior dorsal part of valve. Ventral margin serrated. Intestine without convolutions.

Head: (Fig. 6) with eye bulging, head-shield not reticulate, dorsal margin not serrated. Head-pore oval, small. Ventral margin of head smooth. Compound eye located close to margin. Ocellus smaller than compound eye, located near tip of rostrum. Labrum bi-segmented, serrated, distal segment round.

First antenna: (Figs 6 and 7) long, with two sensory setae, located ventro-laterally, near each other. Some groups of strong setules border the ventral angulation, and several rows in lateral orientation. Nine terminal aesthetascs, in two groups of different length. Distally, neither setules nor spines.

Second antenna: as in female.

Trunk limb 1: (Figs 9 and 10) as in female, except for the presence of a copulatory clasper on the ODL, plus a long, curved acute seta (Fig. 9). The clasper is characteristic: it widens at its apex, and is provided with an apical cleft, somewhat reminiscent of a pair of castanets.

Trunk limbs 2–5: as in female.

Postabdomen: (Fig. 8) the part between the setae natatoriae and the anus is as in the female, but the post-anal part is tubular, and tapering at its apex. No trace of end-claws present. Ventral margin slightly convex.

2. Macrothrix triserialis Brady, 1886

Macrothrix triserialis was originally described from the Island of Sri Lanka (Brady, 1886), but subsequently cited from most of the tropics, although there is doubt whether all records refer to the same species. Most descriptions and illustrations are indeed incomplete, unclear and fragmentary.

2.1. Redescription of the parthenogenetic female from the terra typica

Material examined: sample 7.730, Ratrapura, Sri Lanka, 18 Aug 1972, leg. C. H. Fernando. Number of specimens analyzed: 4.

Shape: (Fig. 18) ovoid, length 0.451 mm, height 0.302 mm, dorsal margin curved from tip of rostrum to posterior dorsal corner, the latter produced into a spine (arrow). A slight but clear serration with fine latero-vertical lines. Ventral margin serrated. Intestine without convolutions.

Head: (Fig. 18) laterally depressed in rostral region, arched in lateral view, head-shield not reticulated, dorsal margin not serrated. Head-pore small, round. Ventral margin of head smooth, slightly concave. Compound eye located close the margin. Ocellus smaller than compound eye, located near tip of rostrum. Labrum bilobed, with light serration along the edge; distal part a lobe with short hair-setae (Fig. 20).

First antenna: (Fig. 19) rod-like. Dorsal and ventral margin with rows of spines. Two aesthetascs long and standing in a V *vis-à-vis* each other. Seven other terminal aesthetascs of only slightly different length.

Second antenna: (Fig. 21): Basipod with inner apical seta and stout apical spine on outer surface. Swimming setae 0-0-1-3/1-1-3; spines 1-2-1-1/0-0-1. Largest seta bi-segmented, unilaterally setulated, with three big spines (longer than diameter of seta) at mid-length (Fig. 25).

Trunk limb 1: (Figs 22–24) epipodite long and sinuous. ODL and endite lobes indistinguishable from those of *M. rosea*, except perhaps that the setulation of seta 8 is shorter and less abundant.

Trunk limb 2: (Fig. 26) as that of *M. rosea* in all details of structure. Perhaps fourth seta of the gnathobasic filter comb more robust than the other three.

Trunk limb 3: (Fig. 27) exopod as in *M. rosea*, but seta 2 almost twice as long as seta 2. Endopodite identical to that of *M. rosea*.

Trunk limb 4: (Fig. 28) exactly as in *M. rosea*.

Trunk limb 5: (Fig. 29) as in *M. rosea*.

Postabdomen: (Fig. 30) general shape and adornment as in *M. rosea*, but postanal spines longer from the anus onwards. Seta natatoria with long proximal segment, its distal segment **almost completely reduced**.

Male: no males from Sri Lanka were seen, but males were obtained from a sample collected in adjacent continental South India (Kerala State).

2.2. *Macrothrix triserialis* from South India

Monastery pond, Kerala state, India, 14 Feb 1992, leg. K. K. Subbhash Babu.

Parthenogenetic female

Only differences with Sri Lankan specimens are indicated here.

Shape: (Fig. 31) length female 0.787±0.0004 mm (SD for *n*=8); male 0.45±0.0005 mm (*n*=3), shape as in Sri Lankan specimens.

Head: (Fig. 31) as in Sri Lankan material.

First antenna: (Fig. 32) as in Sri Lankan material.

Second antenna: (Fig. 33) Largest seta bi-segmented, unilaterally setulated, with numerous small spines throughout, and two long spines at mid-length.

Trunk limb 1: (Fig. 34) indistinguishable from that of *M. rosea* and *M. triserialis* from Sri Lanka.

Trunk limb 2: (Fig. 35) indistinguishable from that of *M. rosea* and *M. triserialis* from Sri Lanka, but fourth

filter seta of gnathobasic comb not thicker than the other three.

Trunk limb 3: (Fig. 36) two apically ciliated hillocks between setae 1 and 2. Seta 3 only slightly longer than seta 2.

Trunk limb 4: (Fig. 37) as in *M. rosea* and *M. triserialis* from Sri Lanka.

Trunk limb 5: (Fig. 38) pre-epipodite with two hyaline setae (third one broken off?), setulated at their tip. Epipodite massive. Exopod not seen. Endopod reduced to a rounded lobe with short setules at tip. Gnathobase: third seta not seen.

Postabdomen: (Fig. 39) not different from that of *M. rosea* and specimens from Sri Lanka.

Description of the male

Shape: (Fig. 40) oval acute in lateral view. Dorsal margin curved from tip of rostrum to posterior dorsal corner, into a small spine. Slight serration in dorsal margin. Ventral margin serrated with spines of two different lengths. Intestine without convolutions.

Head: (Fig. 40) as in female.

First antenna: (Figs 40, 41, 45 and 46) long, nearly halfway to valve, with two sensory setae, located ventro-laterally and dorsally. Groups of spinules near ventral angulation.

Second antenna: (Fig. 42) as in female.

Trunk limb 1: (Fig. 43) as in female, but with clasper at the ODL, shaped exactly as in *M. rosea* (Figs 43, 47 and 51).

Trunk limbs 2–5: as in female.

Postabdomen: (Figs 40, 44, 48–50) as in *M. rosea*, but the apical tubular part rounded, not pointed. True end-claws absent, but **a pair of soft, triangular leaflet-like appendages** in subdistal position on the dorsal side. Two large sperm ducts open at the ventral tip (Figs 49 and 50). Seta natatoria with long proximal segment, its distal segment reduced.

3. *Macrothrix* cf *triserialis* from Kenya

Material examined: Mirim Ahero Dam, Kisumu, Kenya, 30 Aug 1995, Leg. Mutune David Masai. Number of animals examined: 5.

Description of the parthenogenetic female

Shape: (Fig. 52) oval, length 0.619 mm, height 0.321 mm.

Head: (Fig. 52) laterally depressed in rostral region, evenly arched in lateral view, head-shield not reticulated, dorsal margin not serrated. Head-pore small, round. Ventral margin of head smooth. Labrum bi-segmented, with fine serration along the edge, and distal end with a lobe with few short hair-setae (Fig. 54).

First antenna: (Fig. 55) rod-like. Dorsal and ventral margins with strong spines. One sensory seta located ventro-laterally at proximal end. Seven terminal aesthetascs of different length. Distally an external row of spines.

Second antenna: (Fig. 57) Basipod with inner apical seta (arrow), and a stout apical spine on outer surface. Swimming setae 0-0-1-3/1-1-3; spines 1-1-0-1/0-0-1. Largest seta with a mix of *rosea*- and *triserialis*-like characters: one long spine and a rather large number of short spines (beside numerous spinules) (Fig. 56).

Trunk limb 1: (Figs 58–60) as in *triserialis*, but seta 8 provided with rather short setules.

Trunk limb 2: (Fig. 61) as in *triserialis*, but scraper 5 without enlarged subapical spine.

Trunk limb 3: (Fig. 62) as in *triserialis*, but seta 3 of exopodite twice as long as seta 2, and two hillocks between setae 1 and 2.

Trunk limb 4: (Fig. 63) as in *triserialis*, but both setae of exopodite unequal in length and external scraper seta of endopodite provided with conspicuously long setules along most of its length.

Trunk limb 5: (Fig. 64) as in *triserialis*, but 'exopodite' apically rounded, not seta-like.

Postabdomen: (Fig. 53) as in *M. triserialis*.

Male: not seen.

4. *Macrothrix elegans* Sars, 1901

New synonym: *Macrothrix superaculeata* Smirnov, 1992.

Material: numerous parthenogenetic females from Broa (Lobo) Reservoir, SP, Brazil, 1994, leg H. J. Dumont.

Description of the parthenogenetic female

Shape: (Fig. 65) oval, length 0.538 mm, height 0.317 mm, dorsal margin curved from tip of rostrum to posterior dorsal corner. The latter produced into a spine (arrow), with fine serration along the valve. Ventral margin serrated. Intestine without convolutions.

Head: (Fig. 65) laterally depressed in rostral region, arched in lateral view. Head-shield not reticulated, dorsal margin slightly serrated. Head-pore medium sized, oval. Ventral margin of head smooth. Compound eye located close to the margin. Ocellus smaller than compound eye, located near tip of rostrum. Labrum bi-segmented with transversal lines along the edge, distal part a lobe with short hair-setae (Fig. 67).

First antenna: (Fig. 68) rod-like, with subapical group of robust spines. Rows of spinules all over body. Terminal aesthetascs of different length: two long, V-shaped, and seven short ones.

Second antenna: (Fig. 66) Swimming setae 0-0-1-3/1-1-3; spines 2-2-1-1/2-2-1. Largest seta, with fine spines from basis to tip (Fig. 70), and two strong spines at mid-length.

Trunk limb 1: (Figs 71 and 72) as in typical *triserialis* or *rosea*.

Trunk limb 2: (Fig. 73) as in *triserialis*, but subapical spines of scrapers 4 and, especially, 5 enlarged and fourth filter seta of gnathobasic comb thickened.

Trunk limb 3: (Fig. 74) similar to that of *triserialis* and *rosea*, but seta 3 of the exopodite relatively short, and on the endopodite seta 2′ relatively much longer than seta 1′, and no gap between the seta row on the internal and external endites.

Trunk limb 4: (Fig. 75) exopod with the two hyaline (I–II) setae subequal in length. All other characters as in *triserialis* and *rosea*.

Trunk limb 5: (Fig. 76). Epipodite gigantic. Other characters as in *triserialis* and *rosea*.

Postabdomen: (Fig. 69) with postanal rows of dorsal marginal spines very robust.

Male: not seen, but described by Sars (1901) and very similar to that of *M. triserialis* (see discussion).

5. *Macrothrix* cf. *triserialis* from Mexico

Material examined: Cenote Azul y Cenote Zacate. Quintana Roo, México. 29 jun 1995 and 18 Aug 1987, leg. Eduardo Suárez Morales & Rebeca Gasca Serrano. Four specimens were analyzed.

Description of the parthenogenetic female

Shape: (Fig. 77) oval, length 0.608 mm, height 0.402 mm, dorsal margin curved from tip of rostrum to posterior dorsal corner, the latter produced in a small spine (arrow), with serrations along the valve. Ventral margin serrated. Intestine without convolutions.

Head: (Fig. 77) laterally with a depression in rostral region. Arched in lateral view, head-shield not reticulated. Dorsal margin serrated. Head-pore small and round. Ventral margin of head smooth. Compound eye located close to the margin. Ocellus smaller than compound eye, located near tip of rostrum. Labrum bilobed, bi-segmented, serrated, with transversal lines along the edge (Fig. 79).

First antenna: (Fig. 78) rod like. Ventral and dorsal margins with rows of spines, more robust near tip. One sensory seta located ventro-laterally near base of antennula. Seven short, two long aesthetascs.

Second antenna: (Fig. 80) basipod with inner apical seta (arrow), and a stout apical spine on outer surface (arrow). All segments spinulose. Swimming setae 0-0-1-3/1-1-3; spines 2-2-2-1/0-0-1. Apex of segments two and three of exopodite widened. Largest seta with numerous short pines and about four stronger spines around mid-length.

Trunk limb 1: (Figs 83 and 84) as in *triserialis* and *rosea*, but epipodite small and rounded.

Trunk limb 2: (Fig. 85) scrapers 4–5 without enlarged subapical teeth. Seta 4 of gnathobasic filter-comb moderately larger than three other setae.

Trunk limb 3: (Fig. 86) exopodite with seta 3 at least twice as long as seta 2. Hillocks between setae 1 and 2 weakly indicated. Other characters as in *triserialis* and *rosea*.

Trunk limb 4: (Fig. 87) exopod with two hyaline setae of similar size. Endopod as in *triserialis* and *rosea*.

Trunk limb 5: (Fig. 88) pre-epipodite composed of three hyaline hillocks, with setules at their apex. Epipodite round. Exopod of one seta. Endopod with a rounded setulated lobe. Gnathobase with the three usual setae.

Postabdomen: (Fig. 82) as in *M. triserialis*, with robust postanal spines.

Male: not seen.

6. *Macrothrix tabrizensis* n. sp.

Holotype: an adult parthenogenetic female on a permanent glycerin glass slide sealed with Glyceel mounting medium, accession number 2000.274, deposited at the British Museum (Natural History) London.

Paratypes: two adults parthenogenetic females, accession number 2000.275, 2000.276; one adult ephippial female, accession number: 2000.277 and one male, accession number 2000.278; all on permanent slides, deposited in the British Museum. Two adult parthenogenetic females, one ephippial female and one male on permanent slides, access number respectively: I.G.28.882c, I:G.28.882d, I.G.28.882a, I.G.28.882b, deposited at Royal Belgian Institute of Natural Sciences, Brussels. One adult parthenogenetic female on permanent slide, accession number ECO CHZ 778. One adult ephippial female on permanent slide, access number: ECO CHZ 777; and one male on permanent slide, accession number ECO CHZ 776, deposited at ECOSUR. Carr. Chetumal-Bacalar, Quintana Roo. C. P. 77000. MEXICO. All material (Holotype and Para-

types) obtained from cultures by G. Murugan and H. J. Dumont from dry mud collected at Tabriz, Iran, 18 Nov 1995.

Diagnosis of the female
Dorsal margin of finely serrated valves curved, with dorsal spine, intestine not convoluted. Head laterally depressed, head-shield not reticulated, head-pore oval and small. Labrum bi-segmented, with setose apex. First antenna rod-like, with rows of spines of similar size all along. Nine aesthetascs in two groups (2 and 7) of different length. Second antenna: swimming setae 0-0-1-3/1-1-3; spines 0-1-0-1/0-0-1. Largest seta with fine spines from its basis to tip, and some slightly stronger spines along the body. Exopod of trunk limb one with long apical seta and very short lateral seta. Epipodite small. Seta 5 shortened. Exopodite of trunk limb two with plumose seta not curved inwardly, endopod with scraper 5 strongly developed but subapical spine not enlarged and middle seta of gnathobase erect, not curved outwardly. Third limb: all setae of the back row of the endite approximately of the same size. Exopod of trunk limb four very small. Endopod: posterior row with only three filter setae. Pre-epipodite of trunk limb five composed of three fused hyaline setae. Epipodite big and round. Exopod of two small mushroom-shaped setae. Biggest gnathobasic seta inserted on a hillock. Second biggest with a bulge. Postabdomen: preanal zone with small heel, postanal rim with groups of relatively small spines. Setae natatoriae with well individualised distal segment.

Description of the parthenogenetic female
Shape: (Fig. 89) length 0.759 mm, height 0.504 mm, dorsal margin curved from tip of rostrum to posterior dorsal corner, produced into a spine (arrow), with fine serration along the valve. Ventral margin serrated. Intestine without convolutions.

Head: (Figs 98, 109 and 110) with depression in rostral region, arched in lateral view, head-shield not reticulated, dorsal margin not serrated. Head-pore small, round. Ventral margin smooth. Compound eye located close to the margin. Ocellus smaller than compound eye, located near tip of rostrum. Labrum bi-segmented with transversal lines along the edge, distal end with a setose lobe (Fig. 91).

First antenna: (Figs 90, 112–114) rod-like, with rows of short spines over most of its length. These spines not increasing in size distally. One sensory seta ventro-laterally on proximal end of antennular length. Nine terminal aesthetascs in two groups of different length.

A distal external row of crown spines.

Second antenna: (Fig. 92) Basipod with inner apical seta and outer apical spine. Swimming setae- 1-3/1-1-3; spines 0-1-0-1/0-0-1. Largest seta bi-segmented, with row of fine spines from basis to tip (Fig. 93), and few slightly stronger spines dispersed over the seta.

Trunk limb 1: (Figs 94–96, 119, 121–124) epipodite small and rounded. Exopod with long apical and very short lateral seta. Endite 4 with three setae of decreasing length. Endite 3 with four setae, as in *rosea* and *triserialis*. Endite 2 with three setae, of which setae 3–4 of normal appearance, but seta 5 shortened. Endite 1 with two setae. Fork on endite 1 with one big tooth. Fork on endite 2, with two small teeth (Figs 96 and 125).

Trunk limb 2: (Figs 96, 127–131) exopod digitiform with heavy, not curved inwardly plumose apical seta. Endopod with eight scrapers. Scrapers 1–2 long and setiform. Scraper 3–8 of various length, their apical segment comb-shaped. Scraper 5 the most robust of all, its subapical spines not enlarged. Sensillum behind scrapers 8 and 4 absent. Between scrapers 3 and 4 a very small receptor (Figs 126 and 128). Gnathobase with brushing seta erect, setulose. Filter comb composed of the four usual plumose setae. A pore on the surface of the gnathobase (Figs 127–129).

Trunk limb 3: (Figs 97, 132 and 133) exopod with four plumose setae of subequal length. Seta 3 the longest, seta 4 unilaterally setulated. Endopodite with two external scrapers and two rows of setae. Anterior row apparently of five setae only. Posterior row of six setae, all of which about equal in length. Gnathobase with rows of thick setules. A bottle-shaped receptor near its base. Two setose hyaline papillae on its posterior end.

Trunk limb 4: (Fig. 98) exopod reduced in size, with two small plumose setae, similar in length. Endopod with two rows of appendages. Anterior row with five elements and a gnathobasic horsetail-seta, as in *triserialis*. Posterior row with **three** filter setae (1′–3′).

Trunk limb 5: (Fig. 99) pre-epipodite composed of three fused hyaline setose setae. Epipodite big and round. Exopod composed of two coalesced lobes (I–II), with setules around their tip. Endopod a lobe, round and broad. Gnathobase with three setae. Seta I′ long, swollen at its broad basis, constricted and pointed apically. Distal segment with long setules at its basis and short apical setules. Seta 2 with basal swelling. Seta 3 very short, setulated.

Postabdomen: (Figs 100, 116–118) with heel (arrow).

Dorsal margin bilobed. Basal portion with numerous rows of small spines. Anus with groups of larger spines. Terminal claw small, evenly curved, with lateral small teeth. Seta natatoria with distal segment relatively long, about 1/5 of the whole, carrying few setae.

Diagnosis of the male

Shape oval-acute, dorsal margin curved from tip of rostrum to straight in posterior dorsal corner. No serration along the valves. Intestine without convolutions. Head with a depression in rostral region, arched in lateral view, head-shield not reticulated, dorsal margin not serrated. Labrum bi-segmented, without ornamentation. First antenna long, nearly halfway to body, with two sensory setae. Distally neither setules nor spines. Second antenna as in female. Trunk limb 1 as in female, but a clasper plus a strong curved seta present. Clasper not widened apically. Trunk limb 2–5 as in female. Postabdomen hook-shaped in lateral view. Pre-anal zone with small heel. Dorsal margin not bilobed. Anus bordered with groups of spines. Terminal claw big, slightly curved, without lateral teeth. Seta natatoria with long proximal segment, distal segment shorter (about 1/4 of the total length of the seta).

Description of the male

Shape: (Figs 101 and 134) oval-acute, length 0.384 mm, height 0. 250 mm, dorsal margin curved from tip of rostrum to straight in posterior dorsal corner. No serration along the valve. Intestine without convolutions.

Head: (Figs 101 and 134) with a depression in rostral region, arched in lateral view, head-shield not reticulated, dorsal margin not serrated. Head pore small, oval. Ventral margin of head smooth. Compound eye located close to the margin. Ocellus smaller than compound eye, located near tip of rostrum. Labrum bi-segmented, without ornamentation, distal segment round.

First antenna: (Figs 102, 134, 135 and 137) long, nearly halfway to valves, with two sensory setae, located ventro-laterally and dorsally. Some groups of strong spinules border the ventral angulation, and several rows occur in lateral orientation. Seven terminal aesthetascs, of which one much longer than the others. Distally, neither setules nor spines.

Second antenna: (Figs 103, 104 and 134) as in female. Distal segment of largest seta, with numerous strong spines (Fig. 104).

Trunk limb 1: (Figs 105 and 106) as in female, but

a clasper and curved acute seta attached to the exopod (Fig. 105). Clasper not widened apically.

Trunk limbs 2–5: as in female.

Postabdomen: (Figs 108, 138 and 139) hook-shaped in lateral view. Ventral margin convex. Preanal zone with small heel. Dorsal margin not bilobed. Pre-anal zone with numerous rows of spines similar in length, stronger than I the female. Anus bordered with groups of spines. Terminal claw big, slightly curved, smooth, without lateral teeth. Seta natatoria with long proximal segment, distal segment short (c. 1/4) in comparison with proximal. Distal segment carrying some long setae at its tip.

7. Macrothrix agsensis n. sp.

Holotype: an adult parthenogenetic female on a permanent glycerine glass slide sealed with Glyceel mounting medium, accession number 2000.279, deposited in the British Museum (Natural History), London.

Paratypes: three adult parthenogenetic females on permanent slides, accession number: 2000.280, 2000.281, 2000.279, deposited in the British Museum. Three adult parthenogenetic females on permanent slides, accession number I.G.28.881a, I.G.28.881b, I.G.28.881c, deposited at the Belgian Institute of Natural Sciences, Brussels, Belgium. Three adult parthenogenetic females on permanent slides, deposited at ECOSUR. Carr. Chetumal Bacalar. Chetumal Quintana Roo. C.P. 77000, Mexico, accession number ECO CHZ 771, 772, 773. All the material (Holotype and Paratypes) was collected by M. Silva-Briano, Araceli Adabache-Ortiz, Gustavo Quintero Díaz & Joél Vázquez Díaz. EL salto del burro (El Jagüey), Aguascalientes, México, 31 July 1993.

Diagnosis

Shape: oval in lateral view, dorsal margin curved from tip of rostrum to posterior dorsal corner. No dorsal serration. Intestine without convolutions. **Head** laterally depressed, head-shield not reticulated, head-pore big and oval. Labrum bi-segmented with transversal lines along the edge. First antenna rod-like. Dorsal margin with four rows of long spines, sensory seta located ventro-laterally. Nine terminal aesthetascs in two groups of somewhat different length. **Second antenna:** coxal region with unusually long apical spine on outer surface. Swimming setae 0-0-1-3/1-1-3; spines 0-1(1)-1-1/0-0-1. Largest seta with fine spines from basis to tip. Exopod of trunk limb one with long apical seta and very short lateral seta. Setae

4–5 of endite 2 with five spines along their inner apical rim. Epipodite elongated. Trunk limb two: exopod oval with plumose apical seta, turned inwardly. Endopod with eight scrapers, none of which have enlarged subapical teeth. Filter comb composed of four long setae; seta four much thicker than the others. Trunk limb three: exopod with four plumose setae, and seta 2 slightly longer than seta 3. Endopod with back row of six setae, all of the same length. Anterior row with **five** setae. Exopod of trunk limb four with two short setae. Endopod with naked external scraper. Trunk limb five: exopod composed of one seta, endopod a rounded lobe, epipodite big, elongated, gnathobase composed of three setae of different length. Postabdomen: ovoid in lateral view, pre-anal zone with small heel, dorsal margin not bilobed. Anus bordered with several groups of spines. Terminal claw small, with lateral teeth. Seta natatoria with long proximal segment, distal segment shorter but not reduced, carrying some setae at its tip.

Description of the parthenogenetic female

Shape: (Figs 140 and 153) oval, length 0.807 mm, height 0.515 mm, dorsal margin curved from tip of rostrum to posterior dorsal corner, without dorsal serration, but with slight reticulation (Fig. 154). Intestine without convolutions.

Head: (Figs 140 and 153) laterally depressed in rostral region, arched in lateral view, head-shield not reticulated, dorsal margin not serrated. Head-pore big, oval. Ventral margin of head smooth. Compound eye located close to the margin. Ocellus smaller than compound eye, located near tip of rostrum. Labrum bi-segmented with transversal lines along the edge, and apex a lobe with few short hairs (Fig. 142).

First antenna: (Figs 141 and 156) rod-like. Dorsal margin with four rows of long, strong spines. One sensory seta ventro-laterally near proximal end. Nine terminal aesthetascs of different length. Two longer ones inserted in a V-shape. Distally, an external row of small spines.

Second antenna: (Fig. 143) coxal region with two inner basal sensory setae. Basipod with inner apical seta (arrow), and a stout apical spine in outer surface (arrow). Swimming setae 0-0-1-3/1-1-3; spines 0-1-(1)-1-1/0-0-1. Largest seta bi-segmented, unilaterally setulated, with fine spines from basis to tip (Fig. 144).

Trunk limb 1: (Figs 146–148, 158–162) epipodite elongated. Exopod with long apical seta, and a very short lateral seta. Endite 4 with three setae of decreasing length. Endite 3 with four setae. Seta 1 long with

12

short unilateral setules. Seta 2 smaller and finer than 1, with long setules at its tip. Seta 8 with apical tuft of setules. Endite 1 as is *triserialis* and *rosea*. Endite 2 with setae 4–5 modified: their inner apical half lined with about five thick, short setae. Fork on endite 1 with two teeth, on endite 2 with one tooth (Fig. 148).

Trunk limb 2: (Figs 149, 163, 164 and 166) exopod oval with a plumose seta, inserted apically and curved inwardly over scrapers 1–2. Endopod with eight scrapers. Scrapers 1–2 setiform. Scrapers 3–5 robust, their subapical teeth not enlarged. Scrapers 6–8 with fine teeth. No sensillum behind scrapers 8 or 4. Gnathobase with filter comb composed of four plumose setae, of which seta 4 the more robust one.

Trunk limb 3: (Figs 150, 165, 167–169) exopod with four plumose setae. Setae 1–2 similar in length, seta 3 slightly shorter. Seta 4 short and serrate. Endopod: two external scapers and two rows of appendages. Anterior row with five setae. Posterior row with six setae, all of similar length. Gnathobase blunt, with rows of long and thick setules, a receptor and two setose hyaline papillae. A pore at the base of the gnathobase (Figs 125–127).

Trunk limb 4: (Fig. 151) exopod with two plumose setae of similar length. Endopod with two row of setules. Anterior row with the usual five elements. Posterior row with **five** setae (1′–5′). Gnathobase with horsetail seta (6). **Trunk limb 5:** (Figs 152 and 171) pre-epipodite composed of three lobes, setulated at their tip. Epipodite big. Exopod one seta (1), with short setules at its terminal part. Endopod rounded, setulated at its apex. Gnathobase with three setae. Seta 1 very long, bi-segmented, unilaterally setulated. Seta 2 long, fully setulated. Seta 3 small, fully setulated. A probable pore on the gnathobase.

Postabdomen: (Figs 145, 155 and 157) Ventral margin slightly convex with two rows of small spines. Preanal zone with small heel. Dorsal margin not bilobed. Pre-anal portion with numerous rows of small spines, all similar in length. Anus too bordered with several groups of spines. Terminal claw, small, evenly curved, with lateral teeth. Seta natatoria with long proximal segment, distal segment shorter than proximal, carrying long setae at its tip.

Male unknown.

*8. **Macrothrix smirnovi** Ciros-Pérez & Elías-Gutierrez, 1997*

Material examined: Presa La Araña, Sierra Fría, Ags. 25 Sep 1995. Charco en carr. Tanque Los Jimenez, Ags. 17 Aug 1991, México; 20 specimens analysed, leg. M. Silva-Briano.

Description of the parthenogenetic female
Shape: (Fig. 172) oval, length 0.901 mm, height 0.624 mm, dorsal margin curved from tip of rostrum to posterior dorsal corner. No serration along the valve. Intestine without convolutions.
Head: (Fig. 172) with depression in rostral region, arched in lateral view. Head-shield not reticulate, dorsal margin slightly serrated. Head-pore big, round. Ventral margin of head smooth. Compound eye located close to margin. Ocellus smaller than compound eye, located near tip of rostrum. Labrum bi-segmented; distal part a lobe bearing short hair-setae (Fig. 174).
First antenna: (Fig. 173) rod-like. Dorsal margin with up to eight rows of long spines. One sensory seta on proximal end of antennular length. Seven terminal aesthetascs in two groups of different length. Distally, an external crown of spines.
Second antenna: (Fig. 175) Basipod with inner apical seta and stout apical spine. Swimming setae 0-0-1-3/1-1-3; spines 0-2-1-1/0-0-1. Largest seta, with fine spines at it basis, about five strong spines, decreasing in size, at its median part, and fine spines at its distal tip (Fig. 176).
Trunk limb 1: (Figs 178–180) epipodite big, elongated. Exopod with long, serrated apical seta, and short lateral seta. Near it basis, a small plumose seta, and an oblique row of setules between both. Endite 4 with three serrate setae of decreasing length. Endite 3 with four setae. Setae 8 and 7 similar and size, and slightly smaller and less robust than seta 6. Endite 2 with three, endite 1 with two feathered setae 8–9. Forks on endites 1–2 with one big tooth each (Fig. 180).
Trunk limb 2: (Fig. 181) exopod with inwardly curved apical seta. Endopod with eight scrapers. Scrapers 1–2 long and setiform. Scraper 3 the longest of the 'true' scrapers, but scraper five has the most robust teeth. Scrapers 6-8 smaller and similar in length, with fine teeth along their distal segment. No sensillum behind scraper 8, but **a hairy seta behind scraper 4**. Gnathobase with conical posteriormost gnathobasic seta 1. Brushing seta 2 not curved inwardly. Seta 3 hooked. Small seta between gnathobase and filter comb, with fine setules at its tip. Filter comb composed of four plumose setae. Seta four the strongest.
Trunk limb 3: (Fig. 182) exopod with four plumose setae. Setae 1–2 about equally long,. Setae 3–4 also similar in length, and shorter than 1–2. Seta 3 plumose, but seta 4 serrate. Endopod with two external

scrapers, and two rows of appendages. Anterior row with **five** appendages, including at least one receptor seta. Posterior row with six setae, all of similar length. Gnathobase with rows of thick setules, an elongated receptor, and two setose papillae.

Trunk limb 4: (Fig. 183) exopod with two plumose setae, similar in length. Endopod with two rows of setae. Anterior part with a scraper, three flaming-torch setae, and a receptor. Posterior row of five plumose setae long. Gnathobase with horsetail seta and support seta. Gnathobasic filter seta absent.

Trunk limb 5: (Fig. 184) pre-epipodite composed of **two** hyaline lobes of different shape, with long setules at their tip. Epipodite big, curved, apically tapering and pointed. Exopod a hyaline lobe with fine setules at its tip. Endopod a rounded lobe. Gnathobase with three setae.

Postabdomen: (Fig. 177) Ventral margin slightly convex. Pre-anal zone with small heel. Dorsal margin bilobed. Pre-anal portion with numerous rows of spines, all similar in length. Anus bordered with groups of spines. Terminal claw small, evenly curved, with lateral teeth. Seta natatoria with long proximal segment, distal segment much shorter than proximal but not rudimentary, carrying some setae.

Description of the male

Shape: (Fig. 185) oval-acute, length 0.346 mm; dorsal margin curved from tip of rostrum to posterior dorsal corner. No serration along the valve. Ventral margin serrated. Intestine without convolutions.

Head: (Fig. 185) laterally without depression in rostral region. Arched in lateral view, head-shield not reticulate, dorsal margin not serrated. Head-pore not seen. Ventral margin of head smooth. Compound eye located close to the margin. Ocellus smaller than compound eye, located near tip of rostrum. Labrum bi-segmented, with small distal segment rounded.

First antenna: (Fig. 186) long, extending nearly halfway of valve, with two sensory setae located ventro-laterally, close each other. Groups of strong spines border the ventral angulation. Nine terminal aesthetascs in two groups of of different length. Distally neither setules nor spines.

Second antenna: (Fig. 185) as in female.

Trunk limb 1: as in female, but exopod with clasper and long curved seta (Fig. 187).

Trunk limbs 2–5: as in female.

Postabdomen: (Fig. 188) similar to that of the female.

Discussion

It is convenient to begin by an attempt to determine a set of characters common to all the animals described above that provide a contrast with other groups within the genus. Unfortunately, only the *laticornis*-group and *M. tripectinata* have been described in sufficient detail to be usable. *M. tripectinata* is a mesasiatic representative of the *M. hirsuticornis*-group as re-defined by Silva-Briano (1998). In spite of this limited information, trunk limbs 2–5 turn out to be richly endowed with diagnostic characters: the endopodite of limb two has eight scrapers and no receptor or ciliated seta behind scraper 8 in the *rosea*-group, while it has a receptor in the *laticornis*-group, and a seta in *M. tripectinata* (as in other representatives of the *M. hirsuticornis*-group). The exopodite of limb three has four plumose setae in the *rosea*-group and in the *laticornis*-group, against five in *M. tripectinata-hirsuticornis*. The anterior row of setae of the endopodite is composed of five or six setae in *rosea*-group, but of seven setae, of which the three externalmost are modified into aesthetascs and the remaining four are mushroom-shaped in *laticornis* and *tripectinata*. The exopodite of limb four has two plumose setae in the *rosea*-group, three in *laticornis* and *tripecinata*. Limb five has a complex pre-epipodite and a single exopodital seta in the *rosea*-group, a simple pre-epipodite and two exopodital setae in the *laticornis*-group, and one exopodital seta in *M. tripectinata*. In addition, as Kotov (1999) advocated, the first antenna also permits to separate lineages: a first antenna with nine equally long aesthetascs is primitive (*M. tripectinata*), one in which the aesthetascs form two groups is evolved (*rosea* and *laticornis*). It is, therefore, possible to unequivocally create a *rosea*-group, in which *smirnovi* stands out as the only species that has a setose seta behind scraper 4 of limb two, a character which was suggested to provide a transition between *Macrothrix* on the one hand and *Wlassicsia* and *Bunops* on the other hand (Silva-Briano & Dumont, 2001). However, there are many important differences between *Macrothrix* and the latter genera, suggesting that *M. smirnovi* had better continued to be considered a member of the *rosea*-group sl. Examples include the total absence of gnathobasic filter-combs on limbs three and four. Such combs are present and well developed in *Wlassicsia* and *Bunops*, the endite of trunk limb two has three doublings, and pointed, scale-like hillocks behind scrapers 2–4, the exopodite of their limb three has five setae and limb four is with three setae, additional setae occur on the endite 1 of limb one, the Fryer's forks are incompletely modified, etc.

Having thus defined the *rosea*-group s.l. (see Conclusion for a full statement of diagnostic characters), we will now attempt to bring order to the *rosea-triserialis* species complex, sorting out which characters may be useful to delimit species here. Clearly, *rosea* and *triserialis* are nearly related, and in the absence of males, attempts to separate them may fail. In this respect, we note that Korinek misidentified some Czech material as *rosea*, having in hand a species with a 'normal' (i.e. female-like) male postabdomen. Yet, good figures of the blunt-ending postabdomen without end-claws of true *rosea* are available in identification manuals, e.g. Flössner (1972). What Korinek, who kindly allowed us to re-examine his material, had in front of him turned out to be a mixture of three species, including *Wlassicsia pannonica*.

Table 1 provides an overview of 23 diagnostic characters used in the present paper, of which 20 apply to females only. Taking *M. triserialis* from its *terra typica* in Sri Lanka as the species of reference, a systematic comparison of all characters listed shows that three differences exist with the population from Kerala, against seven with *M. rosea*. Of these, female characters 2, 4 and 21 appear taxonomically usable. Even so, there might have been reason for doubting the specific identity of *M. rosea* (as done, e.g. by Dumont & Van de Velde, 1977), had males not been available for comparison. Females of *rosea* can now safely be separated from females of *triserialis* by the fact that their first antenna has no apical enlarged spines, the largest seta of the second antenna has a large number of medium-sized spines throughout, and the terminal segment of the seta natatoria is short but not reduced to just the very apex of the seta (a character first identified by Smirnov, 1992). Here, we also accept that the differences between the *triserialis* population from Sri Lanka and from Kerala (only few hundreds of kilometers apart) are interpopulation variation. It follows that at least the following three characters are suspect as taxonomic tools: the size of the fourth gnathobasic seta of trunk limb two, the size of seta 3 of exopodite 3 relative to that of seta 2, and the lack of plumosity on seta 2 (as seen in the topotypical population).

Comparing new type-*triserialis* with the three African and American populations analysed, it appears that the Kenyan sample shows variation in the same characters as the two populations from India, in addition to the fact that scraper 5 of trunk limb 2 is without enlarged subapical teeth. We find it difficult, at this point in time, to weigh this difference, since the population from Mexico shows it too. In addition, the Mexican population has globular epipodites on trunk limbs 1 and 5, and thus stands a little further away from true *triserialis* than the African population. *Macrothrix elegans* from Sao Paulo state is again closer to typical *triserialis*, except that it has an enormous epipodite on trunk limb 5. If we maintain *elegans* as a good species (which is here done mainly for convenience), the population from Mexico may deserve the same status. Because of the absence of males (males of *triserialis* s.l. have now become available from three sources: Sao Paulo, Central Africa and Kerala State in India, but all look similar), such a decision is not taken here. We suggest to first resample both populations, to test for the stability of the differenciating characters, and to apply Sars' technique of culturing animals in the laboratory (as done here with *M. tabrizensis*). Large numbers of individuals, as well as males can be obtained by this technique. Our proposal to synonymise *M. superaculeata* with *M. elegans* is motivated by the fact that both occur in the same general area, and by the variability of the spine formula of antenna 2 (*superaculeata*-like spines occur as well in the Lobo sample as in the sample from Mexico), leaving no 'solid' characters to separate the two.

M. tabrizensis has 18 differences (out of 23 possible, including male characters) in diagnostic traits with topotypical *M. triserialis*, and is therefore accorded full specific status without hesitation. In *M. agsensis*, of which unfortunately no males were available, 11 differences (out of 20 traits) were recorded, and it is consequently also considered to be of full specific rank. For comparison, it may be stated that *M. smirnovi* differs only by 15 characters out of 23 (14 in males). Its status, of a species and even a subgroup is not questioned here, but the comparison shows that it is almost of the same rank as *agsensis* and *tabrizensis*.

Conclusion

We conclude by summing up all characters that currently allow us to characterise a *M. rosea* s.l. group, a *M. rosea-triserialis* subgroup, and a *M. smirnovi* subgroup.

M. rosea-group s.l.: first trunk limb with nine setae on endites 1–3, three on endite 4, and two (one very short, one long) on the exopodite. Fryer's forks with one or two big teeth. Limb two with eight scrapers in a row, without any doublings (except in *smirnovi*-subgroup). Hillocks behind scrapers 2–4 rounded. Exopodite with apical seta curved inwardly over scrapers 1–2 (except in *M. tabrizensis*). Gnathobase with three apical appendages, of which the middle one the largest, usually curved outwardly and plumose, and a fourth, internal seta between the apex and the four gnathobasic setae. Exopodite of limb three with four setae; endopodite with two marginal scrapers, six appendages in the front row, and six appendages in the back row. Some of the frontal appendages modified to tubular receptors, but none mushroom- or aesthetasc-shaped. Exopodite of limb four with two setae; endopodite with a frontal marginal scraper, three flaming torch setae, a receptor, a horsetail seta, and a support seta. A posterior row of plumose setae composed of 3–6 setae. No gnathobasisc filter setae on trunk limbs three and four. Limb five with a complex pre-epipodite, a strongly developed epipodite, reduced ecto- and endopodites, and three gnathobasic setae. Female postabdomen at least with a rudiment of a heel. Male postabdomen showing a morphological series from end-claws totally reduced to postabdomen fully female-like.

M. rosea-triserialis subgroup: showing the characters of the group, with the following restrictions: postabdominal heel always well developed; rows of strong spines between the rim of the anus and the base of the seta natatoria; fifth (to a lesser degree fourth) scraper of limb two with enlarged subapical teeth (except in populations of Mexico and Kenya, and trait therefore open to further evaluation); hind row of setae on endopodite of limb three composed of two long and four short setae; hind row of setae on endopodite of limb four composed of 5+1 setae (five on the endopodite, one on the gnathobase); epipodite of limb five elongated (exception: population from Mexico); pre-epipodite trilobed.

M. smirnovi-subgroup: showing the characters of the group, with the following restrictions: limb two with a setulated seta behind scraper 4; scraper 5 by far the most robust of the eight, but none of its teeth conspicuously longer than the others; middle appendix of gnathobase not plumose; pre-epipodite of limb five bilobed.

Acknowledgements

We are grateful to Dr V. Korinek, Dr C. F. Fernando, and Dr M. David Masai for presenting us with specimens of various *Macrothrix*. M. Silva-Briano particularly thanks Prof. Enrique Olivares Santana, and the members of the Centro de Ciencias basicas de la Universidad Autonoma de Aguascalientes for their support.

References

Brady, G. S., 1886. Notes on Entomostraca collected by Mr A. Haly in Ceylon. J. linn. Soc. Lond. 19: 293–317.

Ciros-Pérez, J & M. Elías-Gutiérrez. 1997. *Macrothrix smirnovi*, a new species (Crustacea: Anomopoda: Macrothricidae) from Mexico, a member of the *M. triserialis*-group. Proc. biol. Soc. Washington 110: 115–127.

Daday, E. Von, 1901. Mikroskopische Süsswassertiere. Dritte Asiatische Forschungsreise des Grafen Eugen Zichy. Budapest & Leipzig. Volume 2 (edited by G. Horvath): 277–470.

Dumont, H. J. & I. Van de Velde, 1977. Cladocères et Conchostracés récoltés par le Professeur Th. Monod dans la moyenne vallée du Niger en décembre 1972 et janvier 1973. Bull. IFAN A39: 75–93.

Flössner, D., 1972. Krebstiere, Crustacea. Kiemen- und Blattfüsser, Branchiopoda. Fischläuse, Branchiura. Tierwelt Deutschlands Fischer, Jena. 60: 500 pp.

Frey, D. G., 1988. Are there tropicopolitan Macrothricid Cladocera? Acta Limnol. Brasil. 2: 513–525.

Fryer, G., 1974. Evolution and adaptative radiation in the Macrothricidae (Crustacea: Cladocera): a study in comparative functional morphology and ecology. Phil. Trans. r. Soc. Lond. Biol. Sci. B 269: 137–274.

Jurine, L., 1820. Histoire des Monocles, qui se trouvent aux environs de Genève. Paris: 260 pp.

Korinek, V., 1984. Cladocera. Hydrobiological survey of the lake Bangweulu Luapula river basin. Scientific results, 13, 2: 117 pp.

Kotov, A., 1999. Redescription of *Macrothrix tripectinata* Weisig, 1934 (Anomopoda, Branchiopoda), with a discussion of some features rarely used in the systematics of the genus. Hydrobiologia 403: 63–80.

Liévin, F., 1848. Die Branchiopoden der Danziger Gegend, ein Beitrag zur Fauna der Provinz Preussen. Neueste Schriften Naturf. Gesell. Danzig. 4: 1–52.

Luyten, M., 1934. Over de Oecologie van de Cladocera van België. Biol. Jrb. Dodonaea 1: 32–179.

Sars, G. O., 1901. Contributions to the knowledge of the freshwater Entomostraca of South America, as shown by artificial hatching from dried material. Arch. Math. Naturvidensk. 23: 1–102.

Silva-Briano, M., 1998. A revision of the macrothricid-like anomopods. Ph.D. Thesis, Univ. Ghent: 388 pp.

Silva-Briano, M., N. Q. Dieu & H. J. Dumont, 1999. Redescription of *Macrothrix laticornis* (Jurine, 1820) and description of two new species of the *M. laticornis*-group. Hydrobiologia 403: 39–61.

Silva-Briano , M. & H. J. Dumont, 2001. *Wlassicsia, Bunops & Onchobunops*, three related genera. Hydrobiologia 442: 1–28.

Smirnov, N. N., 1976. Macrothricidae i Moinidae fauny mira. Fauna USSR NS, vol. 112 (Rakoobraznye) Nauka, Leningrad 1(3): 237 pp.

Smirnov, N. N., 1992. The Macrothricidae of the world. In Dumont, H. J. (ed.), Guides to the Identification of the Macroinvertebrates of the Continental Waters of the World. SPB, the Hague: 143 pp.

16

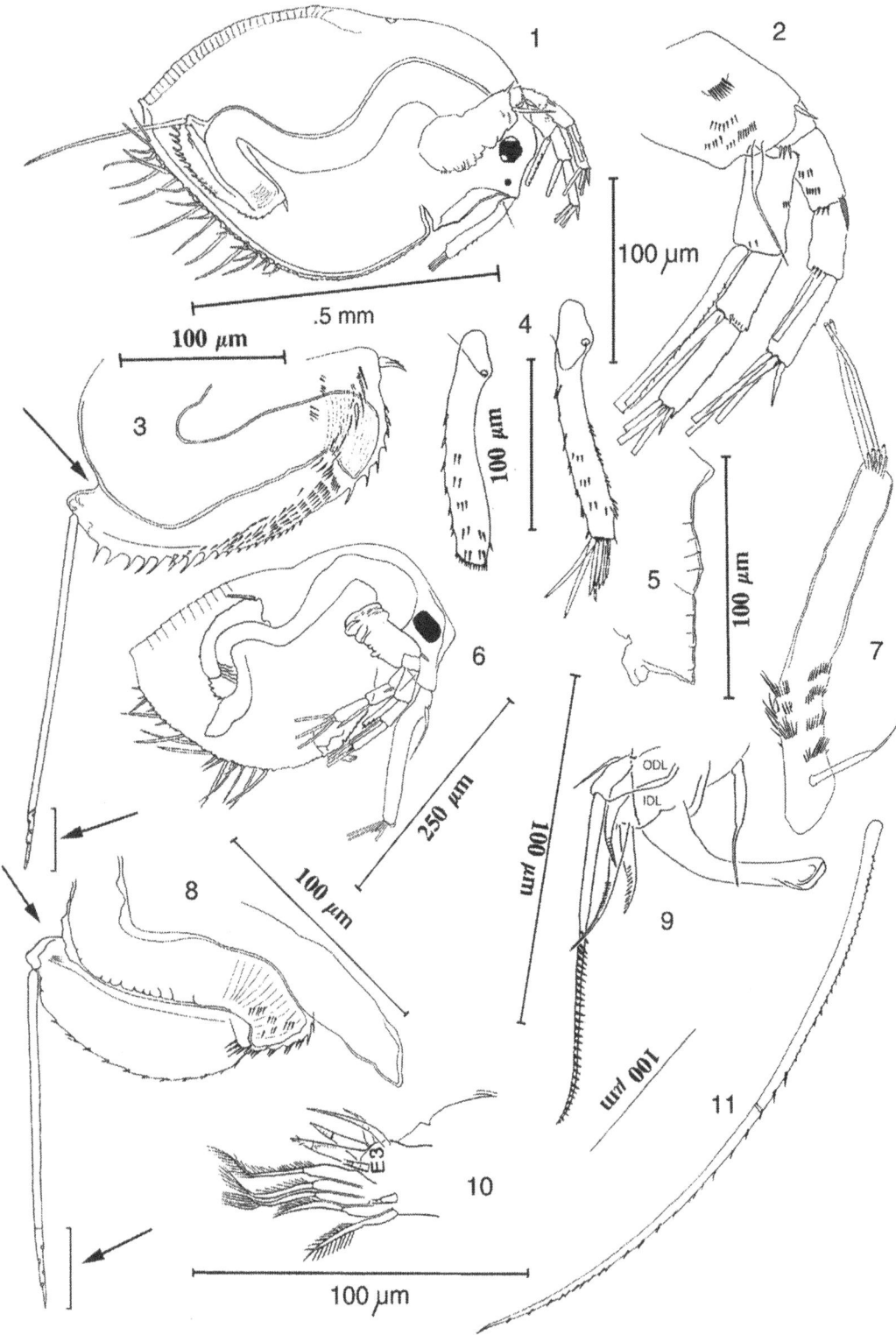

Figures 1–11. Macrothrix rosea, Belgium. *Figure 1.* Habitus of a parthenogenetic female. *Figure 2.* Second antenna of female. *Figure 3.* Postabdomen of female. *Figure 4.* First antenna of female. *Figure 5.* Labrum of female. *Figure 6.* Habitus of a male. *Figure 7.* First antenna of male. *Figure 8.* Postabdomen of male. *Figure 9.* IDL, ODL and copulatory clasper of male. *Figure 10.* Limb one (endites 1–3) of male. *Figure 11.* Longest seta on second antenna in female.

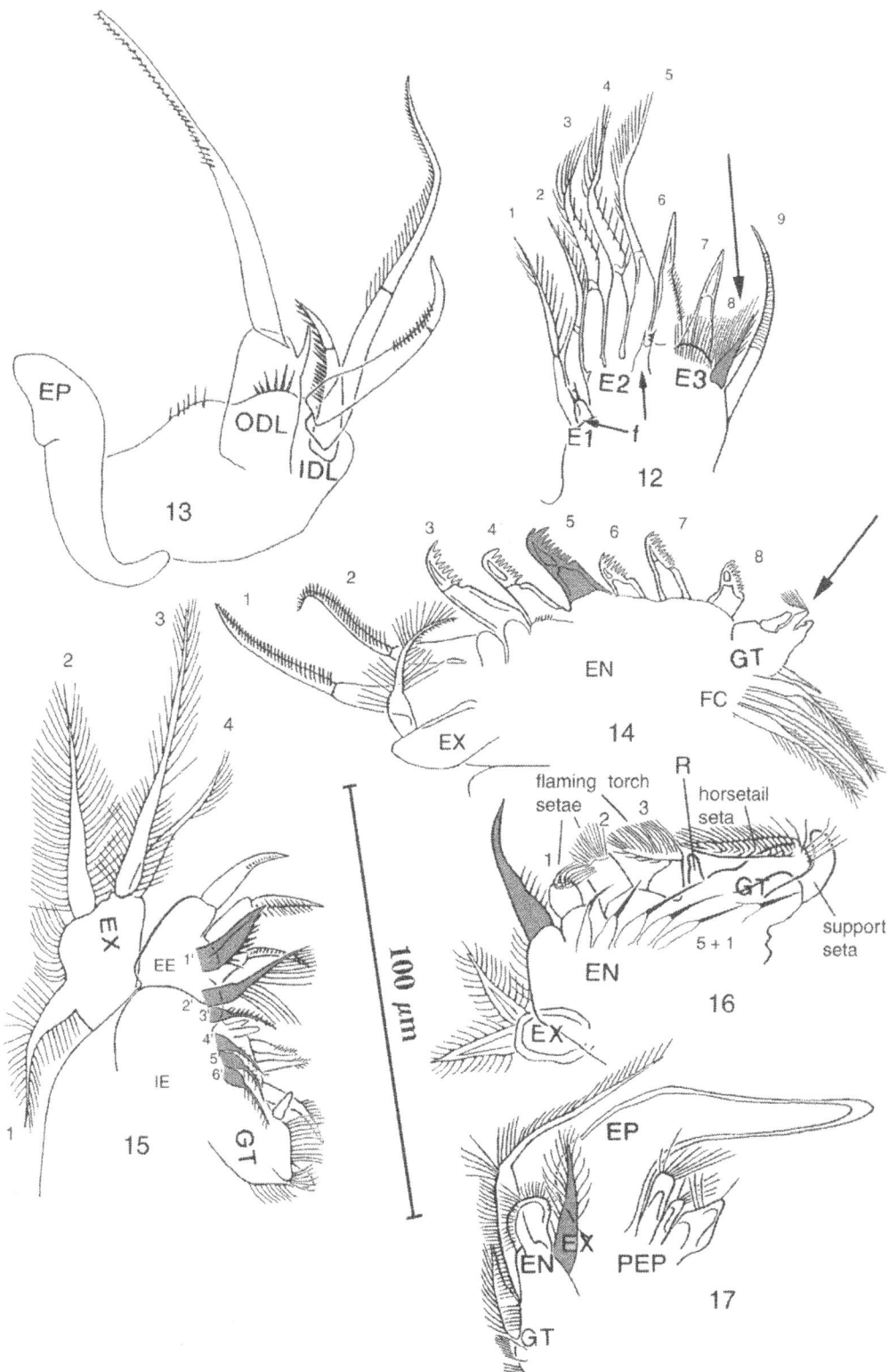

Figures 12–17. Trunk limbs of a female of *Macrothrix rosea*, Belgium. *Figure 12.* Trunk limb one with setae on endites numbered consecutively. *Figure 13.* IDL, ODL and epipodite of limb one. *Figure 14.* Limb two. *Figure 15.* Limb three with rows of setae on exo- and endopodites, provided with a standard numbering. *Figure 16.* Limb four. *Figure 17.* Limb five.

18

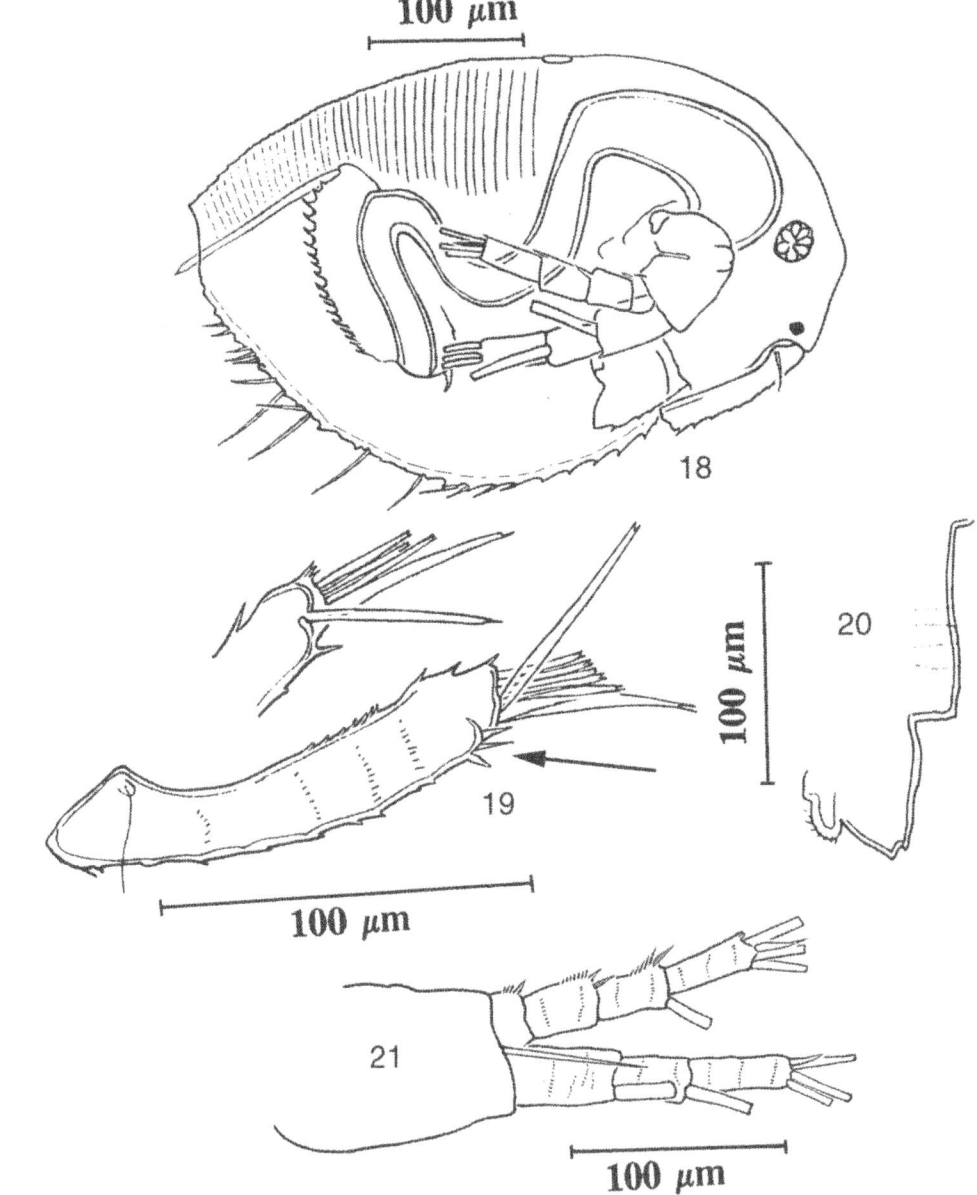

Figures 18–21. Topotypical *Macrothrix triserialis* from Sri Lanka. *Figure 18*. female habitus. *Figure 19*. First antenna. *Figure 20*. Labrum. *Figure 21*. Second antenna.

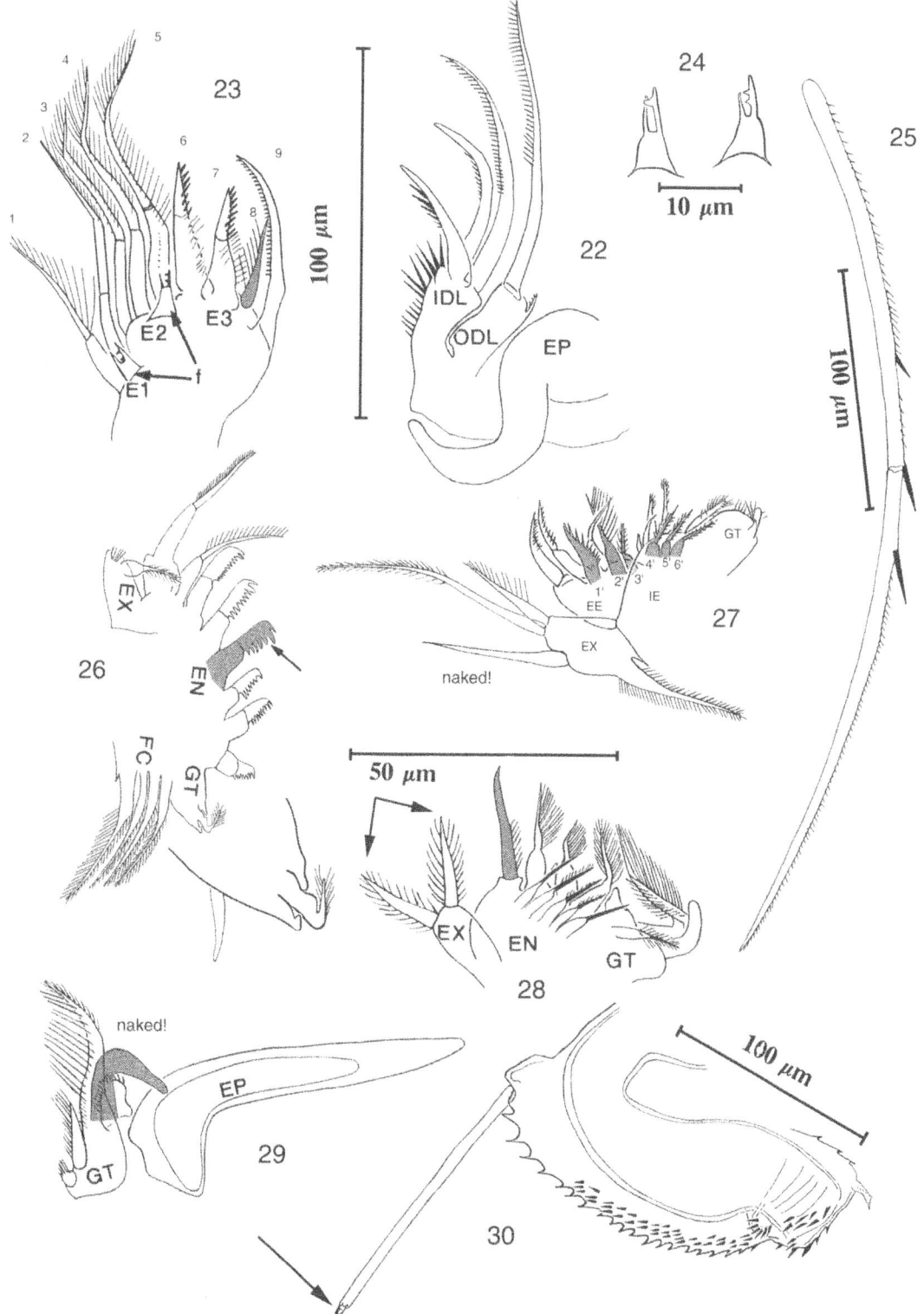

Figures 22–30. Macrothrix triserialis female, Sri Lanka. *Figure 22.* IDL, ODL and epipodite of limb one. *Figure 23.* Endites 1–3 of limb one. *Figure 24.* Fryer's forks. *Figure 25.* Long seta of second antenna. *Figure 26.* Limb two. *Figure 27.* Limb three. *Figure 28.* Limb four. *Figure 29.* Limb five. *Figure 30.* Postabdomen.

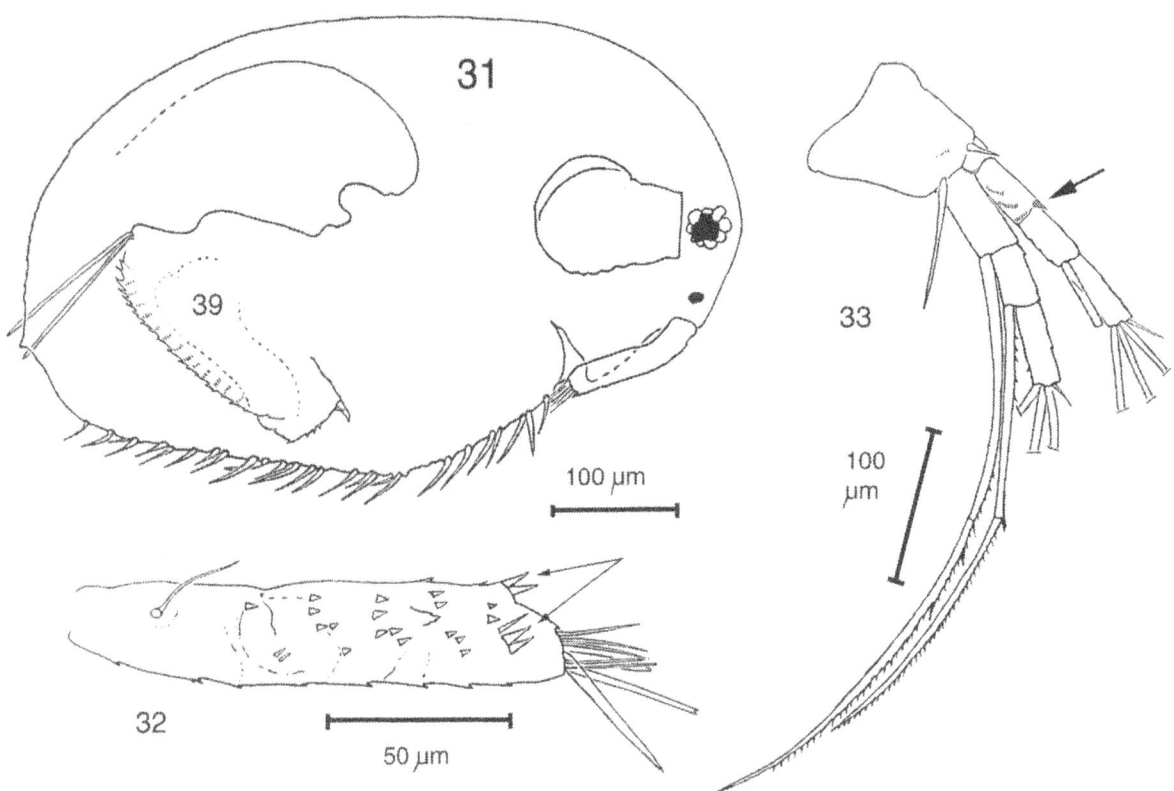

Figures 31–33. *Macrothrix triserialis* from Kerala. *Figure 31.* Female habitus. *Figure 32.* First antenna. *Figure 33.* Second antenna.

Figures 34–38. Macrothrix triserialis from Kerala, female. *Figure 34*. First trunk limb, complete. *Figure 35*. Trunk limb two. *Figure 36*. Trunk limb three. *Figure 37*. Trunk limb four. *Figure 38*. Trunk limb five.

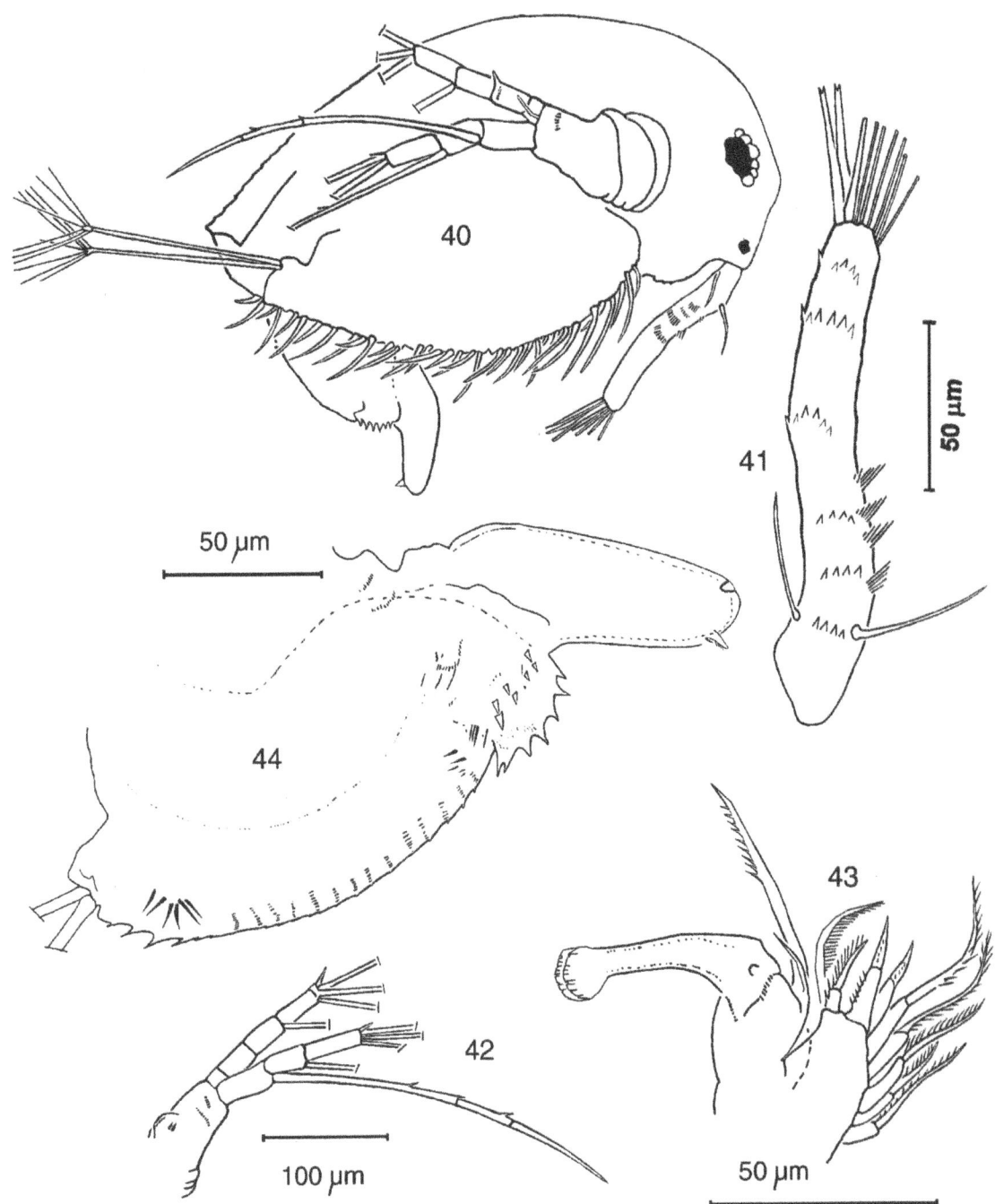

Figures 40–44. *Macrothrix triserialis* from Kerala, male. *Figure 40*. Habitus. *Figure 41*. First antenna. *Figure 42*. Second antenna. *Figure 43*. First trunk limb with copulatory clasper. *Figure 44*. Postabdomen.

23

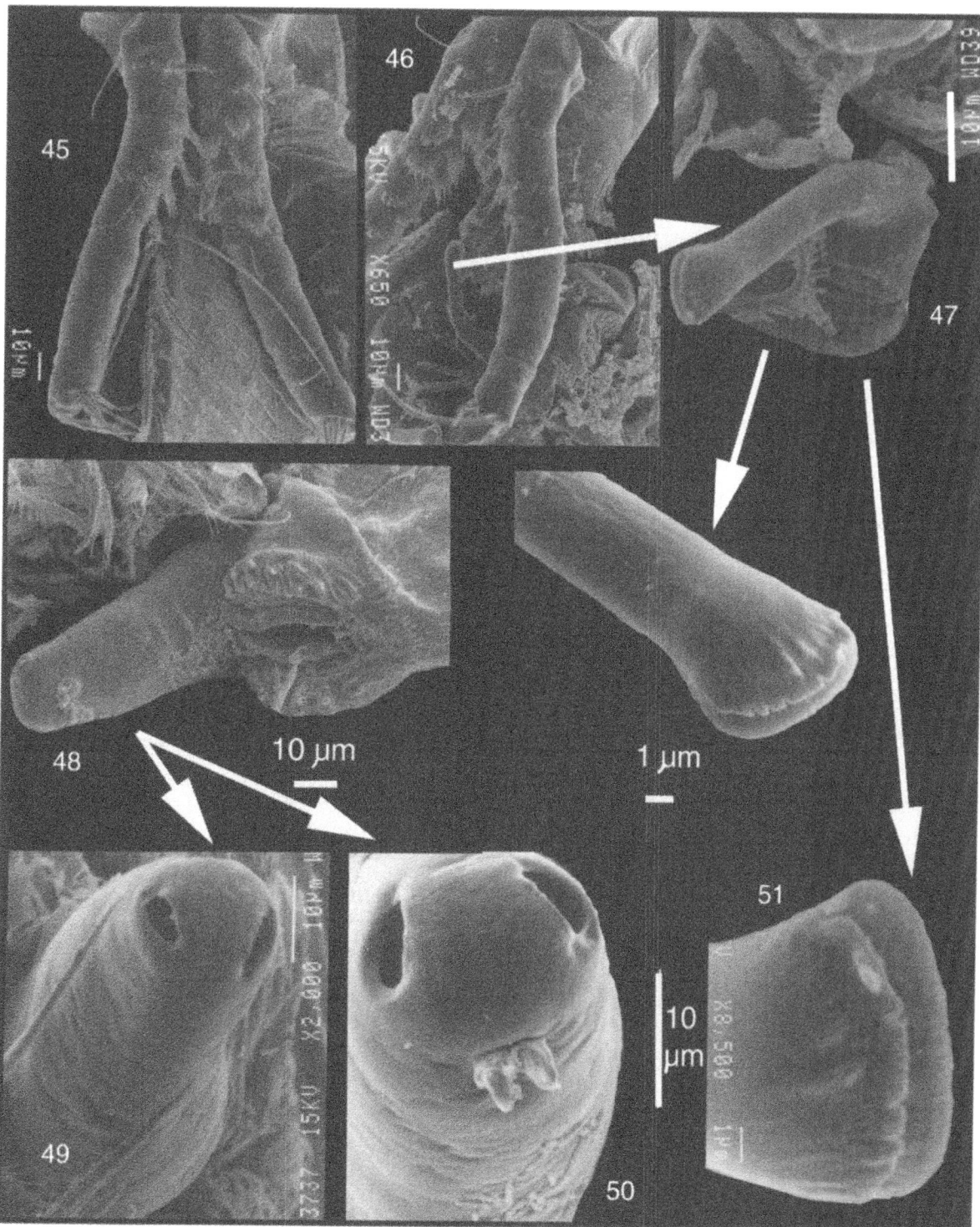

Figures 45–51. SEM's of *Macrothrix triserialis* male, Kerala. *Figure 45.* First antenna of male, showing the spinulation and the sensory setae. *Figure 46.* First antenna of male, showing the sensory setae, and the clasper (arrow). *Figure 47.* Copulatory hook (Clasper) of the male, showing the bifurcate apex. Below, detail of the apex of the clasper, showing small cuts in the edge, making it look like a pair of castanets. *Figure 48.* Postabdomen of the male, showing the anus, the rounded apex, and the two small leaflets below the apex. *Figure 49.* A view of the apex of the postabdomen, showing the openings of the sperm ducts. *Figure 50.* Detail of the same, showing the pseudo-claws below. *Figure 51.* Detail of the bifurcated apex of the copulatory clasper.

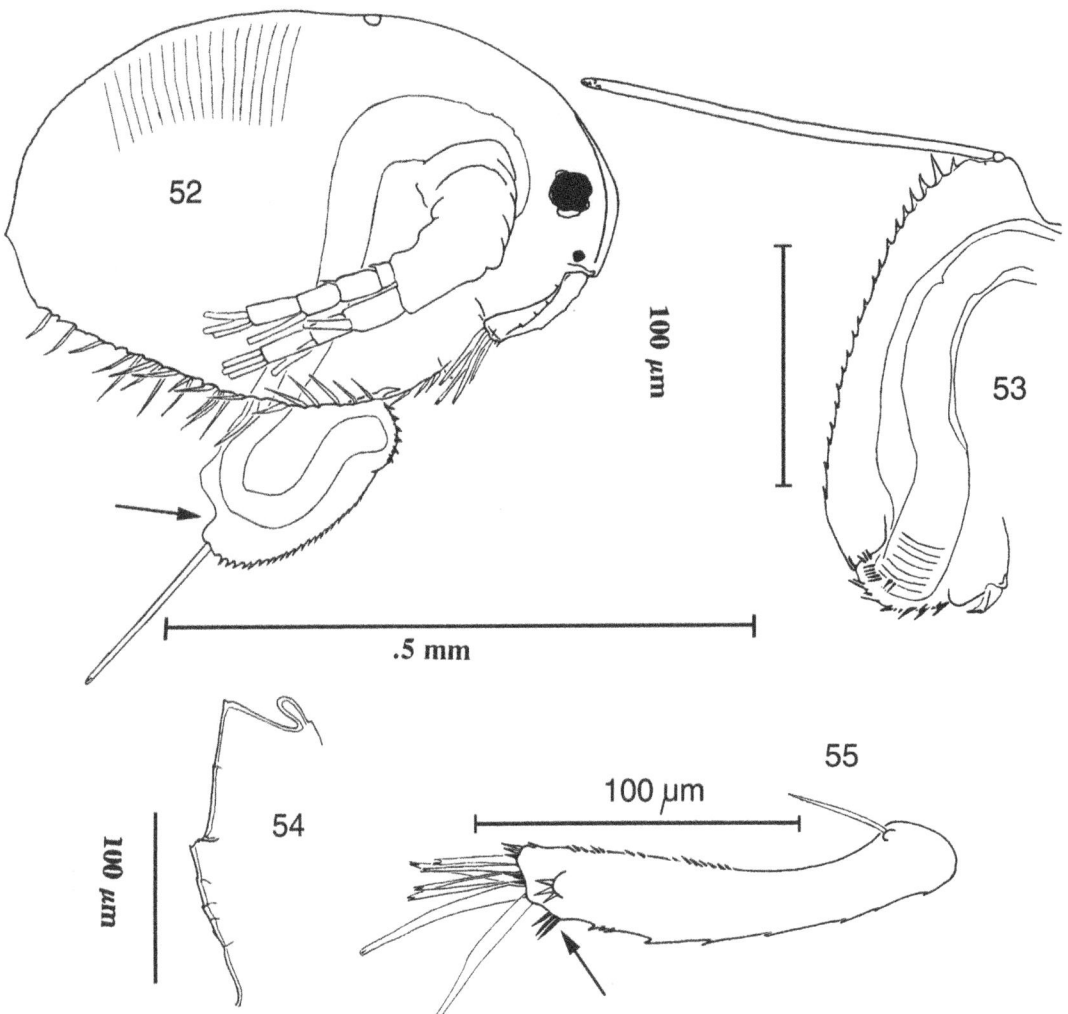

Figures 52–55. *Macrothrix* cf *triserialis* from Kenya. *Figure 52*. Female habitus. *Figure 53*. Postabdomen. *Figure 54*. Labrum. *Figure 55*. First antenna.

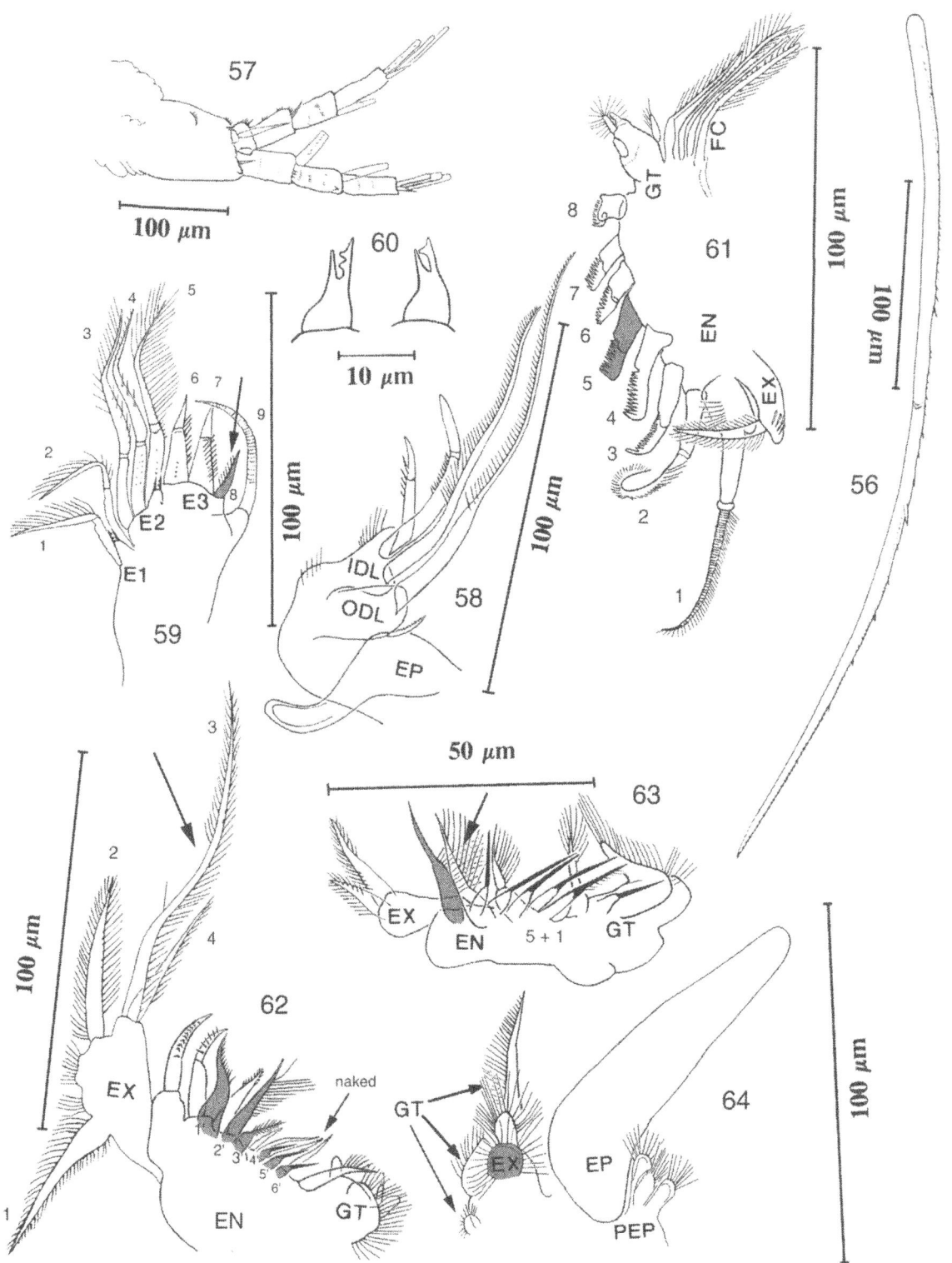

Figures 56–64. *Macrothrix* cf *triserialis* from Kenya. *Figure 56*. Long seta of second antenna. *Figure 57*. Second antenna. *Figure 58*. IDL, ODL and epipodite of first trunk limb. *Figure 59*. Endites 1–3 of first trunk limb. *Figure 60*. Fryer's forks. *Figure 61*. Second trunk limb. *Figure 62*. Third trunk limb. *Figure 63*. Fourth trunk limb. *Figure 64*. Fifth trunk limb.

26

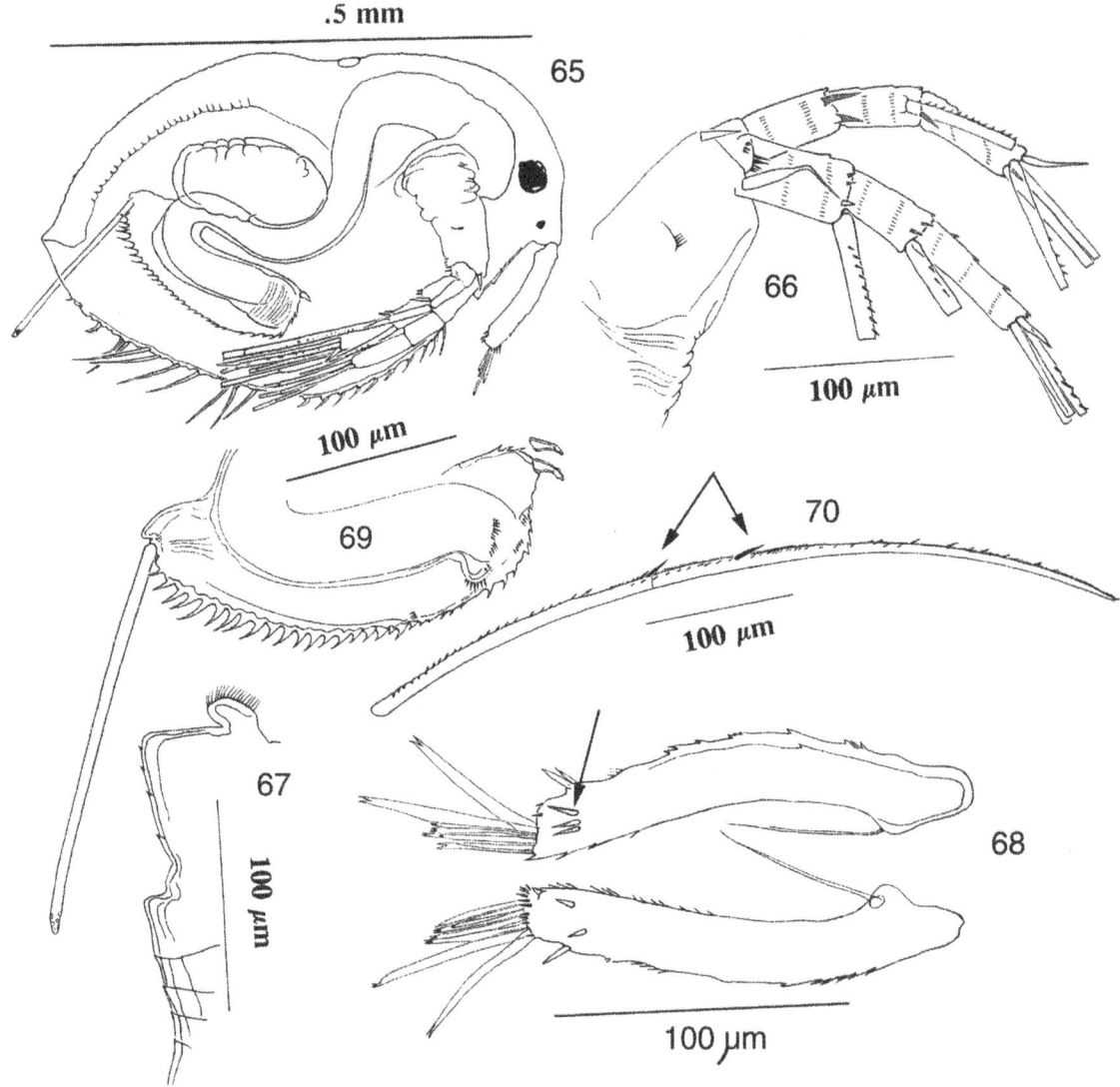

Figures 65–70. Macrothrix elegans from Sao Paulo State, Brasil. *Figure 65*. Female habitus. *Figure 66*. Second antenna. *Figure 67*. Labrum. *Figure 68*. Frist antenna. *Figure 69*. Postabdomen. *Figure 70*. Long seta of second antenna.

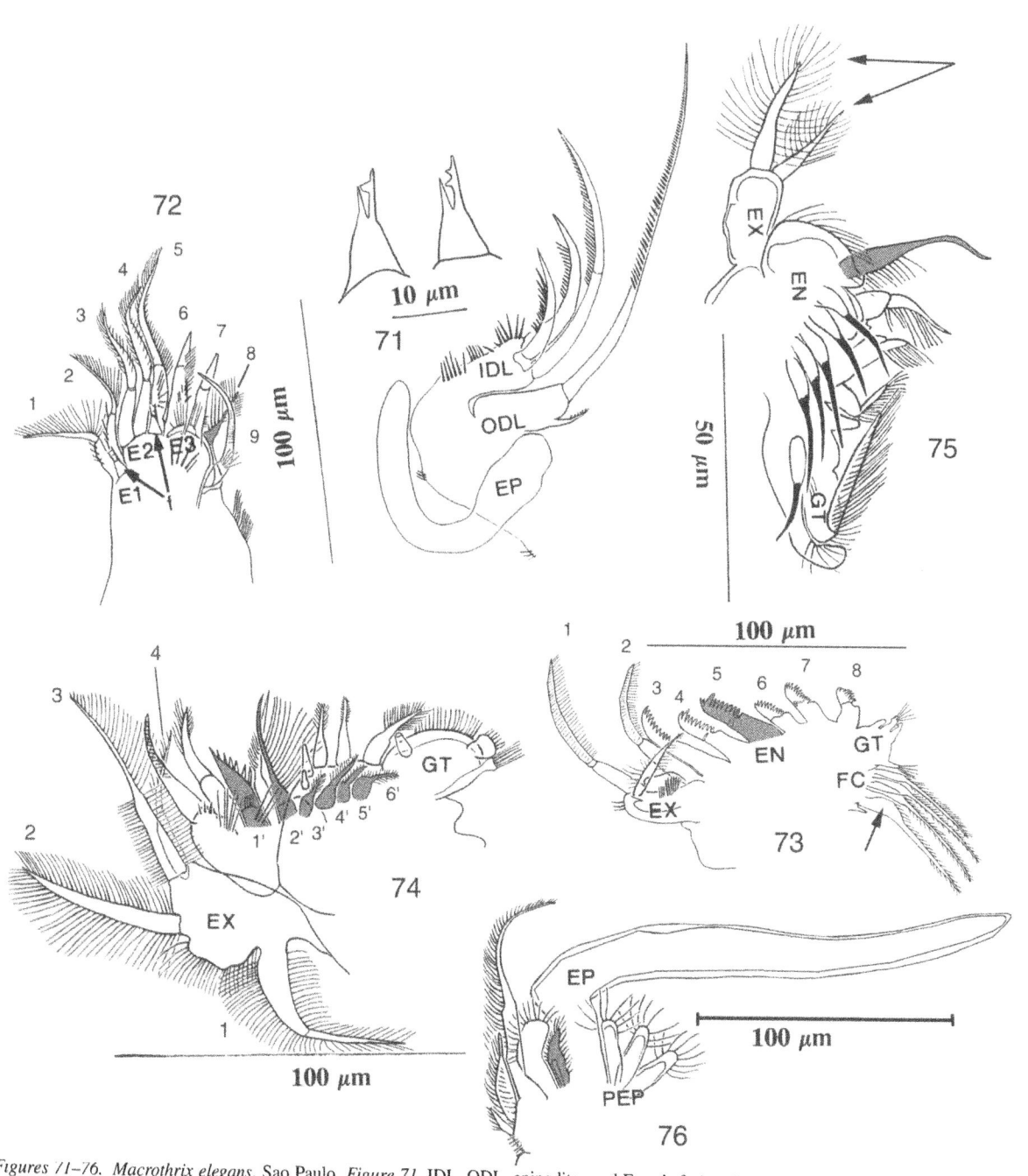

Figures 71–76. Macrothrix elegans, Sao Paulo. *Figure 71*. IDL, ODL, epipodites and Fryer's forks of first trunk limb. *Figure 72*. Endites 1–3 of first trunk limb. *Figure 73*. Second trunk limb. *Figure 74*. Third trunk limb. *Figure 75*. Fourth trunk limb. *Figure 76*. Fifth trunk limb.

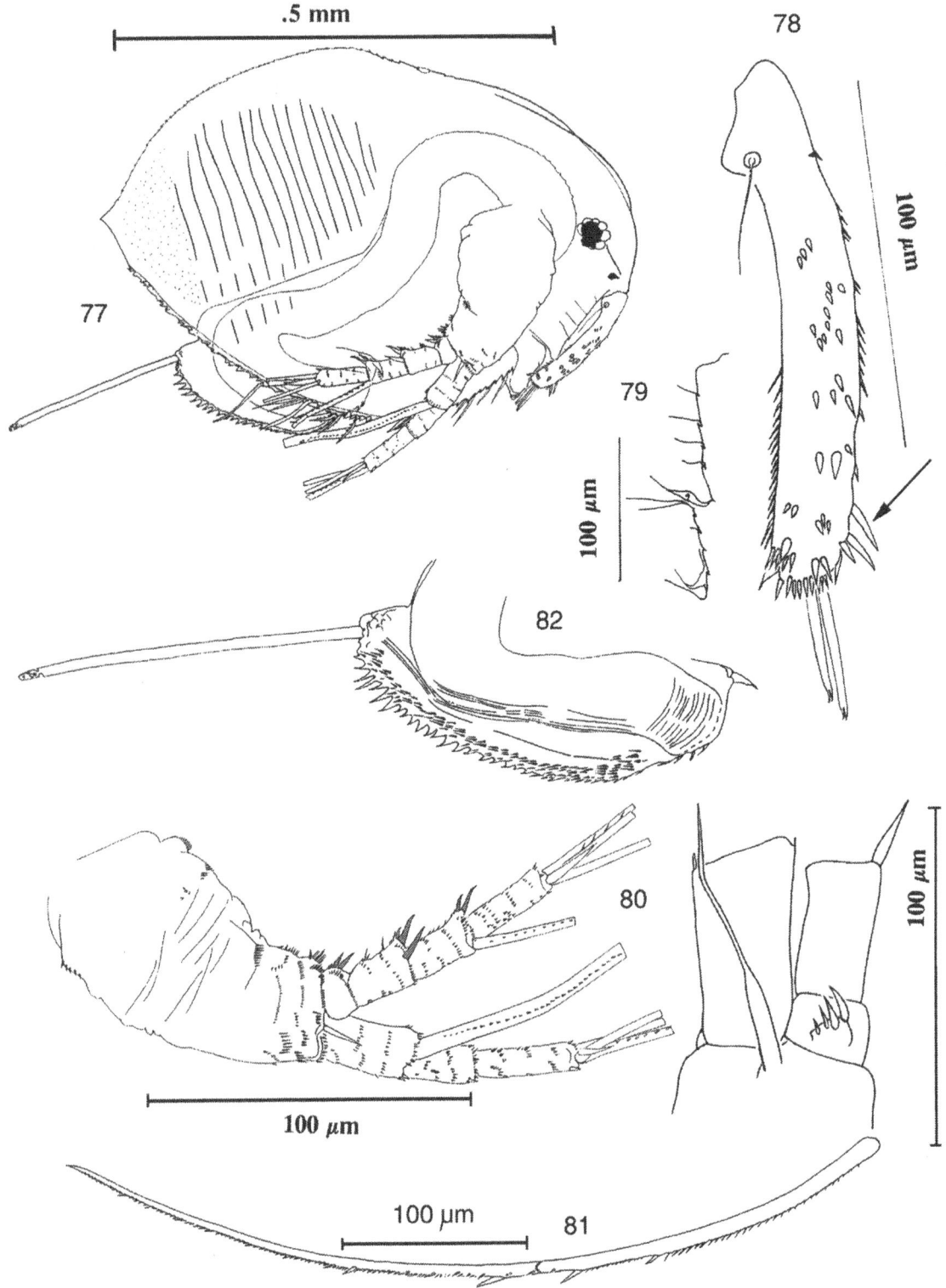

Figures 77–82. Macrothrix cf *triserialis* from Mexico. *Figure 77.* Female habitus. *Figure 78.* First antenna. *Figure 79.* Labrum. *Figure 80.* Second antenna. *Figure 81.* Long seta of second antenna. *Figure 82.* Postabdomen.

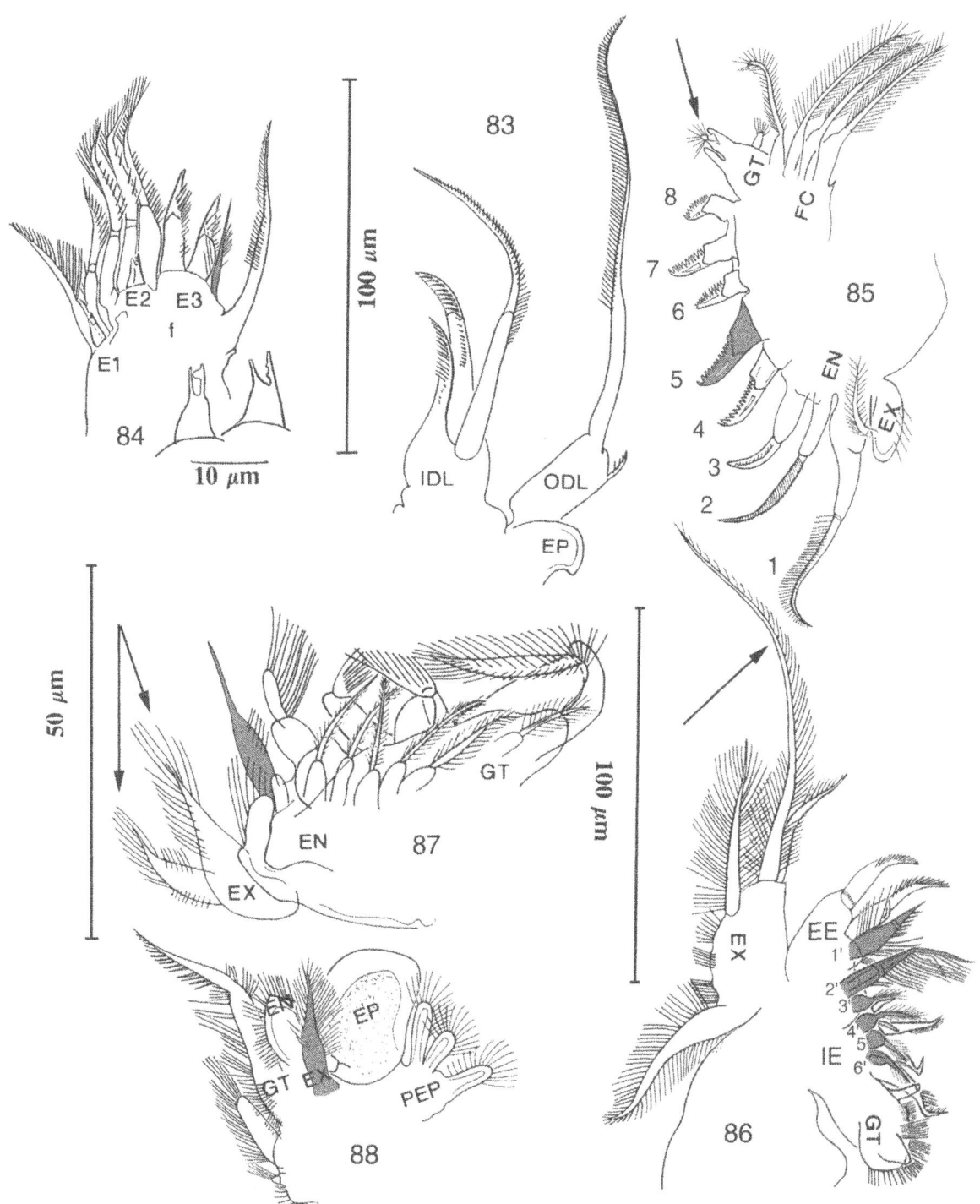

Figures 83–88. Macrothrix cf *triserialis*, Mexico. *Figure 83*. IDL, ODL and epipodite of trunk limb one. *Figure 84*. Endites 1–3 of trunk limb one. *Figure 85*. Second trunk limb. *Figure 86*. Third trunk limb. *Figure 87*. Fourth trunk limb. *Figure 88*. Fifth trunk limb.

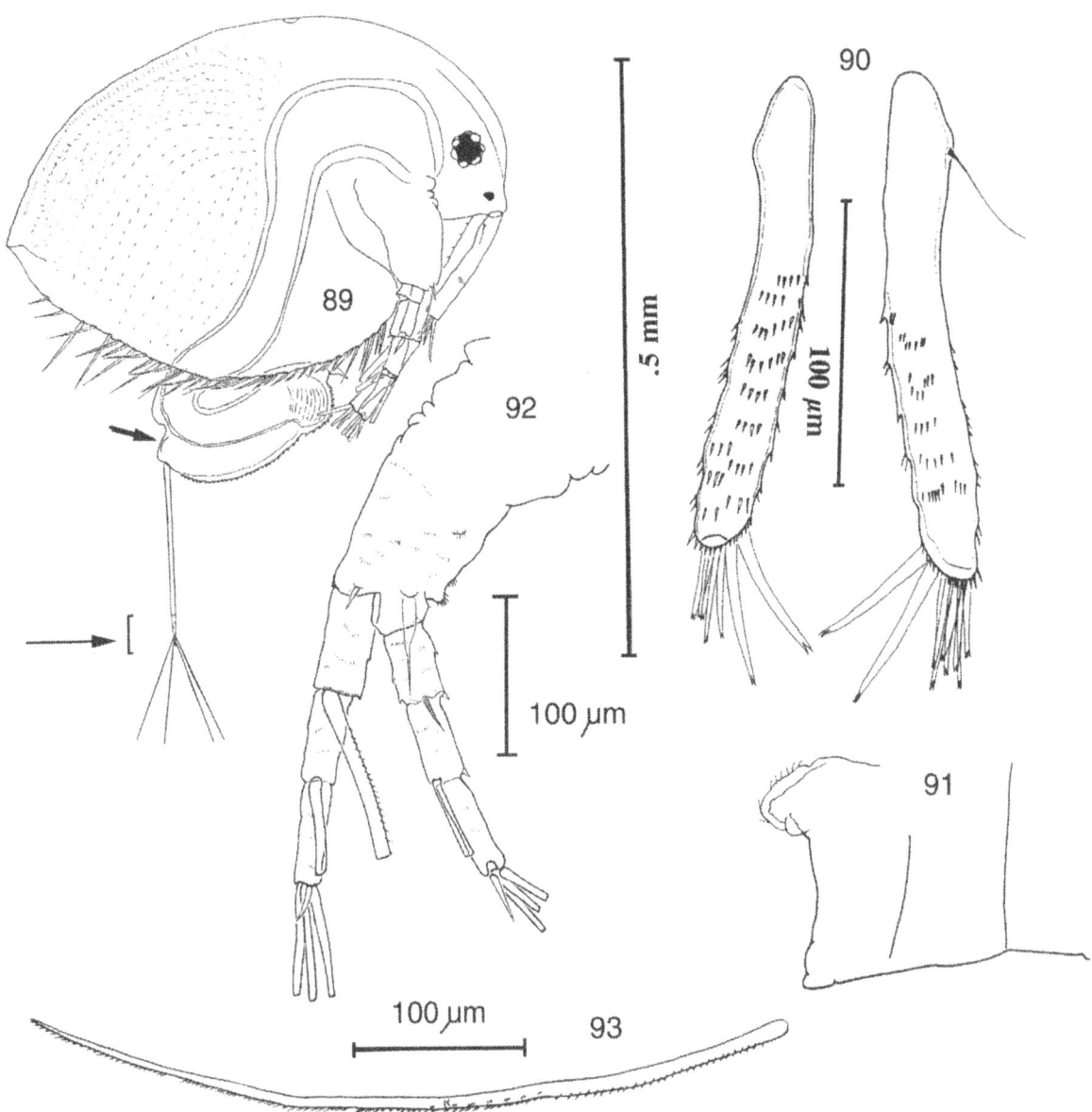

Figures 89–93. *Macrothrix tabrizensis* n. sp. from Iran. *Figure 89.* Female habitus. *Figure 90.* First antenna. *Figure 91.* Labrum. *Figure 92.* Second antenna. *Figure 93.* Long seta of second antenna.

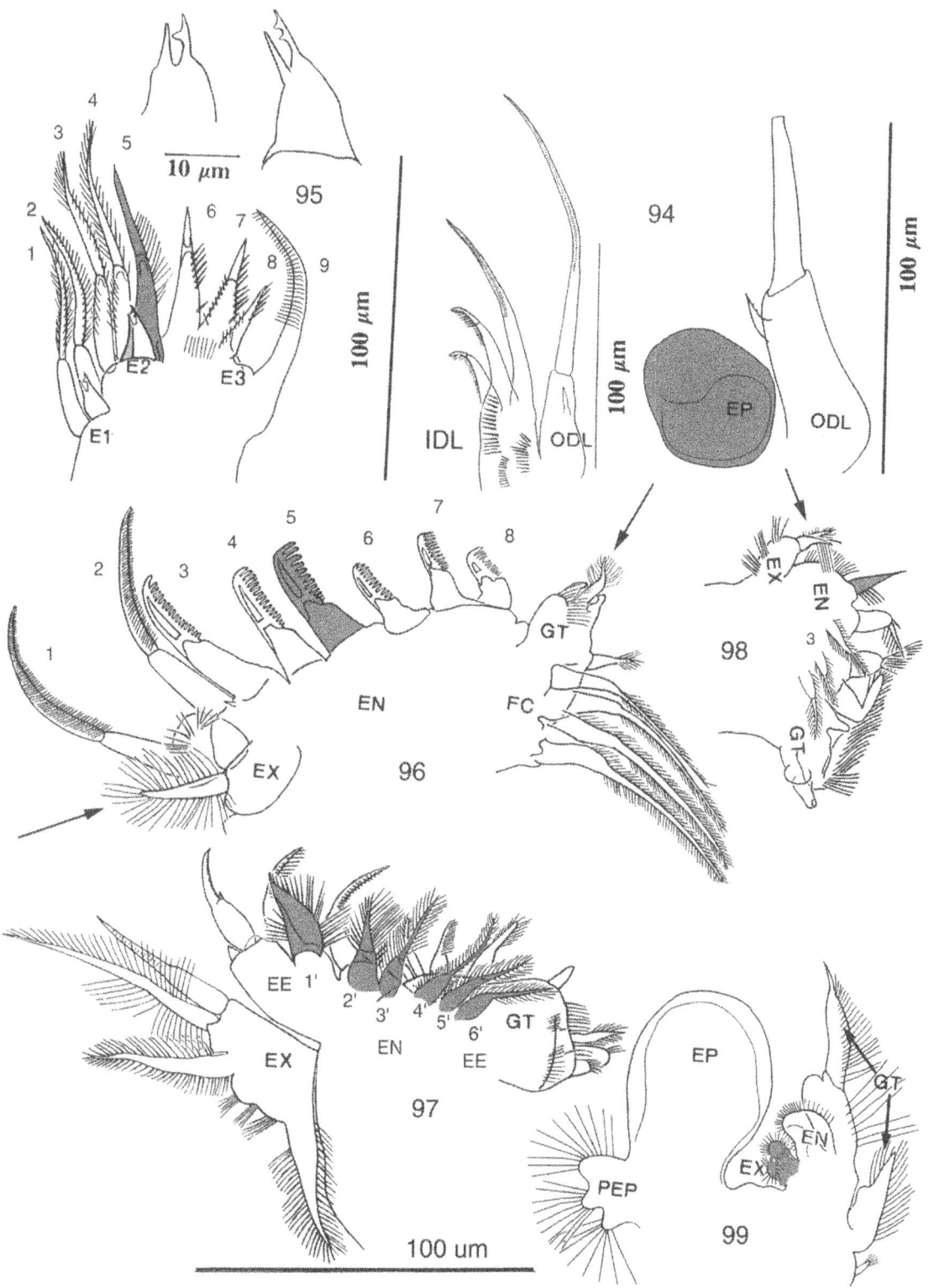

Figures 94–99. Macrothrix tabrizensis n. sp., Iran. *Figure 94*. IDL, ODL and epipodite of trunk limb one. *Figure 95*. Endites 1–3 and Fryer's forks on trunk limb one. *Figure 96*. Second trunk limb. *Figure 97*. Third trunk limb. *Figure 98*. Fourth trunk limb. *Figure 99*. Fifth trunk limb.

32

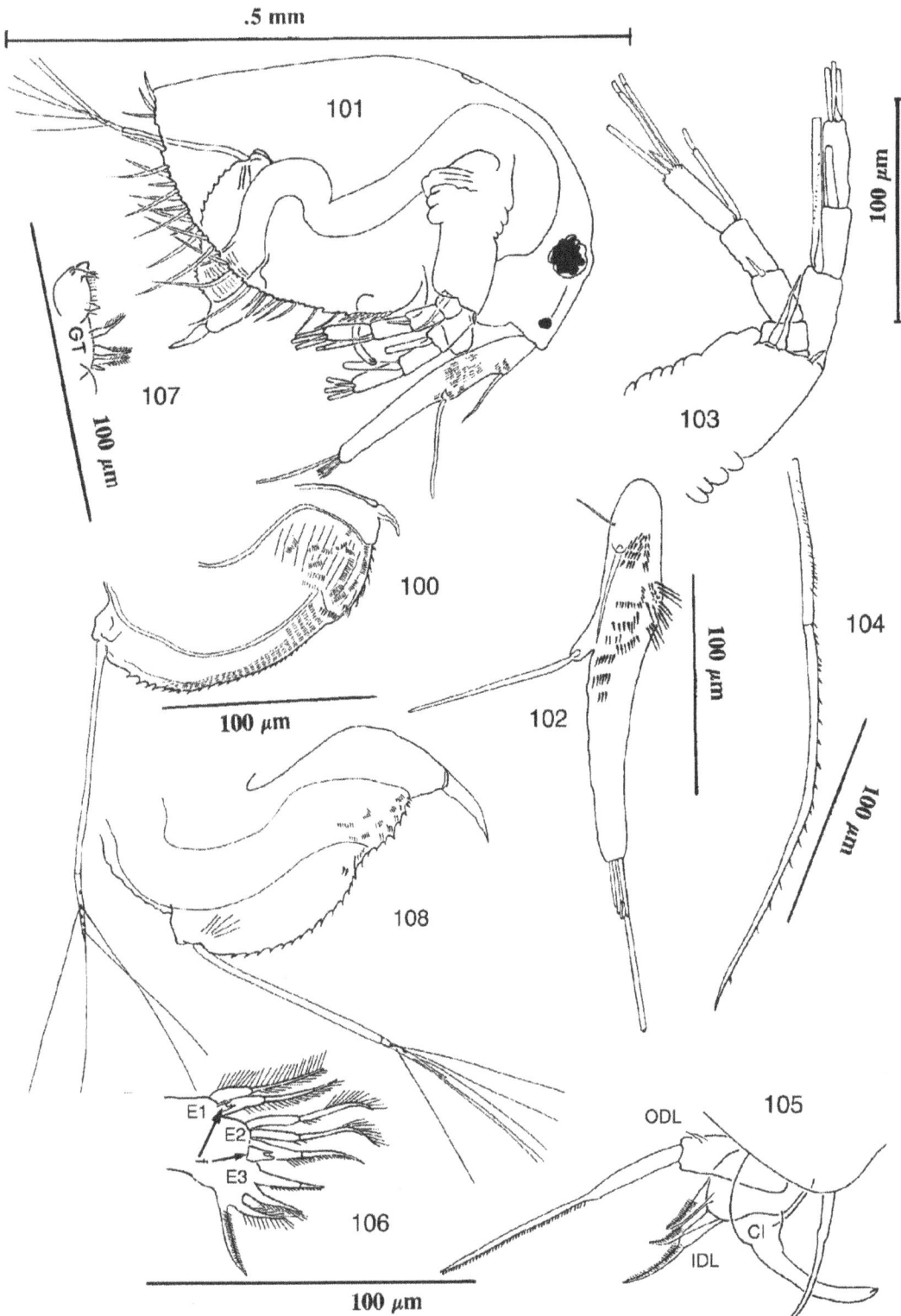

Figures 100–108. Macrothrix tabrizensis n. sp., Iran. *Figure 100.* Female postabdomen. *Figure 101.* Male habitus. *Figure 102.* Male first antenna. *Figure 103.* Female second antenna. *Figure 104.* Long seta of second antenna. *Figure 105.* Male IDL, ODL and clasper. *Figure 106.* Male first trunk limb (endites 1–3). *Figure 107.* Male third trunk limb, gnathobase and adjacent setae. *Figure 108.* Male postabdomen.

33

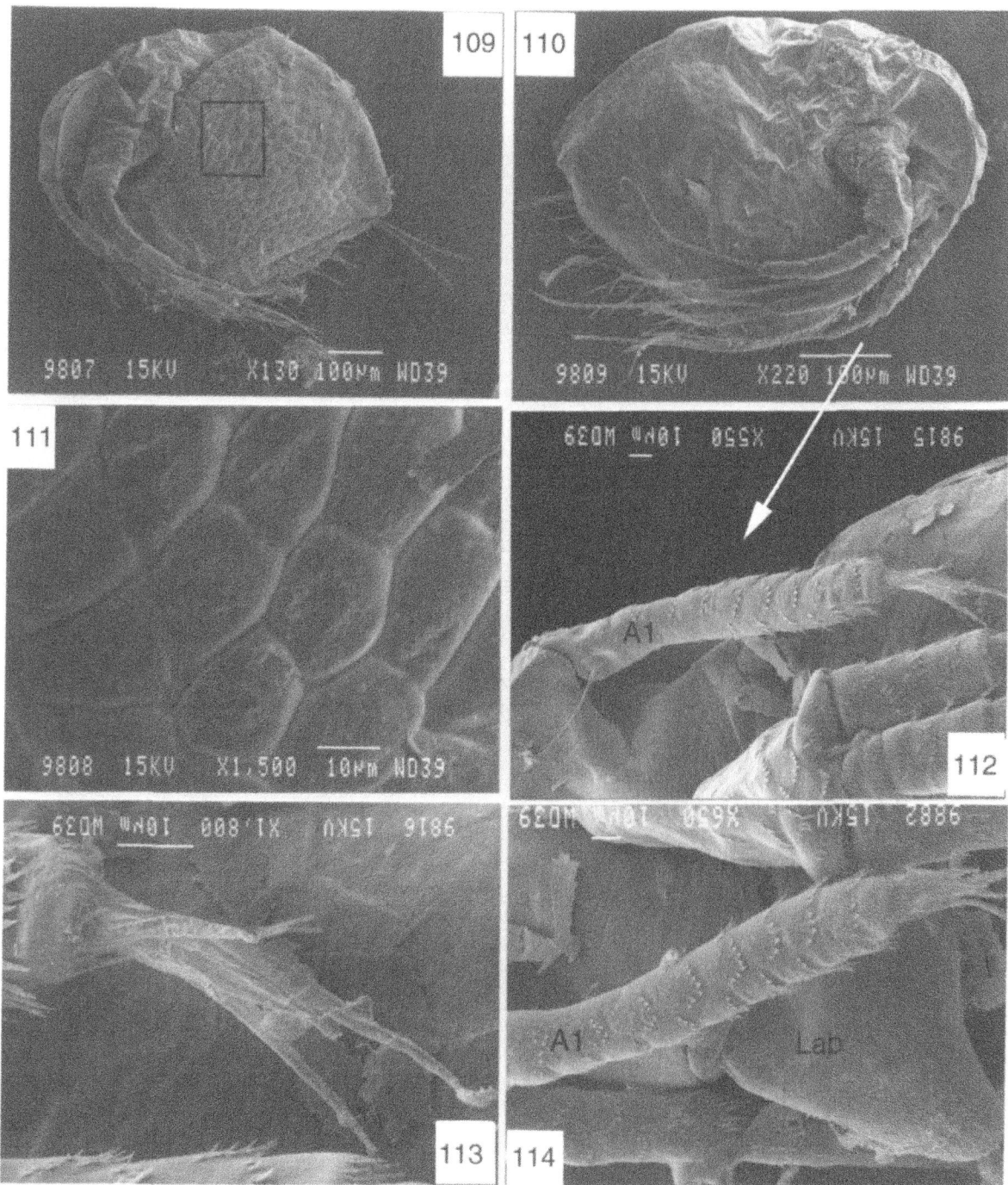

Figures 109–114. *Macrothrix tabrizensis* n. sp. SEM's of gamogenetic female. *Figure 110*. Habitus. *Figure 111*. Ornamentation of surface of ephippium. *Figure 112*. First antenna. *Figure 113*. First antenna, showing long and short aesthetascs. *Figure 114*. First antenna and labrum.

34

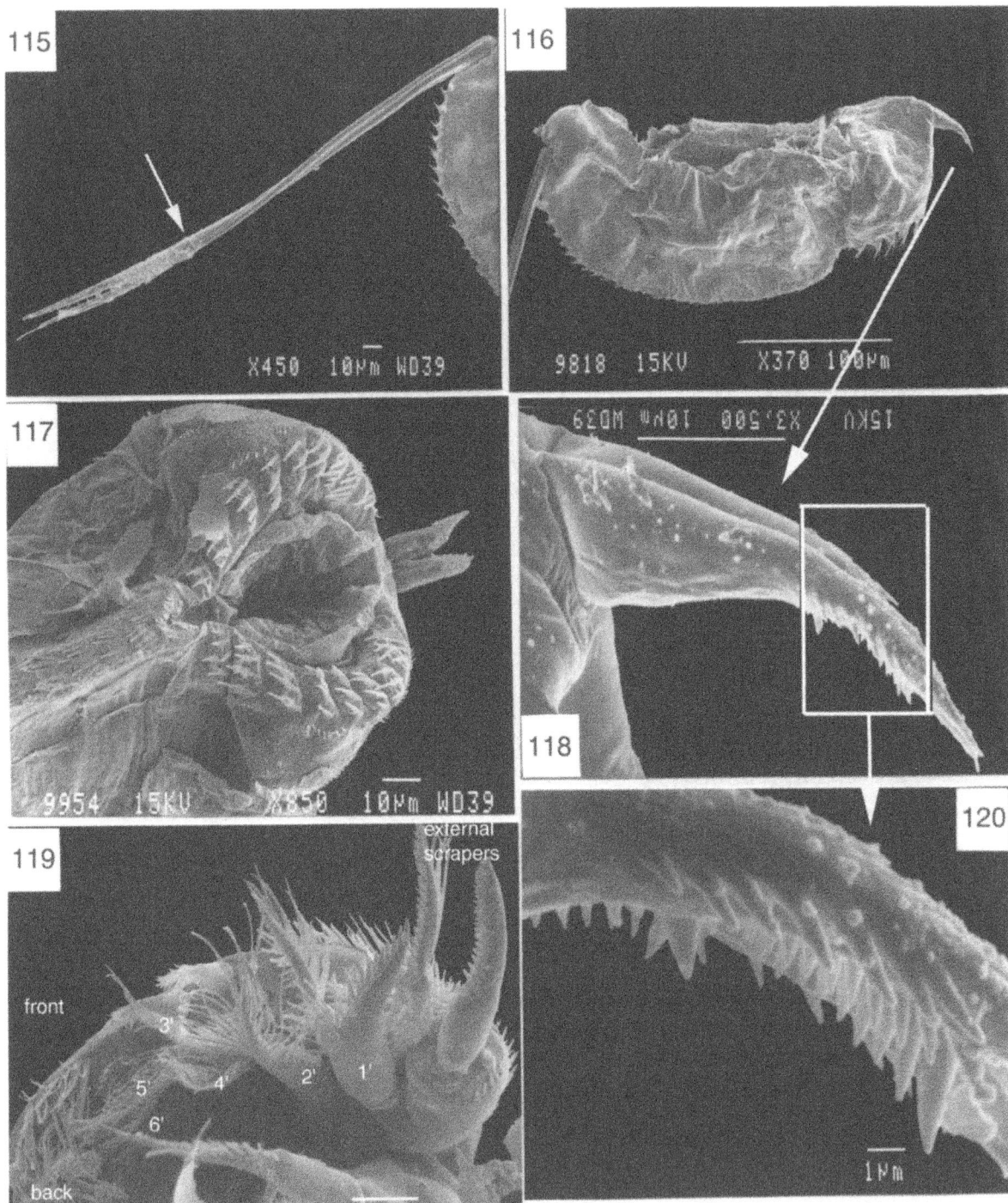

Figures 115–120. Macrothrix tabrizensis n. sp. *Figure 115*. Seta natatoria with articulation of distal segment arrowed. *Figure 116*. Postabdomen. *Figure 117*. Apex of postabdomen, showing anus. *Figure 118*. End-claw, showing lateral row of teeth. *Figure 119*. External zone of endopodite of trunk limb three, showing scrapers and posterior setae. *Figure 120*. Detail of lateral row of teeth of end-claw.

35

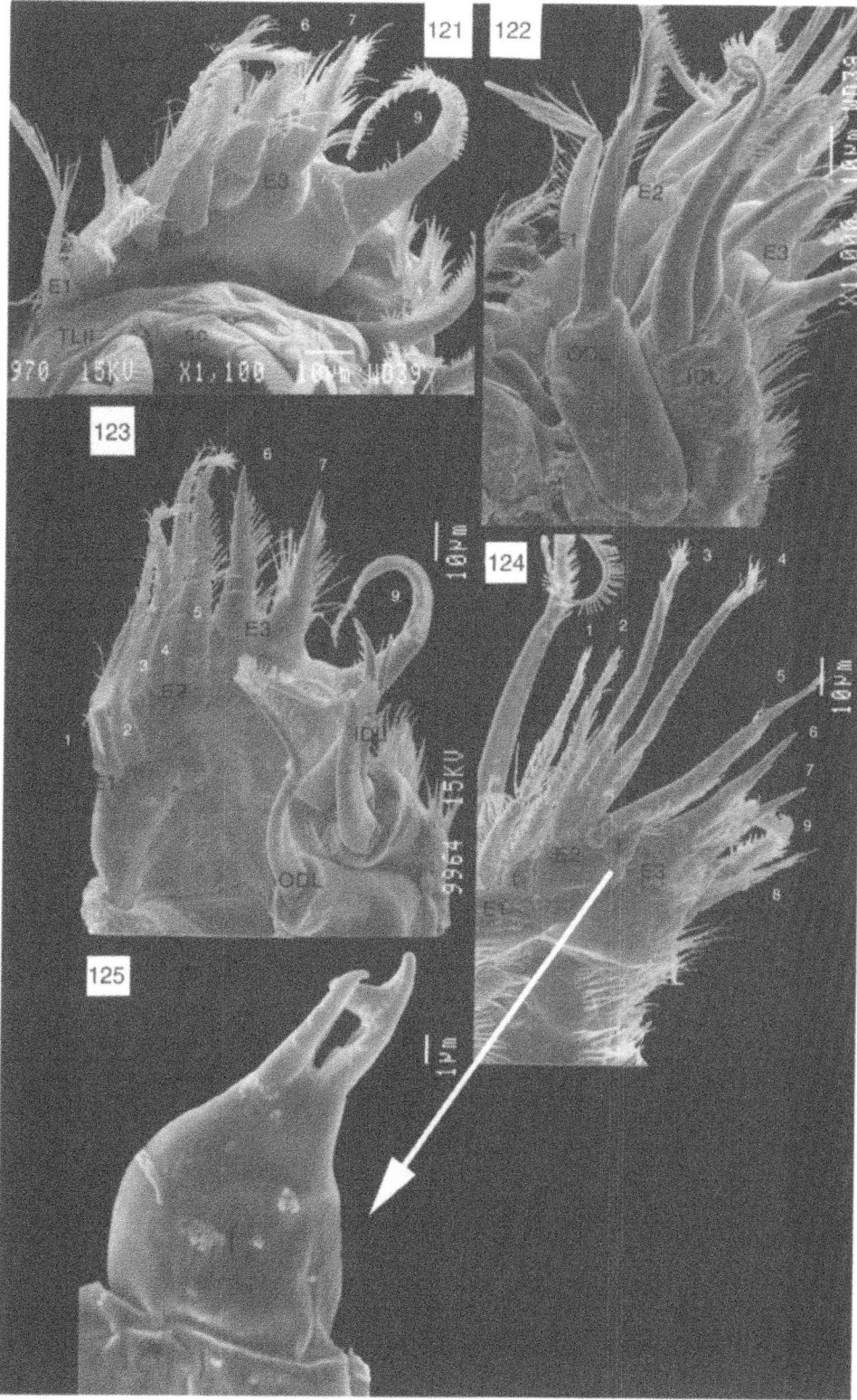

Figures 121–125. Macrothrix tabrizensis n. sp., SEM's of trunk limb one. *Figure 121*. Trunk limb one showing Endites 1–3. Trunk limb two is on the foreground. Seta eight is hidden behind seta seven. *Figure 122*. Trunk limb one, showing ODL, IDL, and endites 1–3 in the background. *Figure 123*. Trunk limb one, showing endites 1–3, ODL and IDL. *Figure 124*. Trunk limb one, showing endites 1–3 and forks (f). Note the shortened seta five. *Figure 125*. Fryer's fork on endite two, with a single big tooth, enlarged.

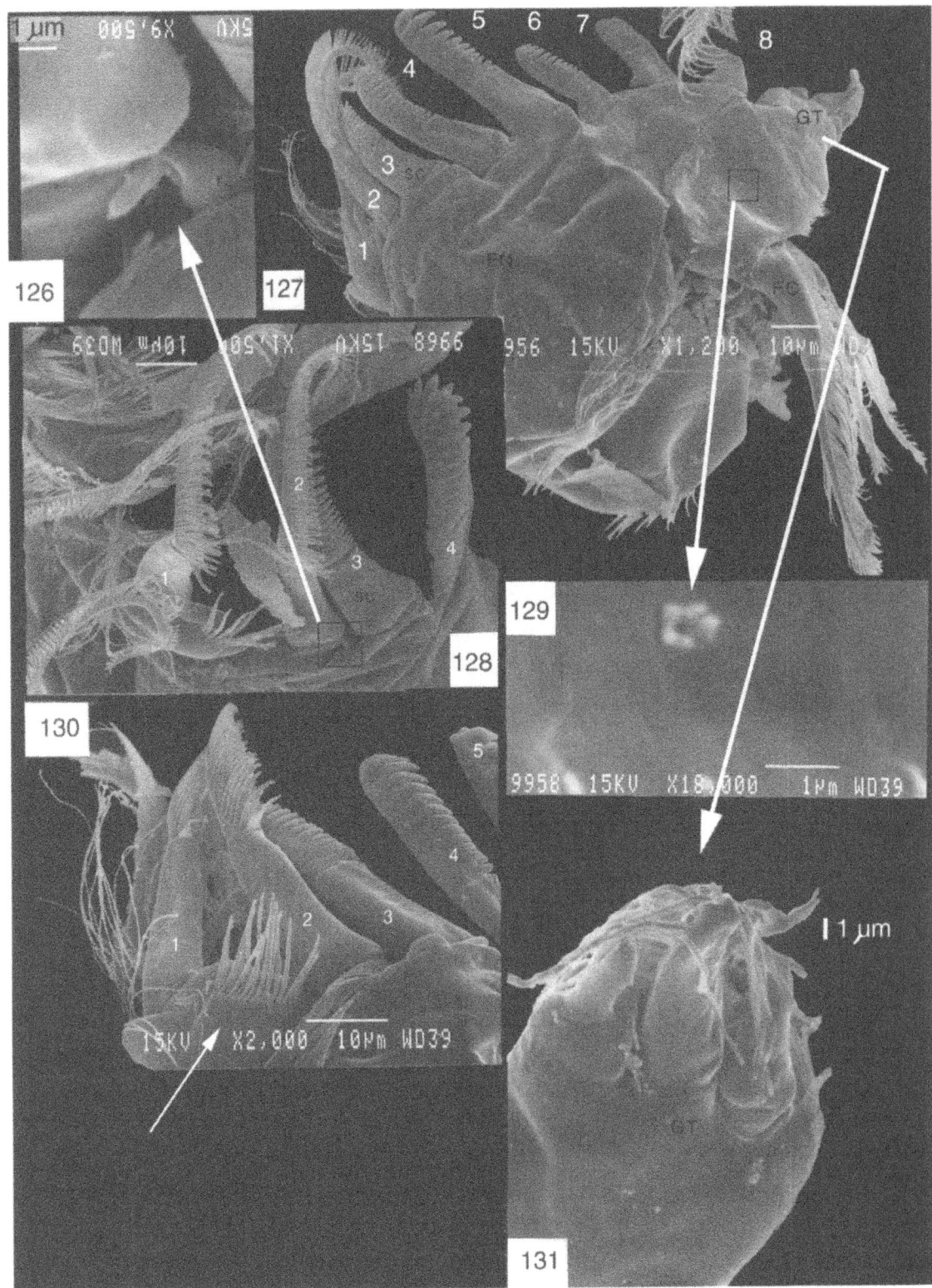

Figures 126–131. Macrothrix tabrizensis n. sp., Iran. SEM's of second trunk limb. *Figure 126.* Small sensory receptor (r) between scrapers 2 and 3 (an unusual location) of second trunk limb. *Figure 127.* Anterior surface of trunk limb two, showing scrapers, Endopod, Gnathobase, and Filter Comb. The square indicates the location of a cuticular pore. *Figure 128.* Trunk limb two, showing scrapers 1–4, and the location of the receptor. *Figure 129.* Pore on gnathobase of trunk limb two, surrounded by a cuticular rosette, detail. *Figure 130.* Alternative view of trunk limb two, showing scraper one, implanted on a remnant of an apically setulated endital 'segment' (arrowed). *Figure 131.* Gnathobase of limb two, showing the three apical appendages.

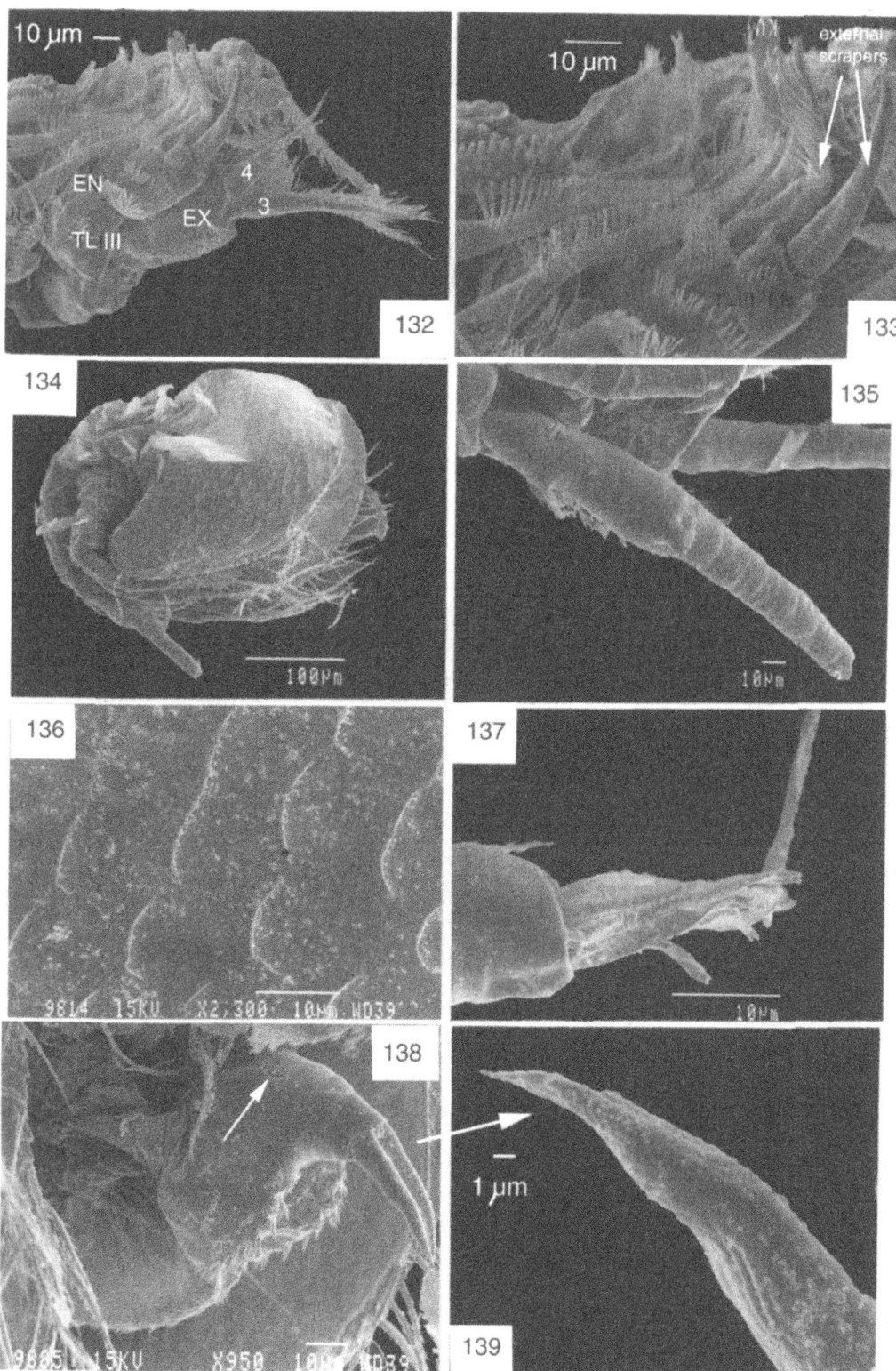

Figures 132–139. *Macrothrix tabrizensis* n. sp., Iran. *Figure 132*. SEM's of trunk limb 3, showing Endopod (EN), and Exopod (EX) with serrate seta four and plumose seta three. *Figure 133*. External scrapers of trunk limb three, and some overlaying scrapers of trunk limb two. *Figure 134*. Male. Habitus. *Figure 135*. Male, first antenna. *Figure 136*. Ornamentation of the male valve. *Figure 137*. Male first antenna, showing bifurcated tips of aesthetascs. *Figure 138*. Male postabdomen, showing the end-claw, the marginal rows of teeth, and a sperm duct (arrow). *Figure 139*. Male postabdomen, detail of glabrous end-claw.

38

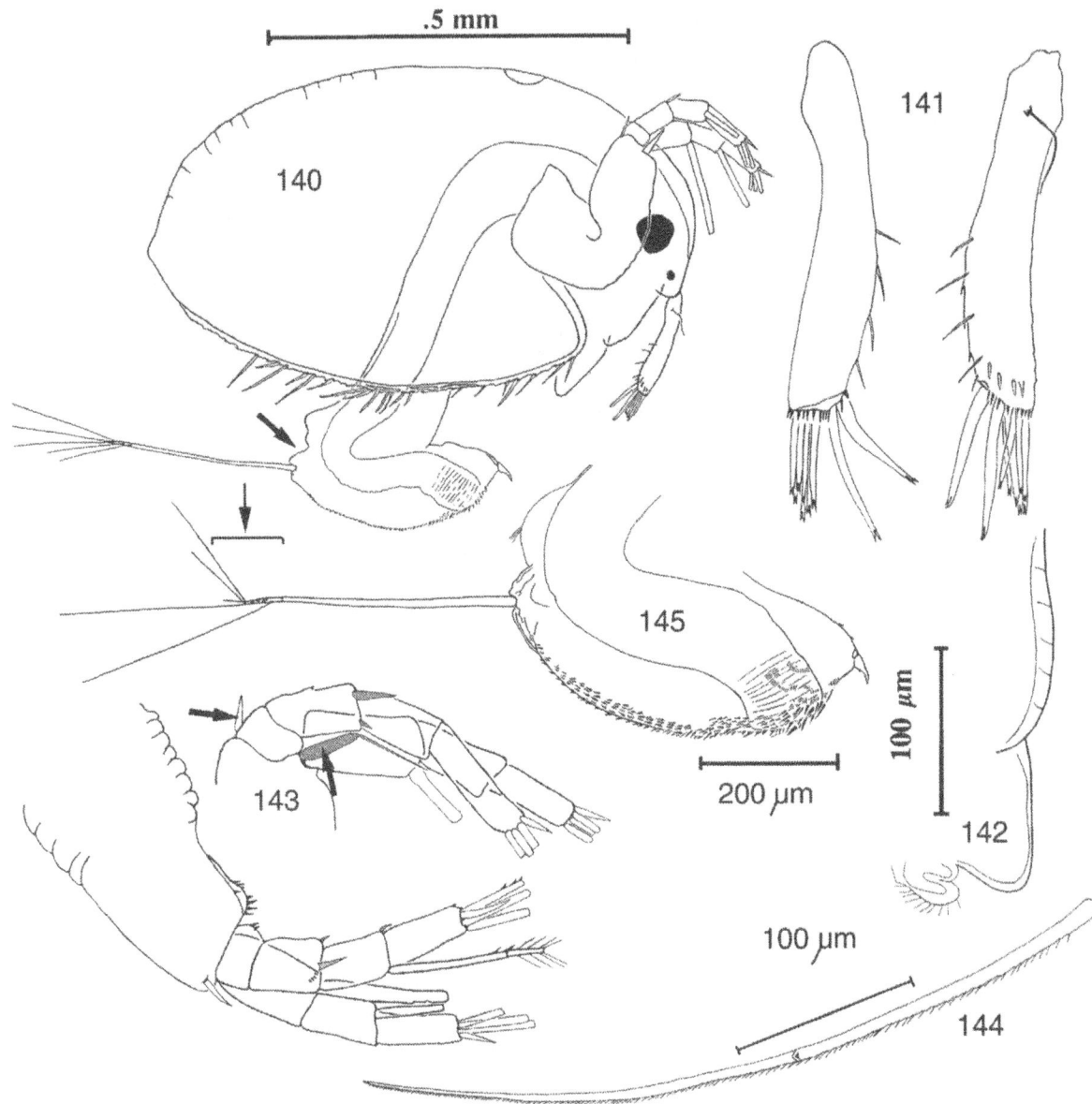

Figures 140–145. *Macrothrix agsensis* n. sp., Mexico. *Figure 140.* Female habitus. *Figure 141.* First antenna. *Figure 142.* Labrum. *Figure 143.* Second antenna. *Figure 144.* Long seta of second antenna. *Figure 145.* Postabdomen.

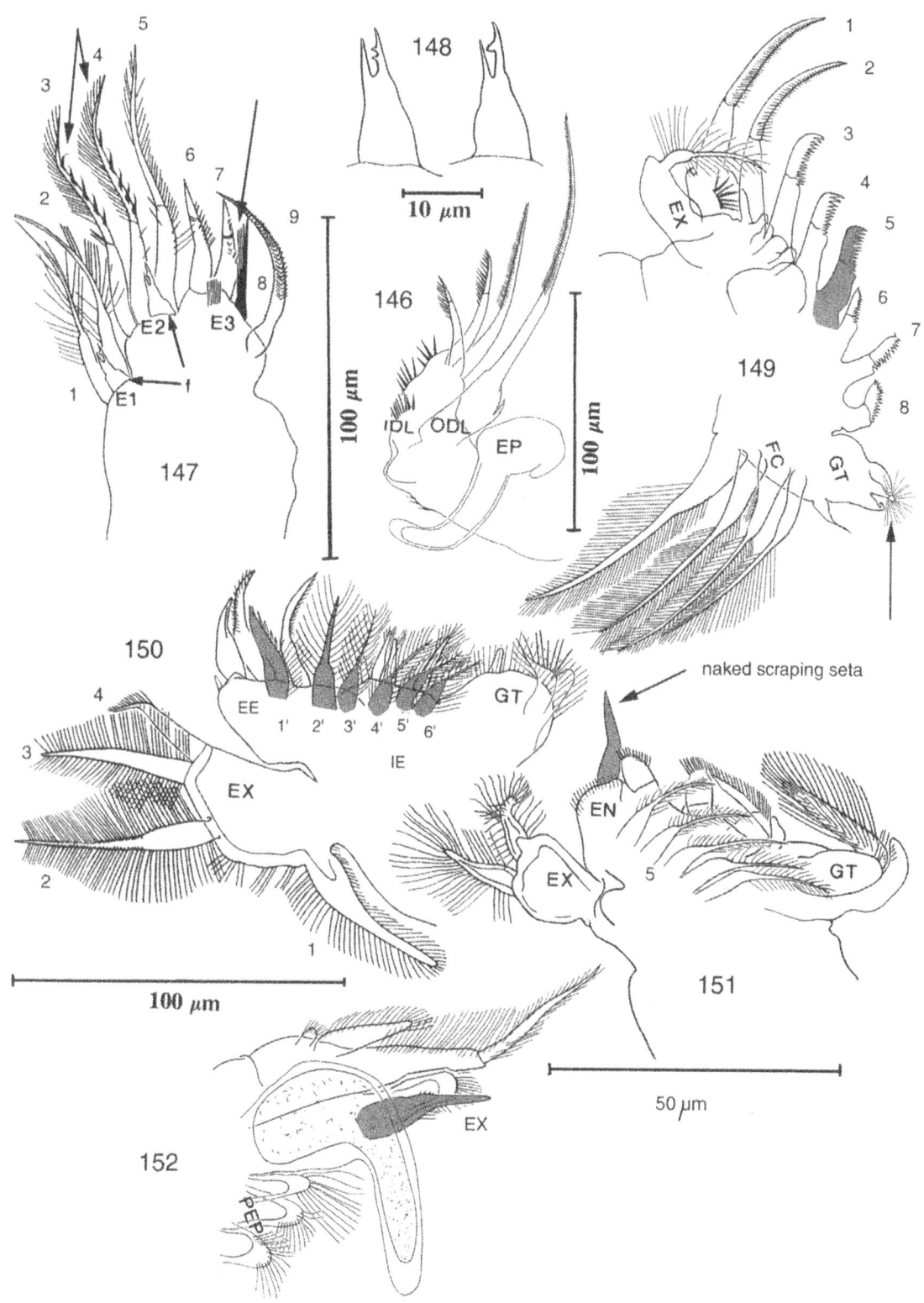

Figures 146–152. Macrothrix agsensis n. sp., Mexico. *Figure 146.* IDL, ODL and epipodite of first trunk limb. *Figure 147.* Endites 1–3 of first trunk limb. *Figure 148.* Fryer's forks. *Figure 149.* Second trunk limb. *Figure 150.* Third trunk limb. *Figure 151.* Fourth trunk limb. *Figure 152.* Fifth trunk limb.

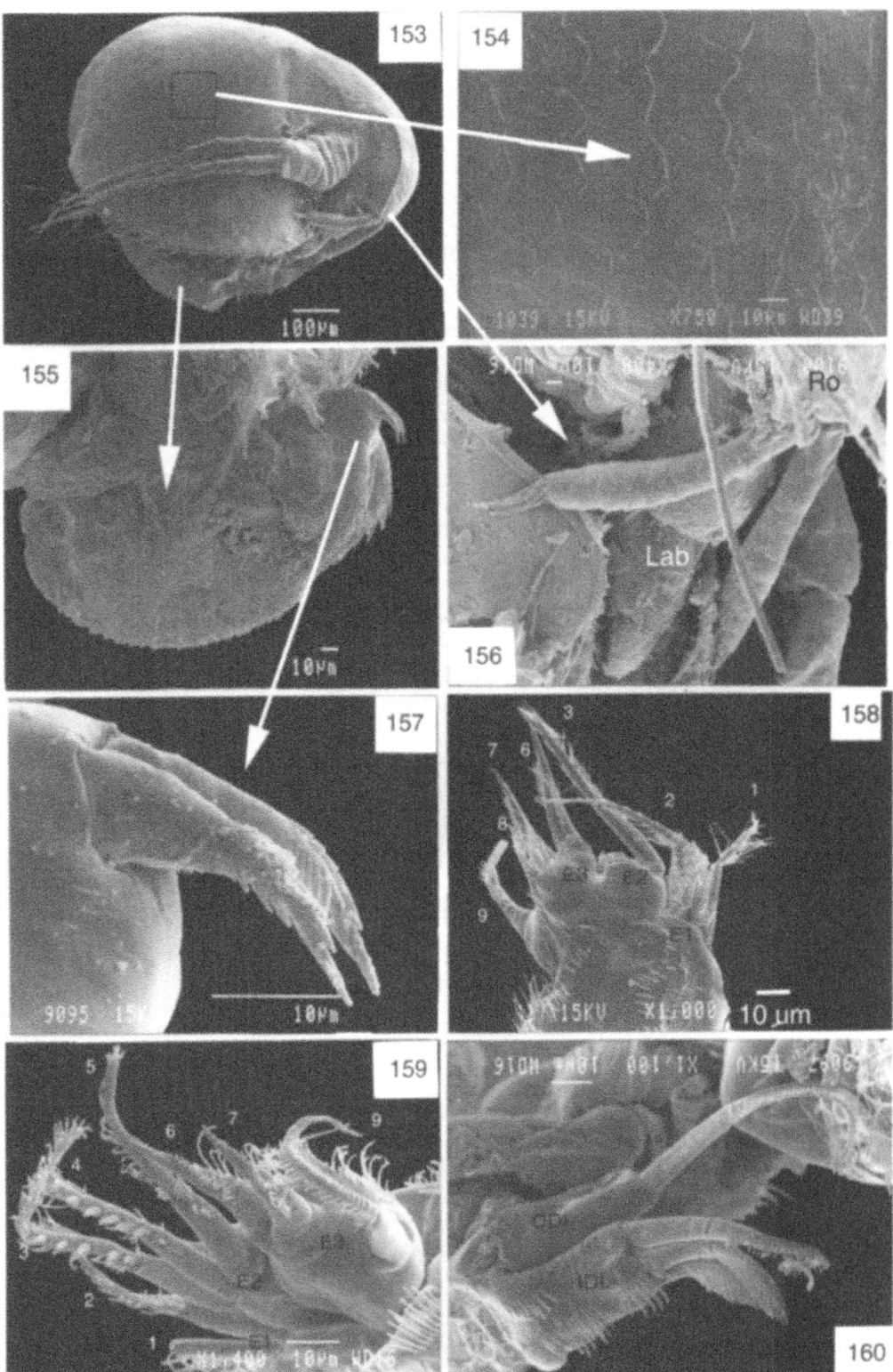

Figures 153–160. Macrothrix agsensis n. sp., Mexico. SEM's. *Figure 153.* Habitus of parthenogenetic female. Small square indicates the ornamentation of the valve in lateral view. *Figure 154.* Ornamentation of the valve. *Figure 155.* Postabdomen, lateral view. *Figure 156.* First antenna. *Figure 157.* Postabdomen, showing the end-claw. *Figure 158.* Trunk limb one, showing endites 1–3, and fork (f) on endite one. Two setae of endite two are broken off. *Figure 159.* Trunk limb one, showing endites 1–3 and the specific spines along setae three and four. *Figure 160.* Trunk limb one, showing ODL and IDL.

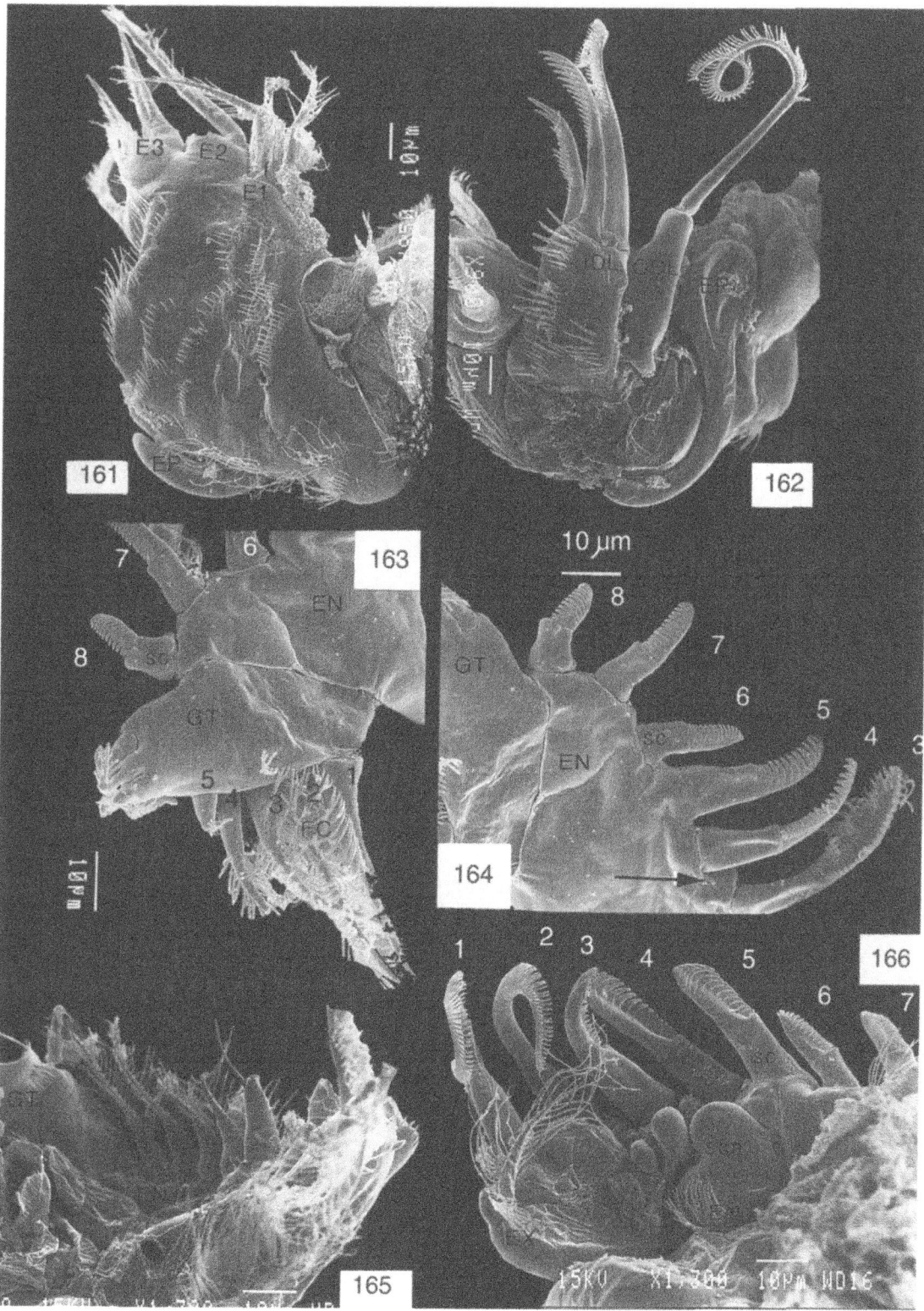

Figures 161–166. Macrothrix agsensis n. sp., Mexico. SEM's of trunk limbs. *Figure 161.* Trunk limb one, showing the spinulated corm, the endites 1–3, Fryer's fork (with one tooth) on endite one, and the long epipodite that lies twisted around the corm. *Figure 162.* Trunk limb one, showing IDL, ODL and the twisted epipodite. *Figure 163.* Trunk limb two, showing the Filter Comb, Gnathobase, and Endopod with scrapers. *Figure 164.* Trunk limb two, mainly showing the row of scrapers. Note a small sensillum between scrapers three and four, but no receptor behind scraper eight, and no doublings behind scrapers four and five. *Figure 165.* Trunk limb three, showing gnathobase and five out of the row of six posterior setae; seta 2′ is partly broken off. *Figure 166.* Posterior view of trunk limb two, showing the exopod and its inwardly curved plumose seta, the endopod with its row of scrapers, and the three rounded hillocks behind scrapers 2–4.

42

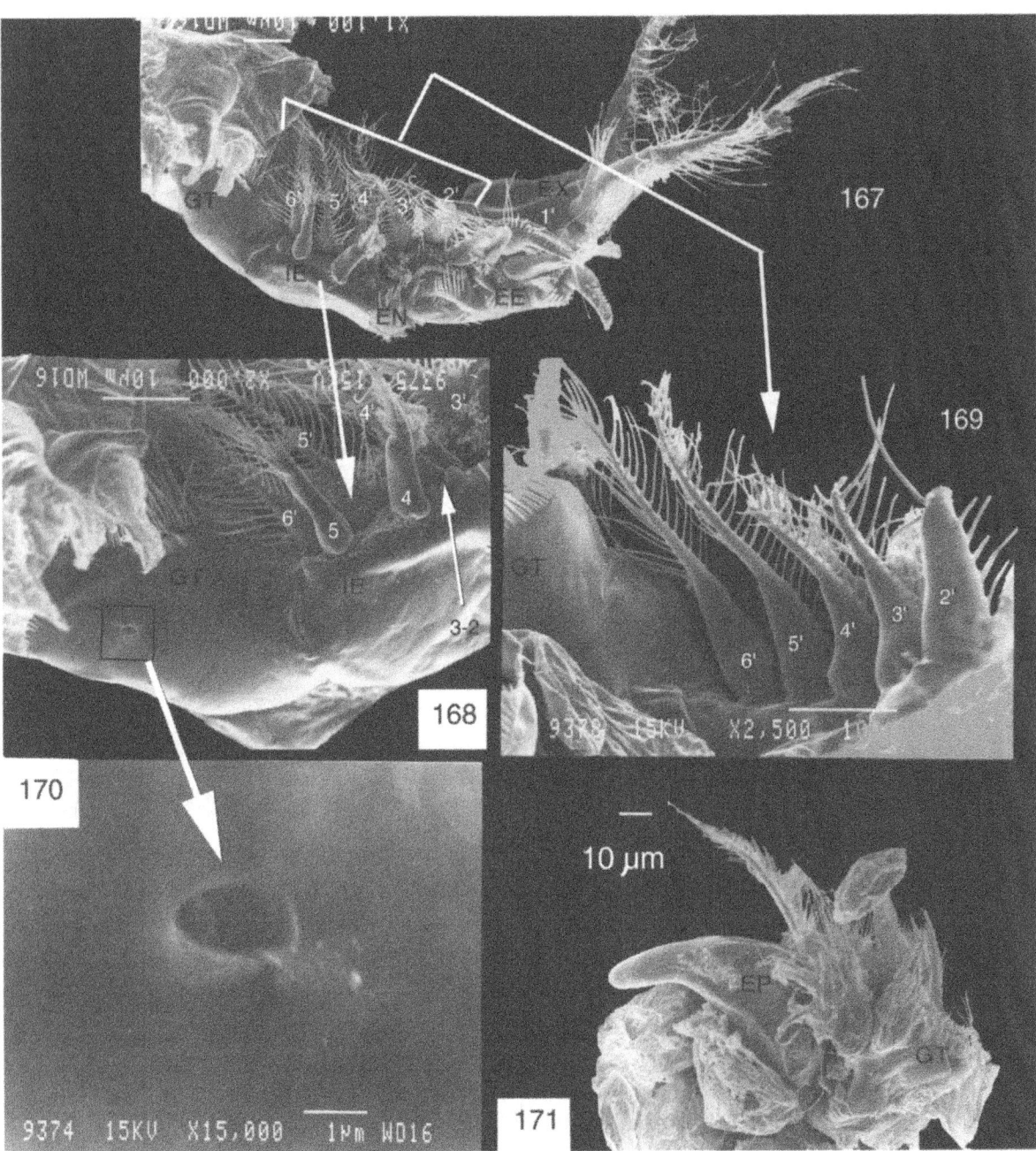

Figures 167–171. Macrothrix agsensis n. sp., Mexico. SEM's of trunk limbs. *Figure 167.* An overview of trunk limb three, showing part of the the gnathobase, and the endopod with internal and external endite (delimitation indicated by an arrow). The exopod is also partly visible. On the endopodite, the two external scrapers, as well as the full line of six posterior setae is visible. Of the anterior setae, the large external curved seta can be seen, and two of the setae on the inner endite. It appears that two small receptors are hidden behind seta four (see arrows on Fig. 168). *Figure 168.* Trunk limb three, showing gnathobase with a pore (inside square) near its basis, and internal endite, with anterior setae 4 and 5 (3–2 possibly hidden behind 4), and posterior plumose setae 3'–6'. *Figure 169.* Trunk limb three, showing the row of posterior setae of limb three, from another specimen (the same as in Fig. 165). *Figure 170.* Pore of gnathobase of trunk limb three, suggesting that some secretion is present. *Figure 171.* Trunk limb five, showing the epipodite and the gnathobase.

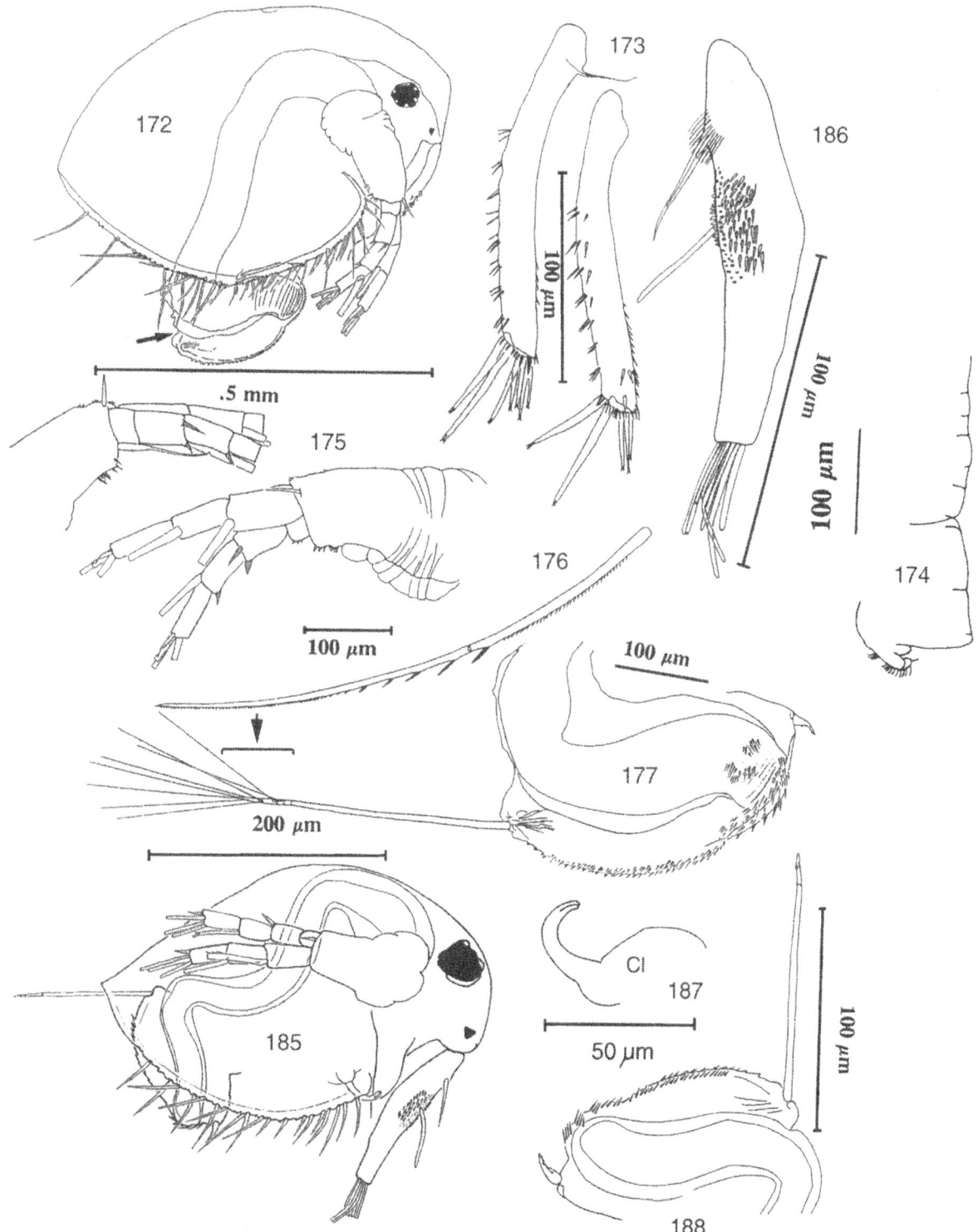

Figures 172–177, 185–188. Macrothrix smirnovi, Mexico. *Figure 172*. Female habitus. *Figure 173*. First antenna. *Figure 174*. Labrum. *Figure 175*. Second antenna of female. *Figure 176*. Long seta of second antenna. *Figure 177*. Female postabdomen. *Figure 185*. Male habitus. *Figure 186*. First antenna of male. *Figure 187*. Copulatory clasper of male. *Figure 188*. Male postabdomen.

44

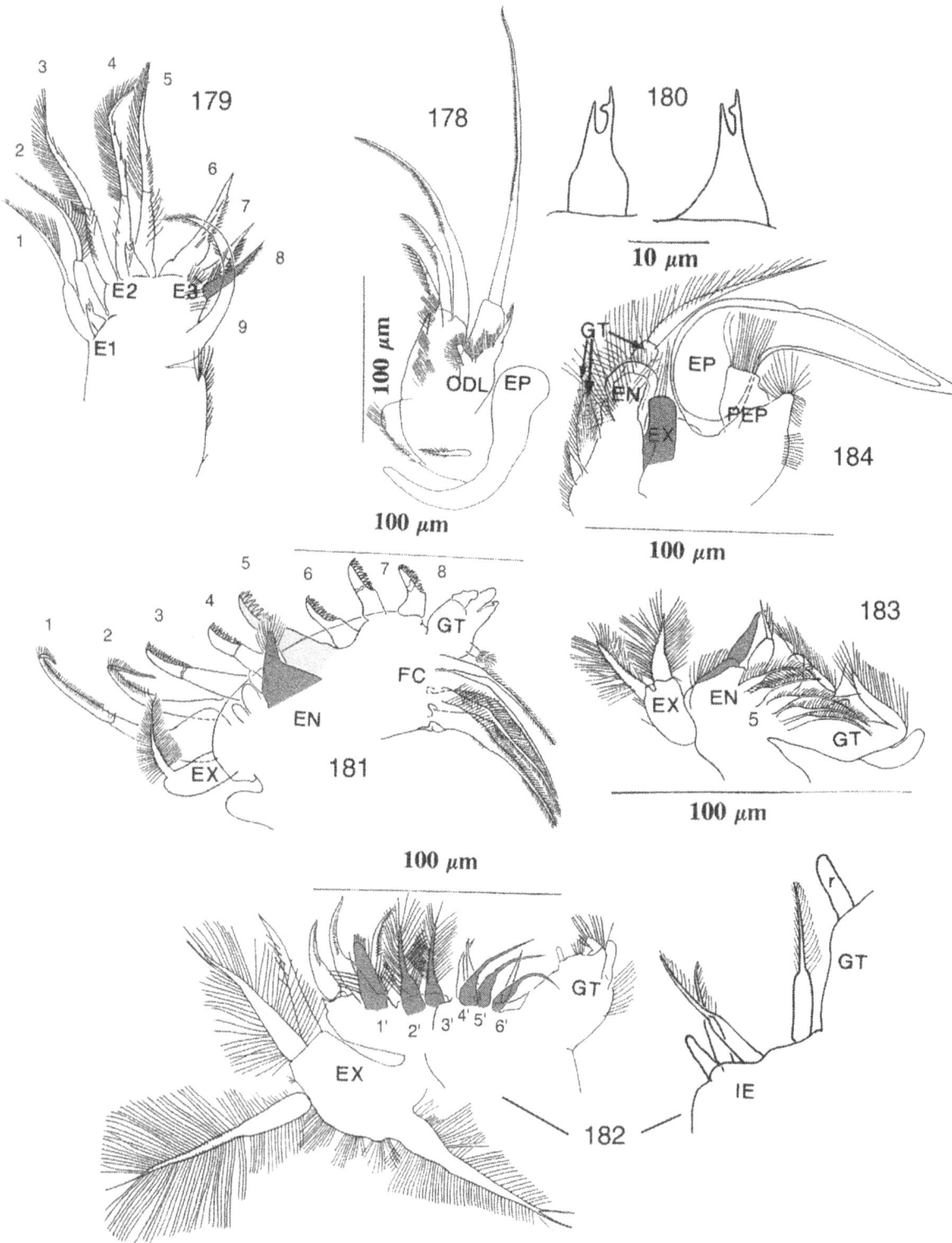

Figures 178–184. Macrothrix smirnovi, Mexico. *Figure 178*. IDL, ODL and epipodite of first trunk limb. *Figure 179*. Endites 1–3 of first trunk limb. *Figure 180*. Fryer's forks. *Figure 181*. Second trunk limb. *Figure 182*. Third trunk limb. *Figure 183*. Fourth trunk limb. *Figure 184*. Fifth trunk limb.

Hydrobiologia **467**: 45–56, 2002.
J. Alcocer & S.S.S. Sarma (eds), Advances in Mexican Limnology: Basic and Applied Aspects.
© 2002 *Kluwer Academic Publishers.*

Branchinecta oterosanvicentei n. sp. (Branchiopoda: Anostraca), a new fairy shrimp from the Chihuahuan desert, with a proposal for the conservation of the Branchinectidae of Mexico

Hortencia Obregón-Barboza[1], Alejandro M. Maeda-Martínez[1], Humberto García-Velazco[2] & Henri J. Dumont[3]
[1]*Centro de Investigaciones Biológicas del Noroeste, S.C., Apdo. Postal 128, La Paz, B.C.S., C.P. 23000, Mexico*
[2]*Centro de Bachillerato Tecnológico Agropecuario 198, Maneadero, B.C., C.P. 22900, Mexico*
[3]*Laboratory of Animal Ecology, University of Ghent, K.L. Ledeganckstraat 35, B-9000 Gent, Belgium*

Key words: Coahuila, Crustacea, geographical distribution, new species

Abstract

Branchinecta oterosanvicentei, new species, a fairy shrimp endemic to the south of Coahuila, Mexico is described and figured. Differential features between the new species and the related *B. lindahli* are discussed on the basis of a SEM micrographs. The main diagnostic characters of *Branchinecta oterosanvicentei* are: (1) a pulvinus covered by scales localized on the middle of the median side of the proximal article of antenna, and (2) a network of prominent cortical crests on the cyst surface. Of seven *Branchinecta* species occurring in Mexico, four (*B. belki*, *B. mexicana*, *B. oterosanvicentei*, and *B. sandiegonensis*) have a restricted geographic range. Because, in addition, extreme fluctuations in the number of mature individuals per population occur, the survival of all of these species is threatened, and measures should be taken to protect them. *Branchinecta mackini* has a wide distribution in North America, but even this form is rare in Mexico.

Introduction

The genus *Branchinecta* Verrill, 1869 occurs on all continents except Australia (Belk, 1982). Of 38 species recognized as valid (Belk & Brtek, 1995; Belk, 2000; Belk & Fugate, 2000), five occur in the old world, 31 in the new world, one is circumpolar (both the old and new world), and one lives in the Antarctic and subantarctic. Thus, the genus *Branchinecta* reaches its major species-richness on the American continent, with 19 species in North America, and 13 in South America. An apparent gap of distribution records across Central America, from Central Mexico in the north, to Peru in the south, requires, further study.

Belk & Brtek (1995, 1997) have stressed the need for zoogeographic studies of the Anostraca; almost 25% of the species are known only from their type localities. This generalization is typically applicable to the Branchinectidae of Mexico. Of the 19 North American *Branchinecta*, only six have been reported from this country (Maeda-Martínez et al., 1997):

B. belki Maeda-Martínez et al., 1992, *B. lindahli* Packard, 1883, *B. mackini* Dexter, 1956, *B. mexicana* Maeda-Martínez et al., 1993, *B. packardi* Pearse, 1912, and *B. sandiegonensis* Fugate, 1993. Three of these occupy restricted geographical ranges, with *B. belki* and *B. mexicana* recorded only from their type localities (Maeda-Martínez et al., 1992, 1993), and *B. sandiegonensis* considered endemic to a limited part of northern Baja California (Norte) in Mexico, and of southern California, in the U.S.A. (Fugate, 1993). After a careful reexamination of material deposited at the Universidad Juárez del Estado de Durango, México, we conclude that there exists a seventh undescribed species, which is morphologicaly very similar to *Branchinecta lindahli*. This new species also shows a limited distribution; it is endemic to the Chihuahuan desert in southern Coahuila along with *Branchinecta belki*.

Scientific interest in the conservation of fairy shrimps has increased in the last several years (see Eng et al., 1990; Simovich & Fugate, 1992; Hamer

& Brendonck, 1997; Petrov & Petrov, 1997; Belk, 1998; Fugate, 1998; Simovich, 1998). Specialists have included 26 anostracan species in the 1996 IUCN red list of threatened animals. Public interest and direct protection for these crustaceans are also increasing. For example, four fairy shrimps are currently included in the endangered species list of the United States by the U.S. Fish and Wildlife Service (Simovich, 1998), and thus these forms are currently legally protected. Two of them, *Branchinecta sandiegonensis* and *Streptocephalus woottoni* Eng et al. (1990), also occur in the state of Baja California (Norte), northwest Mexico. However, in the official list of protected species of flora and fauna published by the Mexican goverment denominated 'Norma Oficial Mexicana NOM-059-ECOL-1994' (Secretaría de Desarrollo Social, 1994), not a single anostracan is included.

Two of us (AMM, HGV) made extensive surveys for large branchiopods in Mexico in 1996, with an aim at obtaining zoogeographical data. Thus, the objectives of the present paper are (1) to describe the new species discovered, and (2) to record our findings on both the distribution, and the conservation status of the Branchinectidae of Mexico.

Materials and methods

Material of *Branchinecta oterosanvicentei* n. sp. was obtained from three different sites on six dates between 1980 and 1985, resulting in a total of nine collection lots. All lots were originally fixed in 70% isopropyl alcohol. For scanning electron micrographs, specimens and cysts of *Branchinecta oterosanvicentei* n. sp., *B. lindahli*, and *B. belki* were critical-point dried, coated with gold (9 nm, Balzers Union SCD 040), and analyzed under a JEOL JSM 840-SEM at 10 kV. Material of *Branchinecta lindahli* was obtained from two sources; one is a lot (Escuela Superior de Biología, Universidad Juárez del Estado de Durango, UJED 003) from Ejido Esteban Cantú, Punta Banda, Ensenada, Baja California (Norte) collected on 18 December 1985 by H. García-Velazco, and the other one was a lot from a laboratory culture using cysts from Morril, Nebraska, U.S.A., collected by Dr M. Fugate. Cysts of *Branchinecta belki* were obtained from female paratypes collected from the same type locality as the new species. For terminology of cyst shell structures we follow Gilchrist (1978). During the surveys in Mexico in 1996, A.M. Maeda-Martínez, and H. García-Velazco sampled ephemeral ponds in

the northwestern region of the country from 26 August to 14 September, and in the center and southern regions from 02 to 31 October. In some localities, the following water characteristics were measured: temperature (thermometer Brannan, UK, −5 °–110 °C), total disolved solids (TDS) (conductivity-meter HATCH, U.S.A.) and pH (Orion 230-A).

The threatened species categories used in this work correpond to the IUCN Red List Categories proposed by the IUCN Species Survival Commission (IUCN, 1994).

Results

Branchinecta oterosanvicentei new species
(Figs 1 A–F, 2 A–C, 3 A,B, 4 A,B, 5 A–C, 6 A–D)
Branchinecta lindahli: Maeda-Martínez (1991)
Branchinecta cf. lindahli: Maeda-Martínez et al. (1992, 1993, 1997)

Distribution and type locality: *Branchinecta oterosanvicentei* n. sp. occurs in three ephemeral ponds in the Chihuahuan desert in the southern part of Coahuila, México. These ponds are located in the semiarid bed of the ancient Laguna de Mayrán which belongs to the endorheic drainage of the rivers Nazas and Parras (Fig. 7). Two ponds are situated in the extreme east of Laguna de Mayrán, while the other one is situated some 115 km away, in the extreme west of Laguna de Mayrán. The specific localities are: (1) Type locality: an ephemeral pond on the south side of a bridge over Federal highway No. 40, approximately 0.5 km W of El Dorado Ranch, Coahuila 25° 39′ N, 101° 35′ W, altitude c. 1130 meters a.s.l., 80.5 km W of Saltillo, Coahuila, Mexico (code number 30-COAH); (2) an ephemeral pond at El Dorado Ranch, Coahuila, 80 km W Saltillo, Federal highway No. 40, 25° 39′ N, 101° 34′ W (code number 13-COAH); (3) an ephemeral pond at El Refugio bridge, Cerro Bola, Coahuila 70 km E of Torreón, Federal highway No. 40, 25° 35′ N, 102° 45′ W (code number 10-COAH).

Material examined: Holotype ♂ (total length 13.9 mm), allotype (total length 14.3 mm), 20 paratypes ♂♂ (total length 11.6–15.7 mm), and 20 paratypes ♀♀ (total length 11.4–14.8 mm) deposited at Centro de Investigaciones Biológicas del Noroeste, S.C. (CIB 486–488), and at Escuela Superior de Biología, Universidad Juárez del Estado de Durango (UJED 128), collected 25 February 1984 from the type locality (30-COAH). Paratypes are also deposited at the Smith-

47

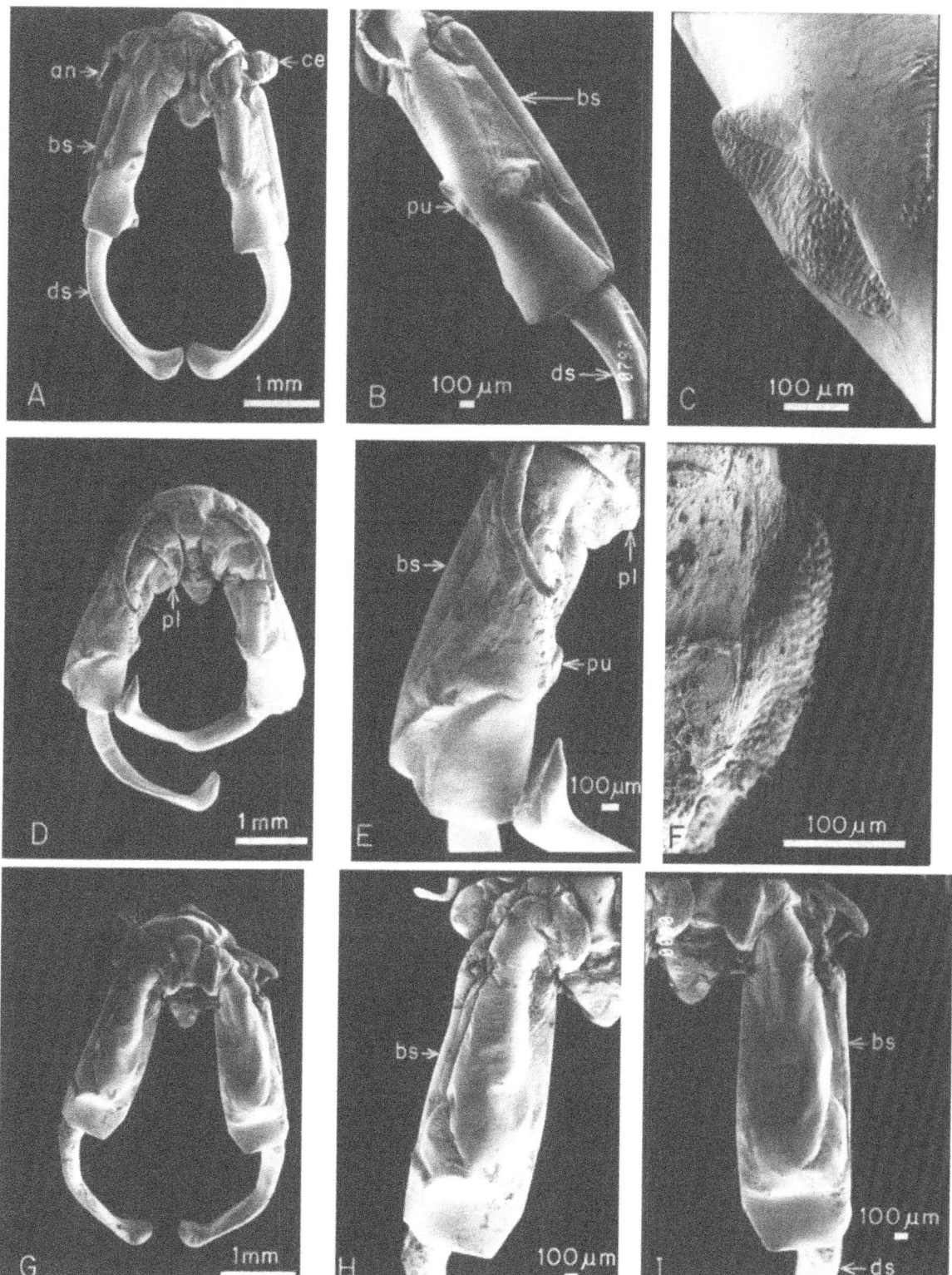

Figure 1. (A–F) *Branchinecta oterosanvicentei* new species (paratypes); (G–I) *Branchinecta lindahli* (Morril, Nebraska, U.S.A.); (A, D, G) head in anterior view; (B, E, H, I) proximal article of antenna in anterior view; (C, F) detail of pulvinus on the median side of the proximal article of antenna in anterior view. an – antennule, bs – proximal article of antenna, ce – compound eye, ds – distal article of antenna, pl – pulvillus on median side of proximal end of antenna, pu – pulvinus.

48

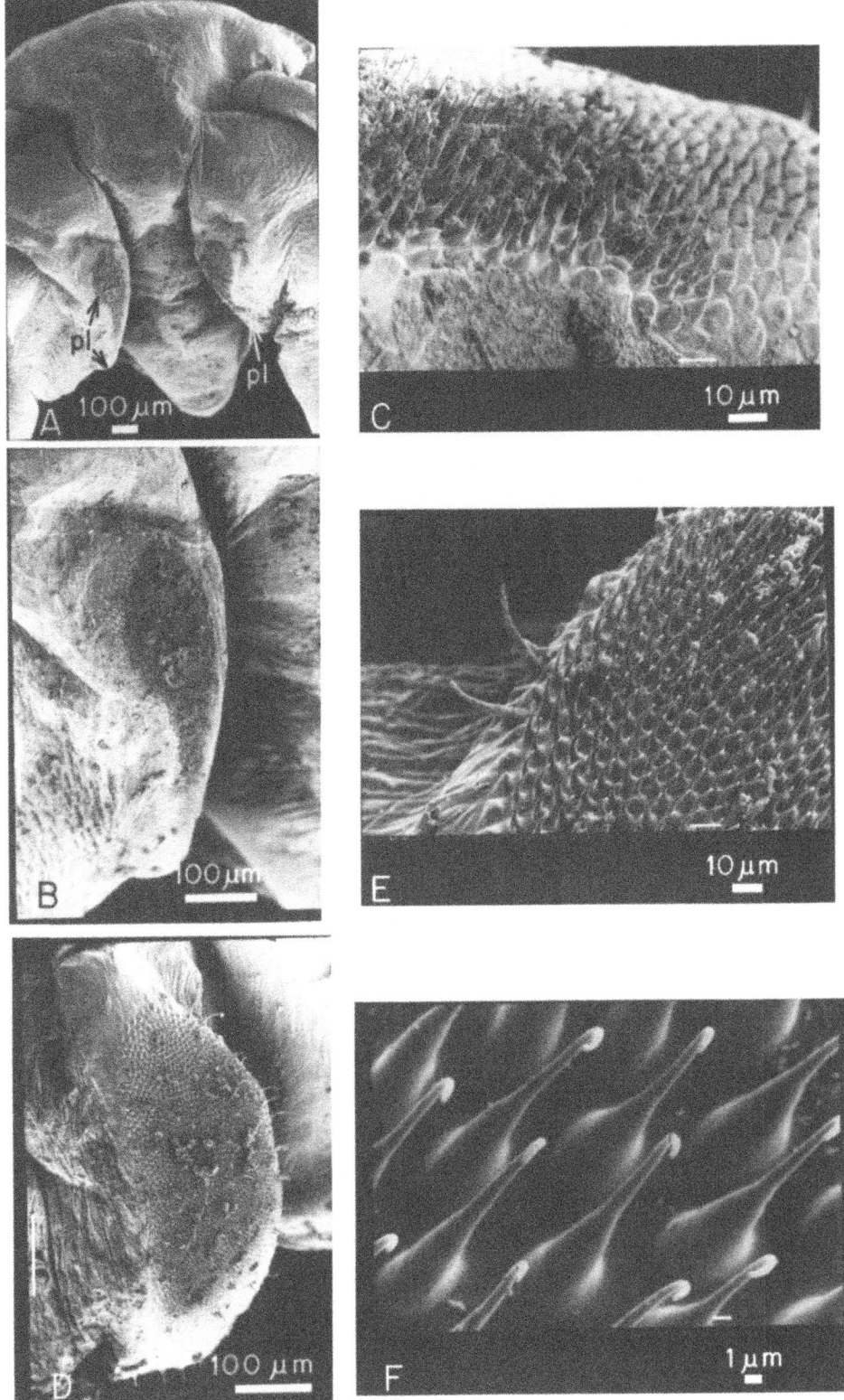

Figure 2. (A–C) *Branchinecta oterosanvicentei*, new species (paratype); (D–F) *Branchinecta lindahli* (Morril, Nebraska, U.S.A.); (A) head in anterior view; (B, D) detail of pulvillus of right antenna; (C) enlarged detail from B; (E) enlarged detail from D; (F) enlarged detail from E. pl – pulvillus.

49

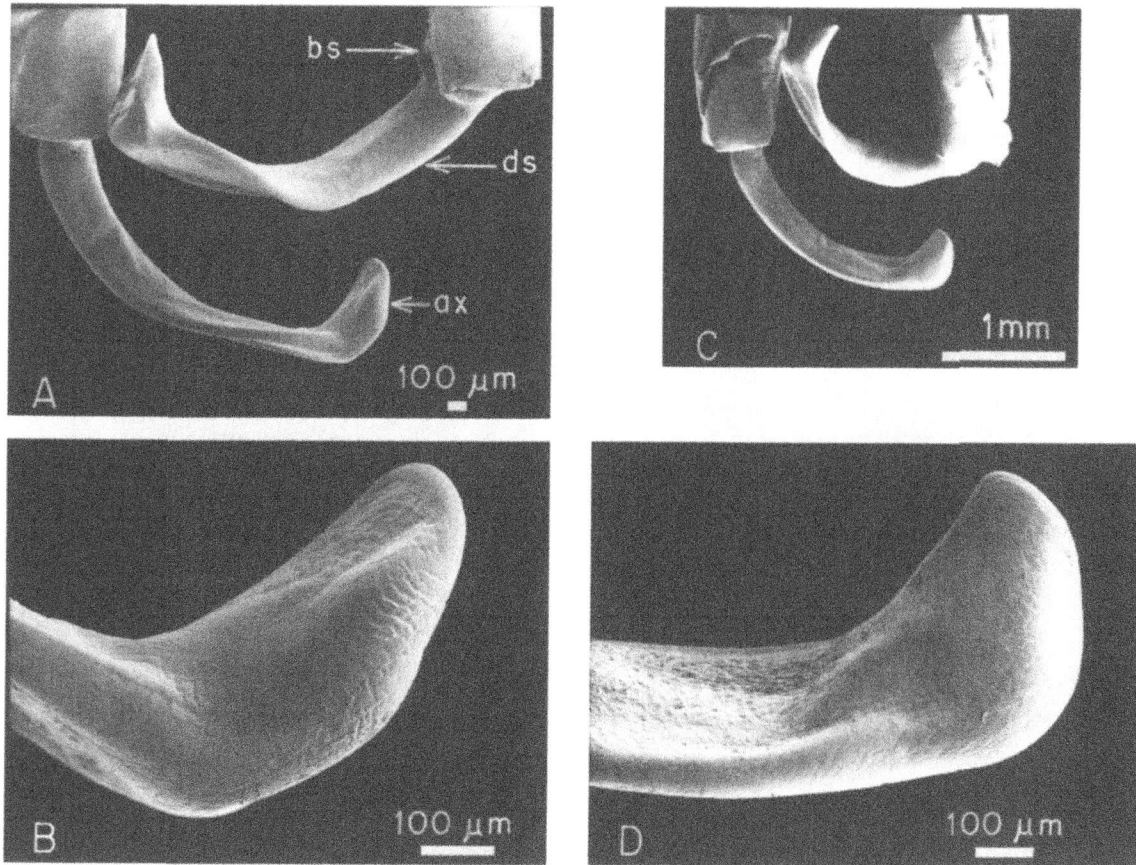

Figure 3. (A–B) *Branchinecta oterosanvicentei*, new species (paratype); (C–D) *Branchinecta lindahli* (Ejido Esteban Cantú, Baja California (Norte), Mexico); (A, C) right and left distal articles of antenna in anterior view; (B, D) detail of apex of right distal article in anterior view. ax – apex of distal article of antenna, bs – proximal article of antenna, ds – distal article of antenna.

sonian Institution (DB 674, Denton Belk's collection). Additional specimens from the type locality 14 ♂♂, 09 ♀♀ UJED 129 collected 30 January 1981. Specimens from locality 13-COAH 02 ♂♂, 01 ♀ UJED 133 collected 06 December 1980, 04 ♂♂ 01 ♀ UJED 134 collected 30 January 1981, and 712 ♂♂, 968 ♀♀ UJED 131. Specimens from locality 10-COAH 205 ♂♂, 112 ♀♀ UJED 005 collected 25 February 1984, 02 ♂♂, 04 ♀♀, UJED 006 collected 27 October 1985, 1 ♂ UJED 189 collected 11 November 1985, and 2 ♂♂ UJED 151 collected 12 November 1985. All the material was collected by A.M. Maeda-Martínez.

Diagnosis: Male: Proximal article of antenna with principal diagnostic characteristic, on the middle of its median side, consisting of pulvinus covered by scales (Fig. 1 A–F). Proximal article of antenna with typical pulvillus on median side near proximal end (Fig. 2 A–C). Pulvillus covered by minute spines and scales (Fig. 2 C). Distal article of antenna compressed, curved inward, with two major inflexions; one local-

ized in median proximal part, and other one at distal end of article, curved most strongly, almost at right angle to preceding parts (Fig. 3 A, B). Tip of distal article a spatule-like projection, concave on ventral side, posterior edge with rasplike surface (Fig. 4 A, B). Cercopods not converging, set with plumose setae along median and lateral borders.

Female: Thoracic segments IV–XI and sometimes first genital segment with, on both sides, dorsolateral rounded lobes covered by cuticular denticles. Brood pouch cylindrical, tip ending under post-thoracic legless segments V or VI. Cercopods as in male. Cyst spherical, diameter about 240 μm. Tertiary shell consisting of outer cortex, and inner alveolar layer with subcortical space between them (Fig. 6 A–C). Outer cortex formed by two parts; (1) external lamellar layer with prominent cortical crests forming net of irregular cells on cyst surface, and (2) outer alveolar layer filled with irregular filaments (Fig. 5 A–C). Internal surface of outer alveolar layer formed into irregular polygons

50

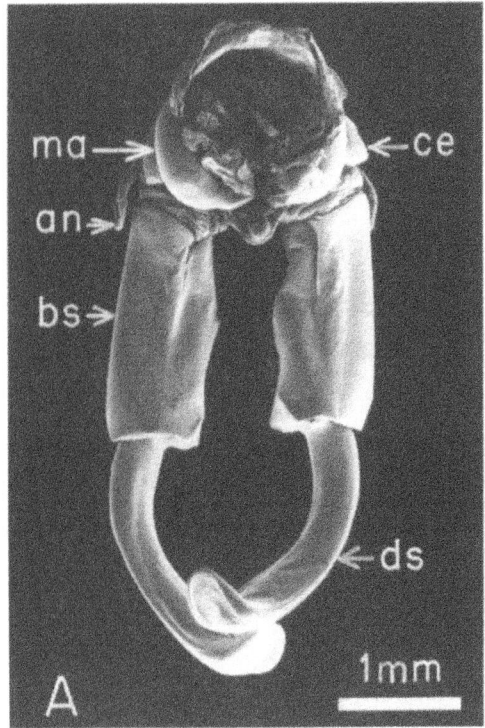 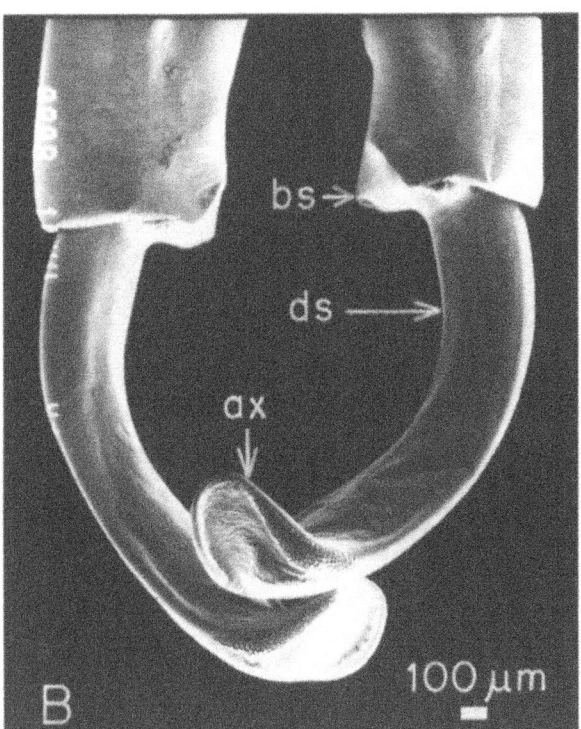

Figure 4. Branchinecta oterosanvicentei, new species (paratype); (A–B) head and antennae in posterior view. an – antennule, ax – apex of antenna, bs – proximal article of antenna, ce – compound eye, ds – distal article of antenna, ma – mandible.

defined by grooves (Fig. 6 B, C). Inner alveolar layer consists of net work of irregular filaments (Fig. 6 A, D).

Differential diagnosis: Branchinecta oterosanvicentei n. sp. is closely related to *B. lindahli.* There are at least two major morphological differences between them. (1) The pulvinus covered by scales localized at the middle of the median side of the proximal article of the antenna, characteristic of *B. oterosanvicentei,* is absent in *B. lindahli* (Fig. 1). According to Lynch (1964), in *B. lindahli* the proximal article of the antenna lacks an apophysis or any other kind of process or denticulate bulge, but often exhibits a slight swelling on the median side below the middle of the article. (2) The morphology of the cyst surface of *B. oterosanvicentei* shows prominent cortical crests while *B. lindahli* cyst exhibits an external surface with no prominent crests (Fig. 5).

Remarks on cyst morphology: In Anostraca, a cyst ornamentation pattern consisting of cells limited by cortical crests is common in species of the Old World genus *Chirocephalus* (Mura, 1992). However, this peculiar cyst morphology has also been reported from one species of *Eubranchipus* from California, U.S.A. (Hill & Shepard, 1997), and from three species of

Branchinecta from South America (i.e. *B. leonensis* César, 1987, *B. palustris* Birabén, 1964, and *B. granulosa* Daday, 1902 (cited as *B. santacrucensis* César, 1987) (César, 1989). Not one of the 13 North American species of Branchinecta studied by César (1989) and Mura (1991) for cyst morphology shows such ornamentation. Thus, *Branchinecta oterosanvicentei* n. sp. exhibiting a cyst surface with prominent cortical crests appears as a unique form among the branchinectids of North America.

Structural morphology of cross-sectioned cyst shells has been studied only in a few anostracan species (see De Walsche et al., 1991, and references therein cited). De Walsche et al. (1991) described structural differences among four species of three genera. They found that the cyst walls are not invariably bilayered as stated by Gilchrist (1978), but the walls can be composed of up to four layers. Recently, Hill & Shepard (1997) found that a thin alveolar layer in the cyst wall of *Eubranchipus* spp. and *Linderiella* spp. readily separates from the cortex. In a similar way, cysts of *Branchinecta oterosanvicentei* n. sp. (Fig. 6) and *B. belki* (Fig. 6) (see also Maeda-Martínez et al., 1992, Fig. 16) show that the outer cortex and the inner alveolar layer are completely separated from each

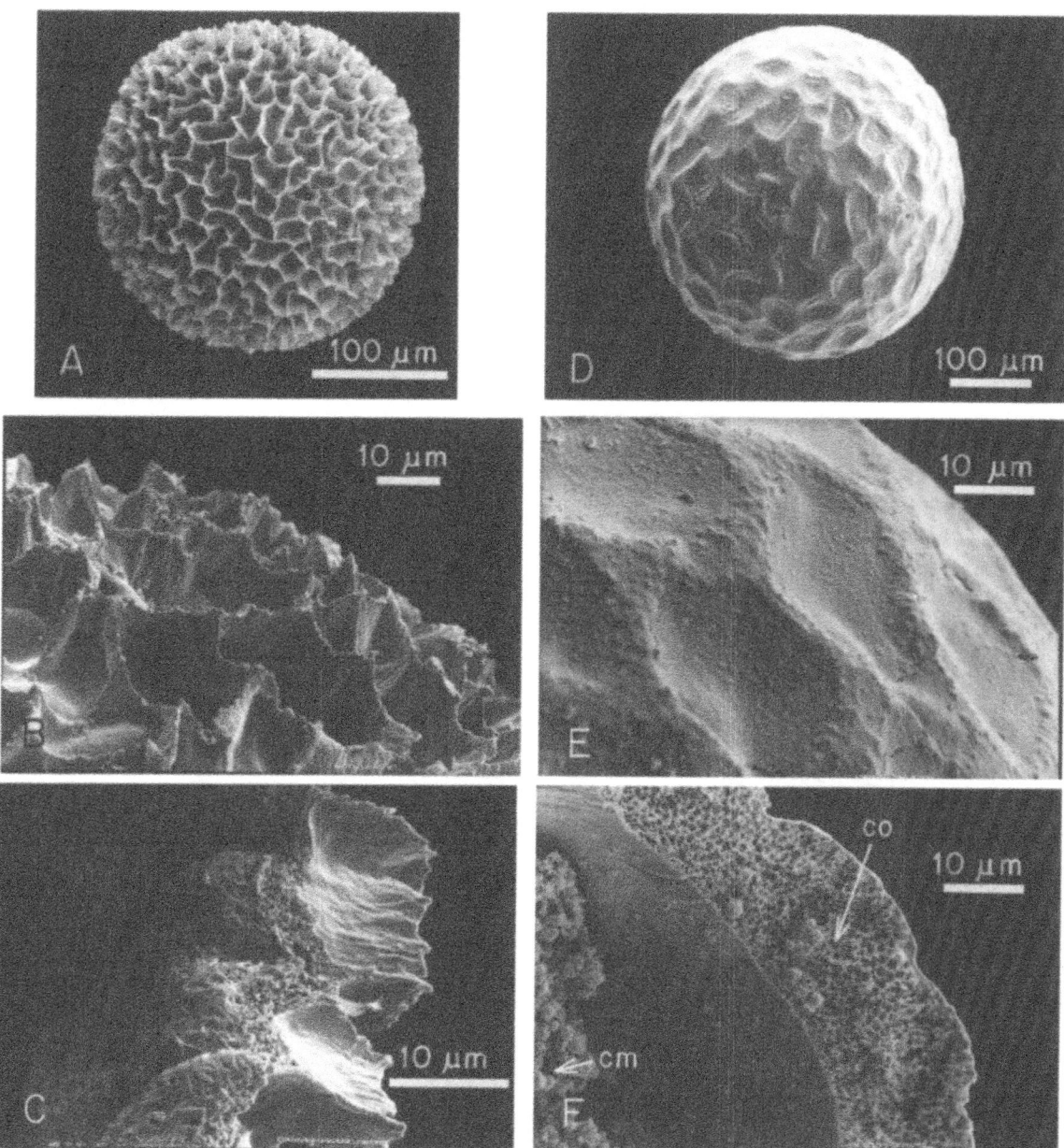

Figure 5. (A–C) Cyst of *Branchinecta oterosanvicentei*, new species (paratype); (D–F) cyst of *Branchinecta lindahli* (Ejido Esteban Cantú, Baja California (Norte), Mexico); (A, D) whole cyst; (B, E), enlarged detail of external surface; (C, D) enlarged detail of cross section of cracked cyst. cm – yolk particles and embrionic cells, co – outer cortex.

other. Therefore, the embryo is covered by a kind of 'double shell'. This feature allows for the embryo covered with the inner alveolar layer to be removed from the outer cortex. Doing this, the inner surface of the outer cortex appears composed of polygonal sections defined by grooves in *B. oterosanvicentei* (Fig. 6 B, C), while in *B. belki* the inner surface is irregular with holes of different sizes (Fig. 6 F). On the basis of available information on cyst structural morphology of

the *Branchinecta* species, three types of shells are observed: (1) shells with a subcortical space present, the outer cortex and inner alveolar layer completely separated from each other; e.g. in *B. oterosanvicentei* n. sp. (Fig. 6 A–C), and *B. belki* (Fig. 6 E, F) (also Maeda-Martínez et al., 1992, Fig. 16), and *B. campestris* Lynch, 1960 (Hill & Shepard, 1997, Fig. 32), (2) shells with a subcortical space present, outer cortex and inner alveolar layer not completely separated

52

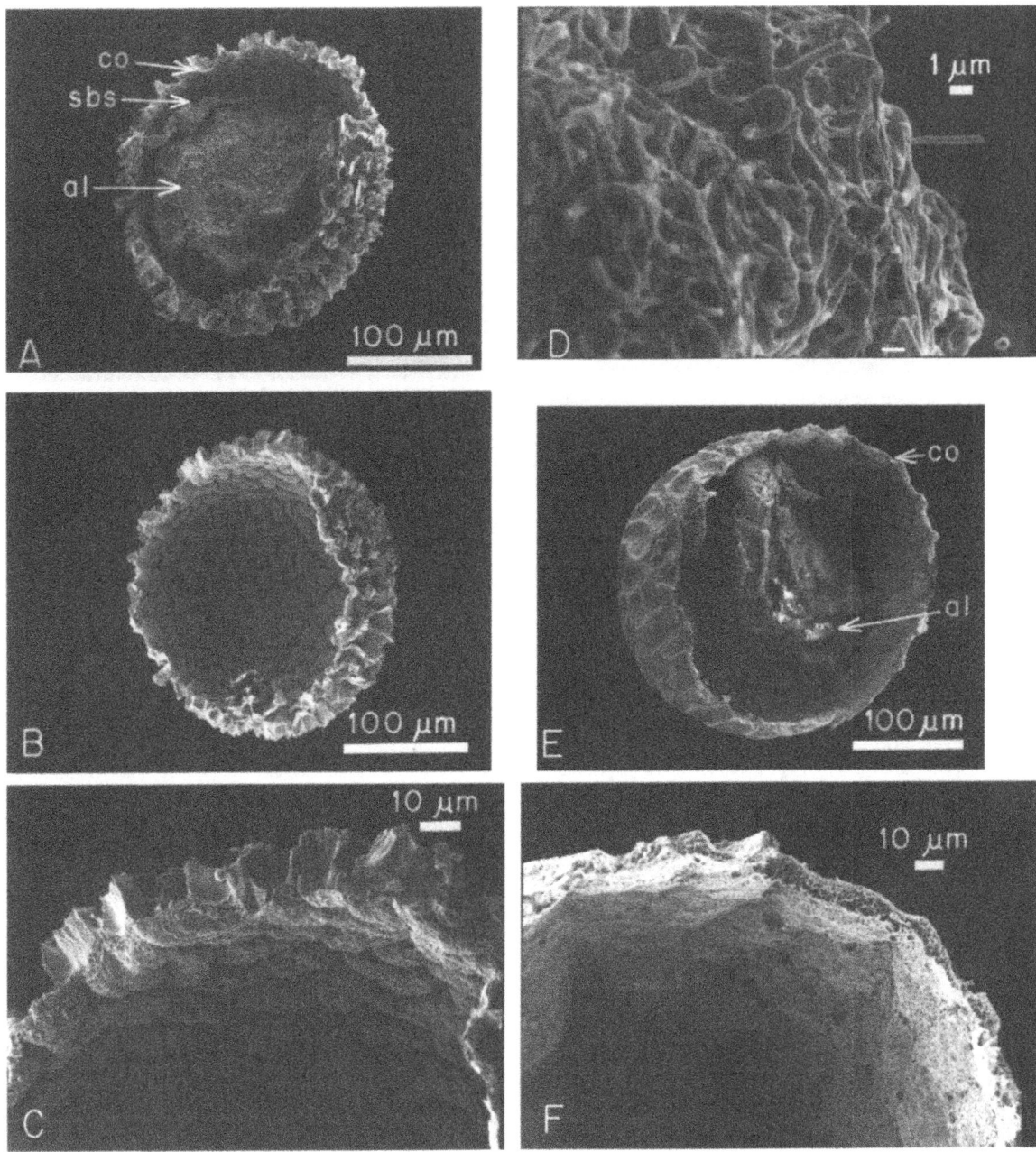

Figure 6. (A–D) Cyst of *Branchinecta oterosanvicentei* new species (paratype); (E, F) cyst of *Branchinecta belki* (paratype); (A, E) detail of cross section of cracked cyst; (B, C, F) detail of cross section of outer cortex (embryo and inner alveolar layer removed); (D) enlarged detail of inner alveolar layer from A. al – inner alveolar layer, co – outer cortex, sbs – subcortical space.

from each other; e.g. in *B. packardi* (Gilchrist, 1978; Maeda-Martínez et al., 1992, Fig. 22), and (3) shells with a subcortical space absent; the shell is composed of a single spongy cortex; e.g. *B. lindahli* (Fig. 5 I) (also Hill & Shepard, 1997, Fig. 26), and *B. mexicana* (Maeda-Martínez et al., 1993, Fig. 4 C–E). It remains, however, to establish the biological significance of

these structural differences in cyst shell morphology of the species. For instance, the presence of a subcortical space in desert inhabitants like *B. belki, B. oterosanvicentei,* and *B. packardi,* and its absence in non-desert inhabitants like *B. mexicana* suggests that the subcortical space (which could give a thermal insulation to the embryo as proposed by Gilchrist (1978)) may be

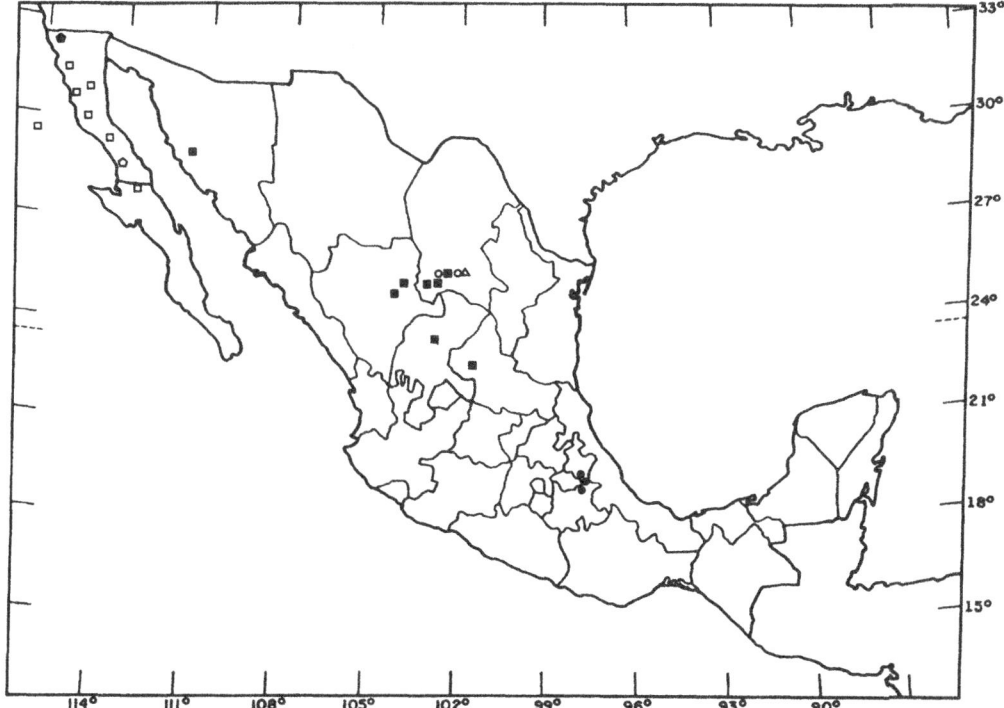

Figure 7. Distribution of *Branchinecta* in Mexico: *B. belki* (open triangle), *B. lindahli* (open squares), *B. mackini* (open pentagon), *B. mexicana* (closed circles), *B. oterosanvicentei* n. sp. (open circles), *B. packardi* (closed squares), and *B. sandiegonensis* (closed pentagon).

an adaptive element to face severe conditions of the habitat.

Etymology: The new species is named in memory of M.Sc. Napoleón Otero-Sanvicente, in recognition of his contributions to our knowledge of the geology and ecology of the basins of the Nazas and Aguanaval Rivers in north-central México.

Conservation status: Branchinecta oterosanvicentei is endemic of the Chihuahuan desert, north of México, and occupies a very limited geographic range (Fig. 7). The species is threatened by agricultural development, extensive cattle raising, and extension of federal highway No. 40. It is currently unprotected. We classify *Branchinecta oterosanvicentei* as endangered (EN) based on the following criteria: (1) Its range is less than 5000 km^2, (2) it exists at no more than five locations, and (3) there are extreme fluctuations in the number of mature individuals.

Branchinecta belki Maeda-Martínez, Obregón-Barboza, and Dumont, 1992

This species is endemic to the Chihuahuan desert, north of Mexico, again with a very limited geographic range. It has been collected from only four sites, localized whitin a radius of 10 km from El Dorado Ranch,

situated in the extreme east of Laguna de Mayrán in southern Coahuila (Maeda-Martínez et al., 1992) (Fig. 7). In 1996, *Branchinecta belki* was listed as endangered (EN) (IUCN, 1996). It is threatened by alteration and destruction of its environment by agricultural development, extensive cattle raising, and extension of federal highway No. 40. No measures of protection exist. Like the preceding case, we rank *Branchinecta belki* as an endangered species (EN) because (1) the extent of its occurrence is less than 5000 km^2, (2) it exists at no more than five locations, and (3) there are extreme fluctuations in the number of mature individuals.

Branchinecta lindahli Packard, 1883

This species has a wide distribution, covering western North America including Canada, the U.S.A., and Mexico (Belk, 1975; Belk & Brtek, 1995). In Mexico, it is a common fairy shrimp in the state of Baja California (Norte), including one record in Isla Guadalupe (Belk & Lindberg, 1979; Belk, 1983; Maeda-Martínez, 1991; Maeda-Martínez et al., 1997). There is, however, only one record of the species in the northern part of Baja California Sur state, almost at the border with Baja California (Norte) (Belk, 1983) (Fig.

7). In September 1996, *Branchinecta lindahli* was found in seven ephemeral ponds on the top of Sierra San Pedro Martir, Baja California (Norte) (CIB 262–268). In one of the ponds (Entronque a la Tasajera, San Telmo de Abajo-Observatorio), at the moment of collection the water was with a temperature of 20.9 °C, and a pH of 6.8. The survival of the species seems to be no immediate issue (Eng et al., 1990); but the known Mexican populations in continental Baja California (Norte) may be threatened by extensions of Federal Highway No. 1. The status of the population in Isla Guadalupe, Baja California (Norte), and the population in Baja California Sur is unknown. Indirect protection could be expected for the population in Isla Guadalupe given that this island has been declared both a 'Reserva Especial de la Biosfera' and 'Area de Protección de Flora y Fauna' by the Mexican government. The same holds true for the populations in San Pedro Martir because this area is included into the 'Parque Nacional Sierra de San Pedro Martir'. Two of the ponds, localized at road posts km 99 and km 98.5 of the road section San Telmo de Abajo-Observatorio Astronómico, finally are located on the campus of the 'Observatorio Astronómico Nacional' of the Universidad Nacional Autónoma de México.

Branchinecta mackini Dexter, 1956

A species with a wide distribution, extending throughout western North America including Canada, the U.S.A., and Mexico (Belk, 1975; Brown et al., 1993; Belk & Brtek, 1995). However, *Branchinecta mackini* is rare in Mexico. It has been recorded from only two localities of Baja California (Norte) (Fig. 7); one location in Laguna Chapala, and another at 45 km S of Laguna Chapala along Federal highway No. 1 (Maeda-Martínez et al., 1997). The species as a whole appears not to be under any immediate threat (Eng et al., 1990), but no protection exists for it in Mexico. Of special interest to conservation is the ephemeral lake Laguna Chapala, because besides the occurrence of *Branchinecta mackini*, this is the habitat of the single known population of the notostracan genus *Lepidurus* in Mexico (*L. lemmoni* Holmes, 1894) (Maeda-Martínez et al., 1997).

Branchinecta mexicana Maeda-Martínez, Obregón-Barboza & Dumont, 1993

An endemic of central Mexico, it was included in the IUCN (1996) list as critically endangered (CR). The species was known from only the type locality at Rafael Avila Camacho in the state of Puebla, just on the border with the state of Tlaxcala. *Branchinecta mexicana* exhibits an erratic occurrence. During the survey in October 1996, the ephemeral pond of the type locality was visited and no anostracans were found. Instead, conchostracans of the genus *Lynceus* were found inhabiting the pond. Fortunately, the survey turned up three new locations for *Branchinecta mexicana* in the same state of Puebla (Fig. 7). Two of them are a group of roadside ephemeral ponds along Federal Highway No. 140, one on highway section Zacatepec-El Seco, municipality of San Salvador El Seco (CIB 351–354), and the other one on section Acatzingo-El Seco, municipality of General Felipe Angeles (CIB 360). At the time of collection, the ranges of the water characteristics in these ponds were: temperature 17.2–23.7 °C, TDS 0.19–0.31 g l^{-1}, and pH 8.0–8.2. In one of the ponds *Branchinecta mexicana* was collected along with *Streptocephalus mackini* Moore, 1966. The third location is a mesohaline temporary pond localized in the extreme north of Laguna de Totolcingo at the west side of Federal Highway No. 129 Zacatepec-Oriental, municipality of Oriental (CIB 357). At the time of collection the water had a temperature of 24.2 °C, TDS 4.8 g l^{-1}, and pH 9.5. During the last 20 years Laguna de Totolcingo, which belongs to the Oriental endorheic basin, has changed from a perennial saline lake to an ephemeral water body (Alcocer et al., 1997). The Oriental basin is suffering dramatic changes due to human activities, where the desiccation process of the area has impacted directly on the aquatic biota (Alcocer et al., 1998). *Branchinecta mexicana* is threatened by alteration and destruction of its biotopes by urban development and extension of Federal Highways. No protection exists for the species. We assess *Branchinecta mexicana* as an endangered species (EN) following the criteria: (1) Extent of occurrence less than 5000 km^2, (2) it is known to exist at no more than five locations, and (3) there are extreme fluctuations in the number of mature individuals.

Branchinecta packardi Pearse, 1912

A species with a wide distribution in North America, including Canada, the U.S.A., and Mexico (Belk, 1975; Belk & Brtek, 1995). In Mexico, *Branchinecta packardi* has been reported from 13 localities in five states, five in Coahuila, four in Durango, one in San Luis Potosí, two in Sonora, and one in Zacatecas (Strenth & Littleton, 1990; Maeda-Martínez, 1991; Campoy-Favela & Quijada-Mascareñas, 1993) (Fig. 7). Given its broad range, the species appears under no inmediate threat, even though no protection exists for it in México. Of the 13 known localities, four are located in urban areas of the cities Torreón, Coahuila; Gómez Palacio, Durango; and Hermosillo, Sonora. These population are threatened by habitat destruction. In fact, the temporary pond at Torreón has already been lost because of the expansion of the local international airport facilities.

Branchinecta sandiegonensis Fugate, 1993
Endemic from southern California (Santa Barbara and San Diego counties), U.S.A., and northern Baja California (Norte), Mexico (Fugate, 1993). In 1996, it was included in the IUCN (1996) list as endangered (EN). Recently, the U.S. Fish and Wildlife Service also listed *Branchinecta sandiegonensis* as an endangered species (Simovich, 1998). In Mexico, *Branchinecta sandiegonensis* is known from two localities; one at Valle de las Palmas, about 23 km south of Tecate, and the second in Bajamar, about 20 km north of Ensenada (Brown et al., 1993; Fugate, 1993) (Fig. 7). H. García-Velazco sampled the pools in Bajamar during the winter seasons of 1995–1997, but no specimens were obtained. This suggests an erratic occurrence for this species. Based upon these data, we continue to consider *Branchinecta sandiegonensis* a endangered species (EN).

Discussion

Studies aimed at establishing the status of species in nature in terms of the number of viable populations and the extent of their geographic range are important not only for biogeography, but also for conservation. A first step, of course, is to know the total number of species that inhabit a given territory, say Mexico. In spite of our efforts of the past few years, the inventory of the large branchiopods of this country remains significantly incomplete. There are regions that have not been studied up to now; for example the northern parts of Sonora, Chihuahua, and Coahuila. On the other hand, the acquisition of appropriate field samples from remote areas is costly and difficult. This follows from the fact that most large branchiopods are limited to life in temporary stagnant water. The occurrence of a reproductive population (needed for taxonomic identification) at a given time is hard to be predicted, because it depends on multiple factors such as the (more or less random) inundation of the water body (wet phase) and the successful hatching of the dormant eggs (cysts). Indeed, during the wet phase of the hab-

itat, the presence of adults of each species depends, among other factors, on a particular range of water temperature (Belk, 1977). The influence of this factor may result in a seasonally restricted occurrence and in a succession of species. Therefore, to obtain a reasonably complete inventory, a seasonal sampling protocol of repeated visits (monitoring) to representative water bodies is required.

According to Wilcove et al. (1998), the most common threats to biodiversity can be divided in five categories: habitat destruction (degradation or loss), the introduction of non-native (alien) species, overharvesting, pollution, and disease. Habitat destruction is recognized as the greatest threat to biodiversity (Wilcove et al., 1998), and it certainly is to the crustaceans of temporary pools (Belk, 1998; Fugate, 1998). Given that a large number of branchiopod species are known from only a single location (Belk & Brtek, 1997), the degradation or loss of a unique temporary pool may cause the extinction of one or more species.

The seven *Branchinecta* species occurring in Mexico may serve to illustrate this point. Four (*B. belki*, *B. mexicana*, *B. oterosanvicentei* and *B. sandiegonensis*) have limited geographic ranges. Due to their restricted distribution pattern, and extreme fluctuations in the number of mature individuals, we consider these four to be acutely endangered. Even *Branchinecta mackini* with its wide distribution in North America, is a rare species in Mexico. Based on this information, it would seem timely for Mexican authorities to give attention to the data presented here. They suggest a critical situation in the country and reveal the high risk for an irrevocable loss of large branchiopod diversity in a short time, unless appropriate protection measures are taken.

Acknowledgements

We thank Drs D. Belk and M. Fugate for valuable suggestions and critical review of the manuscript. We are grateful to Dr M. Fugate for making available the cyst sample of *B. lindahli*, Mrs S. Wellekens for her assistance with the scanning electron microscopy, Mr Martín Prieto-Salazar for helping during field collection in 1996, and Mr Gerardo Hernández-García for the preparation of the map. This study forms part of the project 'Biología y Cultivo de Branquiópodos' (ABM-13) supported by the Centro de Investigaciones Biológicas del Noroeste, S.C. (CIBNOR) and the

56

Comisión Nacional para el Conocimiento y Uso de la Biodiversidad (CONABIO-H068).

References

Alcocer, J., A. Lugo, E. Escobar & M. Sánchez, 1997. The macrobenthic fauna of a former perennial and now episodically filled Mexican saline lake. Int. J. Salt Lake Res. 5: 261–274.

Alcocer, J., A. Lugo, M. R. Sánchez, M. Chávez & E. Escobar, 1998. Threats to the saline lakes of the Oriental Basin, Mexico, by human activities. Verl. int. Ver. Limnol. 26: 1383–1386.

Belk, D., 1975. Key to the Anostraca (fairy shrimps) of North America. S. West. Nat. 20: 91–103.

Belk, D., 1977. Zoogeography of the Arizona fairy shrimps (Crustacea: Anostraca). J. Ariz. Acad. Sci. 12: 70–78.

Belk, D., 1982. Branchiopoda. In Parker, S. P. (ed.), Synopsis and Classification of Living Organisms. McGraw Hill: 174–184.

Belk, D., 1983. New fairy shrimp distribution records among collections at the California Academy of Sciences. S. West. Nat. 28: 380–381.

Belk, D., 1998. Global status and trends in ephemeral pool invertebrate conservation: Implications for Californian fairy shrimp. In Witham, C. W., E. Bauder, D. Belk, W. Ferren & R. Ornduff (eds), Ecology, Conservation, and Management of Vernal Pool Ecosystems. California Native Plant Society, Sacramento, CA: 147–150.

Belk, D., 2000. Branchinecta readingi new species name for a well-known fairy shrimp from east of the North American Continental Divide. J. Crust. Biol. 20: 566–570.

Belk, D. & M. Fugate, 2000. Two new Branchinecta (Crustacea: Anostraca) from the southwestern United States. S. West. Nat. 45: 111–117.

Belk, D. & D. R. Lindberg, 1979. First freshwater animal reported for Isla de Guadalupe represents a southern range extension for Branchinecta lindahli (Crustacea: Anostraca). S. West. Nat. 24: 390–391.

Belk, D. & J. Brtek, 1995. Checklist of the Anostraca. Hydrobiologia 298: 315–353.

Belk, D. & J. Brtek, 1997. Supplement to 'Checklist of the Anostraca'. Hydrobiologia 359: 243–245.

Brown, J. W., H. A. Wier & D. Belk, 1993. New records of fairy shrimps (Crustacea: Anostraca) from Baja California, Mexico. S. West. Nat. 38: 389–390.

Campoy-Favela, J. & A. Quijada-Mascareñas, 1993. Additional distribution records of fairy shrimps (Anostraca) from Sonora, Mexico. S. West. Nat. 38: 85–86.

César, I. I., 1989. Comparative study on the resting eggs of several anostracans (Crustacea). Key for the determination of the species based upon the egg structure and diameter. Studies on Neotropical Fauna and Environment 24: 169–181.

De Walsche, C., N. Munuswamy & H. J. Dumont, 1991. Structural differences between the cyst walls of Streptocephalus dichotomus (Baird), S. torvicornis (Waga), and Thamnocephalus platyurus Packard (Crustacea: Anostraca), and comparison with other genera and species. Hydrobiologia 212: 195–202.

Eng, L. L., D. Belk & C. H. Eriksen, 1990. Californian Anostraca: Distribution, habitat, and status. J. Crust. Biol. 10: 247–277.

Fugate, M., 1993. Branchinecta sandiegonensis, a new species of fairy shrimp (Crustacea: Anostraca) from western North America. Proc. Biol. Soc. Wash. 106: 296–304.

Fugate, M., 1998. Branchinecta of North America: Population structure and its implications for conservation practice. In

Witham, C. W., E. Bauder, D. Belk, W. Ferren & R. Ornduff (eds), Ecology, Conservation, and Management of Vernal Pool Ecosystems. California Native Plant Society, Sacramento, CA: 140–146.

Gilchrist, B. M., 1978. Scanning electron microscope studies of the egg shell in some Anostraca (Crustacea: Branchiopoda). Cell Tiss. Res. 193: 337–351.

Hamer, M. L. & L. Brendonck, 1997. Distribution, diversity and conservation of Anostraca (Crustacea: Branchiopoda) in southern Africa. Hydrobiologia 359: 1–12.

IUCN, 1994. IUCN Red List Categories. IUCN, Gland, Switzerland, 21 pp.

IUCN, 1996. IUCN Red List of Threatened Animals. IUCN, Gland, Switzerland.

Hill, R. E. & W. D. Shepard, 1997. Observations on the identification of California anostracan cyst. Hydrobiologia 359: 113–123.

Lynch, J. E., 1964. Packard's and Pearse's species of Branchinecta: analysis of a nomenclatural involvement. Am. Midl. Nat. 71: 466–488.

Maeda-Martínez, A. M., 1991. Distribution of species of Anostraca, Notostraca, Spinicaudata, and Laevicaudata in Mexico. Hydrobiologia 212: 209–212.

Maeda-Martínez, A. M., H. Obregón-Barboza, & H. J. Dumont, 1992. Branchinecta belki, n.sp. (Branchiopoda: Anostraca), a new fairy shrimp from Mexico, hybridizing with B. packardi Pearse under laboratory conditions. Hydrobiologia 239: 151–162.

Maeda-Martínez, A. M., H. Obregón-Barboza & H. J. Dumont, 1993. Branchinecta mexicana new species (Branchiopoda: Anostraca), a fairy shrimp from Central Mexico. J. Crust. Biol. 13: 585–593.

Maeda-Martínez, A. M., H. Obregón-Barboza & H. García-Velazco, 1997. New records of large branchiopods (Branchiopoda: Anostraca, Notostraca, and Spinicaudata) in Mexico. Hydrobiologia 359: 63–68.

Mura, G., 1991. SEM morphology of resting eggs in the species of the genus Branchinecta from North America. J. Crust. Biol. 3: 432–436

Mura, G., 1992. Additional remarks on cyst morphometrics in anostracans and its significance. Part II: Egg morphology. Crustaceana 63: 225–246.

Petrov, B. & I. Petrov, 1997. The status of Anostraca, Notostraca and Conchostraca (Crustacea: Branchiopoda) in Yugoslavia. Hydrobiologia 359: 29–35.

Secretaría de Desarrollo Social, 1994. Norma Oficial Mexicana NOM-059-ECOL-1994. Diario Oficial de la Federación CDLXXXVIII:10, México, D.F.

Simovich, M. A., 1998. Crustacean biodiversity and endemism in California's ephemeral wetlands. In Witham, C. W., E. Bauder, D. Belk, W. Ferren & R. Ornduff (eds), Ecology, Conservation, and Management of Vernal Pool Ecosystems. California Native Plant Society, Sacramento, CA: 107–118.

Simovich, M. A. & M. Fugate, 1992. Branchiopod diversity in San Diego County, California, U.S.A. Trans. West. Sect. Wildl. Soc. 28: 6–14.

Strenth, N. D. & T. G. Littleton, 1990. First record of Branchinecta packardi Pearse (Crustacea, Anostraca) from Mexico. Tex. J. Sci. 42: 411–412.

Wilcove, D. S., D. Rothstein, J. Dubow, A. Phillips & E. Losos, 1998. Quantifying threats to imperiled species in the United States. Bioscience 48: 607–616.

Hydrobiologia **467**: 57–62, 2002.
J. Alcocer & S.S.S. Sarma (eds), Advances in Mexican Limnology: Basic and Applied Aspects.
© 2002 *Kluwer Academic Publishers.*

The external micro-anatomy of the cephalon of the asellotan isopod *Craseriella anops*

Elva Escobar[1], Luis Oseguera[2], Gerardo H. Vázquez Nin[2] & Javier Alcocer[3]
[1]*Unidad Académica en Sistemas Oceanográficos y Costeros, Instituto de Ciencias del Mar y Limnología,*
UNAM, A.P. 70-305, México D.F. 04510, Mexico
E-mail: escobri@mar.icmyl.unam.mx
[2]*Facultad de Ciencias, Laboratorio de Microscopía Electrónica, UNAM, Mexico*
[3]*Limnology Laboratory, Environmental Conservation and Improvement Project, UIICSE UNAM- Campus Iztacala,*
Mexico

Key words: anchialine, mechanoreceptor, scales, setae, sensory hair, turbulence, Yucatan

Abstract

The micro-anatomy of the cephalon is described in the troglobic asellotan isopod *Craseriella anops* from the Nohoch Nah Chich anchialine cave system in southeast Mexico. The cephalon is entirely covered by cuticular scales bordered by marginal spines. The anterior end of the cephalon is bordered by a carina that is wider medially. The isopod is eyeless. The distal seventh portion of the cephalon is characterized by the presence of two sutures and six setae. A suture is found on each side of the distal margin of the cephalon. Each suture is bordered by microtrichs. Two simple setae with a sensory hair, articulated on the base by a socket, are found one on each side of each of the sutures. Two additional setae, similar in shape and size, occur medially on the cephalon. A terminal pore is absent on the sensory hairs of all setae. These setae are suggested to be mechanoreceptors that provide directional sensitivity and enhance the sensibility of turbulent motion, viscosity and changes of hydrostatic pressure.

Introduction

Adaptation is a central problem in evolutionary biology and an extraordinary topic for research in subterranean and deep-sea ecology (Christianen, 1992). Crustacea have been recognized as one of the most diversified and abundant groups inhabiting the anchialine cave systems in the tropics, the deep-sea and other interstitial habitats and have been used as models to exemplify how animals adapt to a life in total darkness with short food supply (Kaufman, 1994). Among the adaptations recorded in troglobic fauna are the reduction of size (Holsinger, 1986), the loss of pigment and of visual organs and the development of specialized sensorial structures (George, 1981).

Underwater cave systems remain in total darkness. The absence of light limits photosynthesis, which is the basis of the food web in other aquatic habitats. The food sources sustaining the cave system faunal assemblages depend on the import of debris from the neighboring terrestrial and aquatic environments through the sinkholes. Organic material generally enters in dissolved or particulate form and is subsequently processed by bacteria (Eichem et al., 1993) and detritovores. The relationship between the food resources entering the cave systems and the consumers in the caves are not well documented; neither is the prey–predator relationship.

The cuticle of crustaceans is a dynamic system that has been recognized to perform diverse functions. In habitats where light is absent, structures on the epicuticle may provide a way to link the external environment with the sensory function and behavior. Isopods have been recognized as scavengers and predators that, in shallow habitats in presence of light, can utilize visual cues to locate their prey. An extensive terminology has been developed for describing the nature of crustacean cuticle. Studies on decapods date back to last century and early this century and have extensively described mechanosensory and chemosensory structures. Experimental results have recognized the former to be related to the detection of movement and vibration produced by the prey or by the predators (Wägele, 1993). The latter have been considered to help detect the source and concentration of compounds of potential food sources (Laverack, 1989).

The asellotan isopod *Craseriella anops* (Creaser, 1936), is the dominant invertebrate in the community of anchialine cave systems in southeast Mexico. This species is troglobic and inhabits both the marine and fresh water masses of the cave system. To find the food used for growth, maintenance and reproduction in the isopod requires structural and functional features to facilitate a subterranean lifestyle. It is the purpose of this study to describe the external micro structures on the cephalic cuticle of *C. anops* and discuss the role of the observed cuticular appendages.

Sampling methods

Area of study

Nohoch Nah Chich are limestone caverns located in the eastern portion of the Yucatan Peninsula that drain into the northern Caribbean Sea with a known connection to the sea. The location named Main Entrance (20° 17.9′ N, 87° 23.7′ W), where sampling took place, is located 5 km inland from the coast near the city of Tulum. The cave passages develop at a halocline between 10 and 25 m water depths. Nohoch Nah Chich is an oligotrophic system characterized by the absence of light, constant temperature (26 °C) and low dissolved oxygen content (2 ± 0.2 mg $O_2.L^{-1}$). The system is subject to tidal influence and to seasonal input of freshwater and import of particulate organic matter (Alcocer et al., 1998).

Baited scavenger traps were deployed by means of cave diving. The traps used to collect the isopods were placed below the halocline at positions marked on a permanent line. The distances were 30–100 m from the cave entrance for a period of 12 h. This study used scanning electron microscopy to examine the external cuticular micro structures of the cephalon using a modified technique than the one described by Felgenhauer (1987). The heads of specimens were sectioned and fixed in a 2.5% glutaraldehide solution prepared in the field with cave water and phosphate buffer at 0.1 m and pH 7 in 60 ml glass jars. Post processing in the laboratory included several baths with phosphate solution 0.1 m at pH 7 and post fixation

with 2% OsO_4 followed by new baths and dehydration with ethanol baths and acetone. Specimens were dried to critical point in a Samdri-780 dryer substituting the acetone with CO_2 in a vacuum. Heads were mounted and covered with Au–Pd in a 1:1 ratio using a metal ionizer JEOL-JFC 1100. Scanning electron microscopy observations and microphotographs were taken using a JEOL SEM model JSM-54IOLV at high vacuum.

Scales and setae were described using the classification scheme of Watling (1989) and on those structures not considered we used the setal classification of Fish (1972). Scale morphotype was based on average measurements of length and width as in spider cuticular scales (Townsend & Felgenhauer, 1998). SEM micrographs were used to determine cuticular scale length and width of at least 30 scales from 10 specimens. The number and length of marginal setae were recorded.

Results

C. anops lacks body pigment. Cuticular scales cover the external surface of the cephalon as well as the body, the scales rest parallel to the surface of the cuticle and overlap slightly, they are spatulate. The surface is smooth. The scales are bordered by marginal spines 5 μm in length. The dimensions of the scales range 22 to 28 μm wide and 17 to 30 μm long (Fig. 1). The anterior end of the cephalon is bordered by a carina that widens slightly medially (230 μm) and expands to a triangular rostrum that is ventrally directed. The carina extends beyond the right and left borders of the cephalon (Fig. 2).

The eyes are absent on the lateral, ventral or frontal regions of the cephalon (Fig. 2). Omatidia, lenses or reflecting pigment membranes are absent. Two sutures, one on each side, are located on the distal seventh portion of the cephalon almost parallel to the border at each margin. The sutures extend medially approximately 90 μm and are 2 μm wide. The sutures are bordered by microtrichs (Figs 2 and 3). Two simple, non-annulated, robust setae with one sensory hair are located 20 to 25 μm from each suture. The setae are 1.3 μm wide and 55–70 μm in length. The surface of the setal shaft is smooth. The sensory hair is 15–20% of the total length of the setae. The apical portion of the sensory hair is acute, a terminal pore is absent both on the setae and sensory hair. The setae articulate with the cuticle by a socket (Figs 3–5).

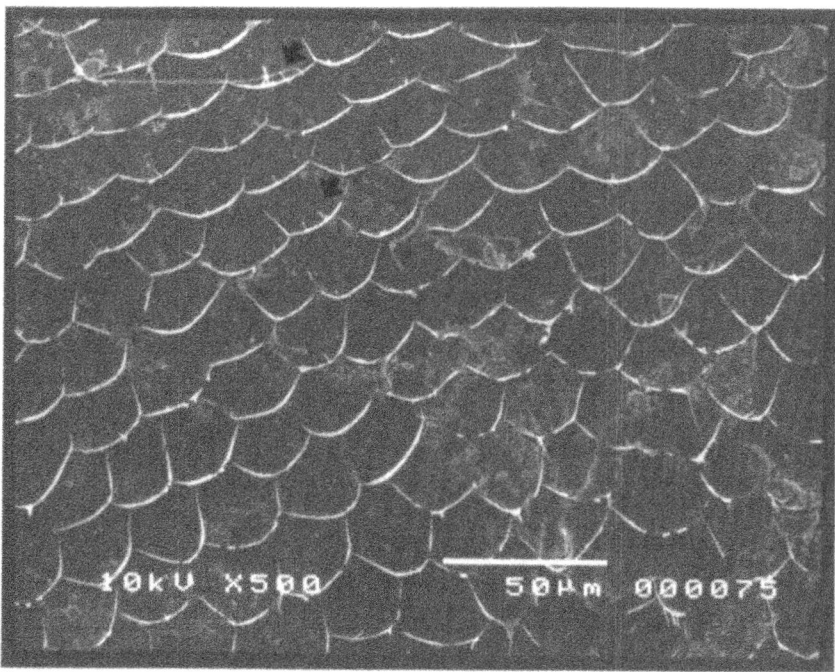

Figure 1. Cuticular scales covering the cephalon. The scales are bordered by marginal spines (⇓).

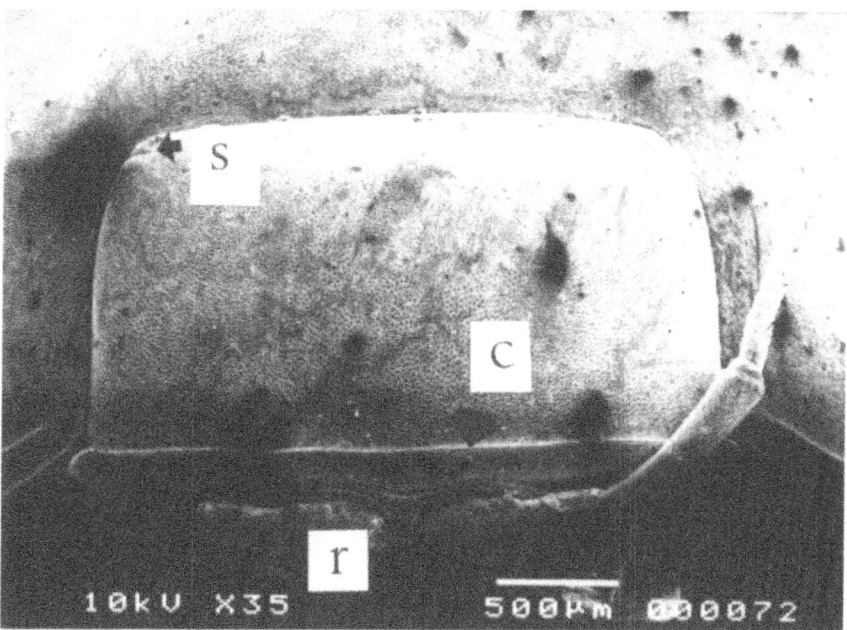

Figure 2. Anterior view of the cephalon showing the carina (c), rostrum (r), suture (s).

Two similar setae as described above are located medially on the cephalon 800 μm from the lateral margins and 28 μm from the distal border. A total of six setae are present on the cephalon. All setae are equal in length, have similar ornamentation, bear a sensory hair and are articulated to the cuticle by a socket inserted in a supra cuticular mode. The setae occur isolated on the cephalon.

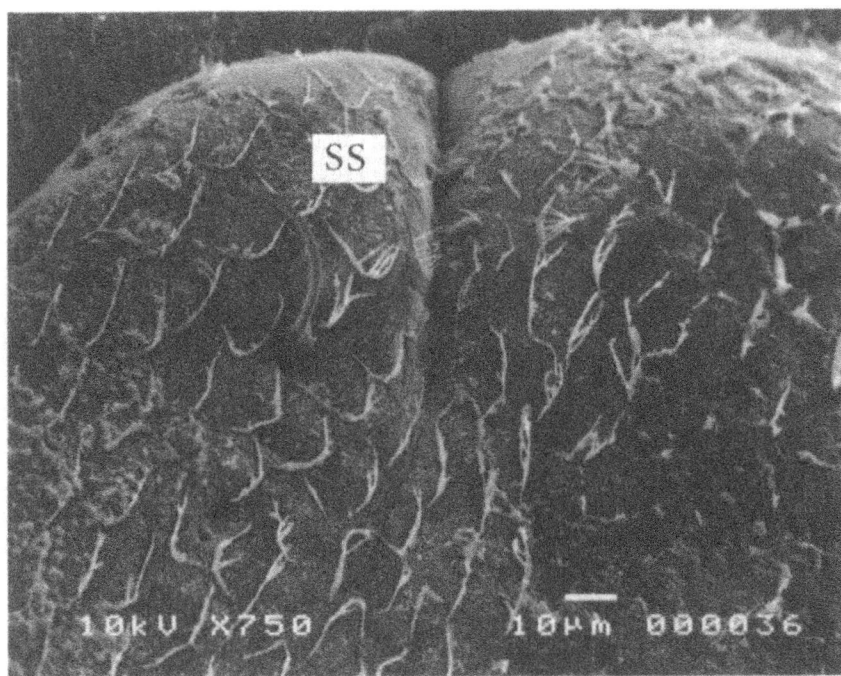

Figure 3. Suture on the left lateral angle of the cephalon. One simple setae (ss), is shown on the left side of the image.

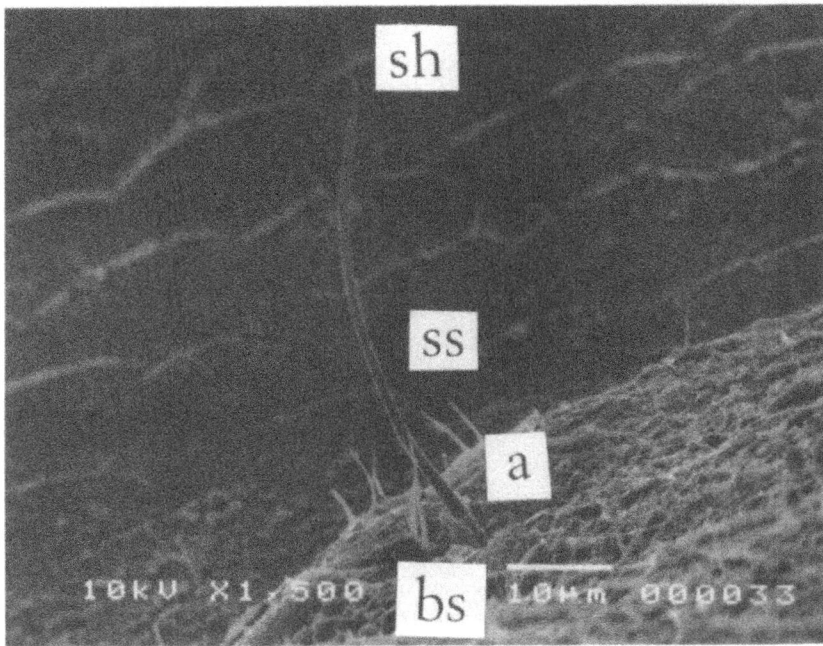

Figure 4. Simple setae (ss) with sensory hair (sh), annulation (a) and basal socket (bs).

Discussion

Constraints imposed by submerged caves, impoverishment and monotony of available food resources coupled with the aphotic nature of the habitat, have led in a surround of the isopod to use chemosensory and mechanoreceptor mechanisms. As other archetypal troglobic organisms, the isopods display morphological characteristics linked to the physical limitations of the cave environment, such as a general lack of

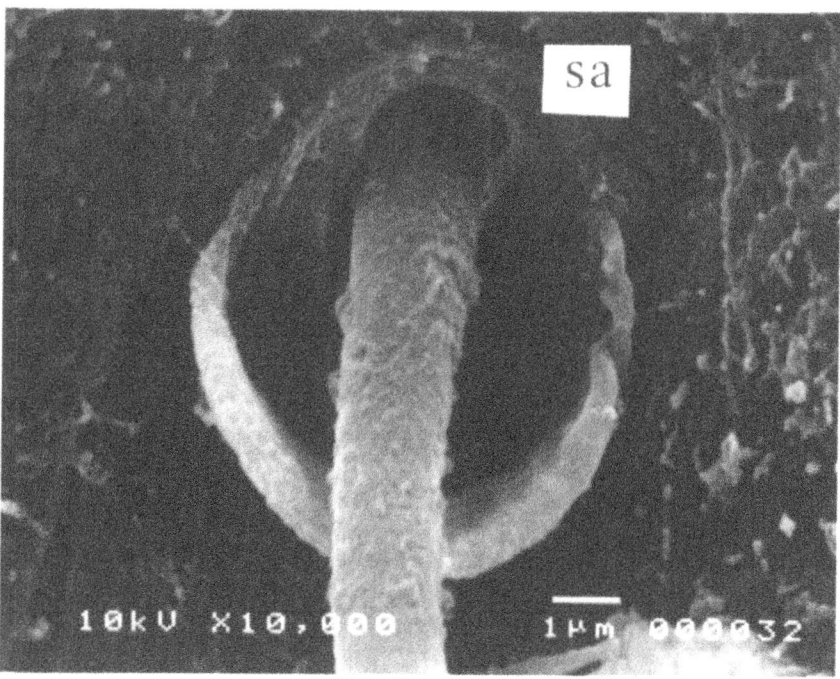

Figure 5. Socket articulation (sa) and shaft of simple seta.

pigmentation, an ocular regression, and hypertrophy of sensory organs (Christiansen, 1992). Adaptive responses for increased efficiency at food finding and avoiding predators can be observed in cephalic cuticular structures of Crustacea.

Cuticular ornamentation and scales can have an important function (Wägele, 1993) and in trying to understand evolutionary trends within Crustacean (Spano & Rapisarda, 1995, 1997). The functions attributed to the polygonal scales include enhancing chemical communication, interspecific defense, mechanical sensibility and thermoregulation as has been recognized in insects (Birch et al., 1990; Scoble, 1995). However it has been suggested that scales, as well as other micro structures, might be the result of differences in phylogeny and not function. Structures as the frontal carina, common to several isopod species, has been suggested to reduce friction and drag during displacement in the deep-sea (Lazier & Mann, 1989). Anophtalmy may be compensated by the complex setal arrangement on the cephalon that can be used as mechanoreceptors that provide the organisms with means to determine direction of vibration in the surrounding medium (De Lattin, 1939); the vibration would then be concentrated in the stalk and passed to underlying nerves. The movement provided by the setal socket may allow the bending of the setae and eli-

cit mechanosensory transduction to the nervous cells (Felgenhauer, 1992).

Mechanoreceptive setae, involved in the detection of currents and directional sensitivity, have a variable external morphology. A way of assessing the function of the seta at the ultra structural level is provided by electro physiological tests (Derby, 1989). Micro-anatomy may describe the innervation of these receptors. However, a common feature of mechanoreceptive setae is the presence of a movable socket at the base of the sensillum (Felgenhauer, 1992). In contrast chemoreceptive setae usually bear an apical pore or very thin cuticle (Laverack, 1989).

Little is known on the biology of troglobic isopods in Mexico, most efforts include the description of new species (Bowman, 1975; Contreras-Balderas & Purata-Velarde, 1982; Rodríguez-Almaraz & Bowman, 1995), their natural history and habitat (Bolivar & Pietain, 1950) and the compiling of faunal lists (Redell, 1981). Few studies have described the physiology of cave Crustacea (Derby, 1989) and lesser number have described the mechanoreceptors in aquatic isopods (Menzies, 1956; Fish, 1972; Wägele, 1993). Further studies are needed in this direction.

Acknowledgements

This study was supported by grants DGAPA UNAM IN-203894, IN-213197 and IN-213798. Logistics and fieldwork support was provided by the Nohoch Diving Team lead by Mr M. Madden.

References

Alcocer, J., A. Lugo, L. E. Marín & E. Escobar, 1998. Hydrochemistry of waters from five cenotes and evaluation of their suitability for drinking water supplies, Northeastern Yucatán, México. Hydrogeol. J. 6: 293–301.

Birch, M. C., G. M. Poppy & T. C. Baker, 1990. Scents and eversible scent structures of male moths. Ann. Rev. Ent. 35: 25–58.

Bolivar, A. & C. Pietain, 1950. Estudio de una *Cirolana* cavernícola nueva de la región de Valles, San Luis Potosí, México. Ciencias 10: 211–218.

Bowman, T. E., 1975. A new genus and species of troglobic isopod from San Luis Potosi, Mexico. Occas. Pap. Mus. Texas Tech Univ. 27: 1–7.

Christiansen, K., 1992. Biological processes in space and time: cave life in the light of modern evolutionary theory. In Camacho, A. I. (ed.), The Natural History of Biospeleology. Monogr. Mus. Nac. Cienc. Nat. C.S.I.C. Madrid: 456–478.

Contreras–Balderas, S. & D. C. Purata- Velarde, 1982. *Specirolana guerrai* sp. nov., cirolánido troglobio anoptalmo de la cueva de la Chorrera, Linares, Nuevo León, México. Association of Mexican Cave Studies Bulletin 8: 1–12.

Creaser, E. P. 1936. Large cave crustaceans of the Yucatan Peninsula. Carnegie Institution Washington 491: 159–164.

De Lattin, G., 1939. Untersuchungen von Isopodenaugen. Zool. Jrb. Anat. 65: 417–468.

Derby, C. D., 1989. Physiology of sensory neurons in morphologically identified cuticular sensilla of Crustaceans. In Felgenhauer, B. T., L. Watling & A. B. Thistle (eds), Functional Morphology of Feeding and Grooming in Crustacea. Crustacean issues. Vol. 6. Netherlands. Balkema: 27–48.

Eichem, A. C. W. K. Dodds, C. M. Tate & C. Edler, 1993. Microbial decomposition of elm and oak leaves in a karst aquifer. Appl. Envir. Microbiol. 59: 3592–3596.

Felgenhauer, B. E., 1992. External Anatomy and Integumentary structures. In: Microscopic Anatomy of Invertebrates. Vol. 10: Decapod Crustacea. Wiley Liss, Inc: 7–43.

Felgenhauer, B. E., 1987. Techniques for preparing crustaceans for scanning electron microscopy. J. Crust. Biol. 7: 71–76.

Fish, S., 1972. The setae of *Eurydice pulchra* (Crustacea: Isopoda). J. Zool. Lond. 166: 163–177.

George, R. Y., 1981. Functional adaptations of deep-sea organisms. In Jornberg, F. J. & W. B. Verndery (eds), Functional Adaptations of Marine Organisms. New York. Academic Press: 280–347.

Holsinger, J. R., 1986. Zoogeographic patterns of North American subterranean amphipod Crustaceans. In Gore, R. H. & K. L. Heck (eds), Biogeography. Crustacean Issues Vol. 4. Balkema, Netherlands: 85–106.

Kaufman, R. S., 1994. Structure and function of chemoreceptors in scavenging lysianassoid amphipods. J. Crust. Biol. 17: 54–71.

Laverack, M. S., 1989. The diversity of chemoreceptors. In Atema, J., R. R. Fay, A. N. Popper & W. N. Tavolga (eds), Sensory Biology of Aquatic Animals. New York. Springer Verlag: 287–312.

Lazier, J. R. N. & K. H. Mann, 1989. Turbulence and diffusive layers around small organisms. Deep-Sea Res. 36: 1721–1733.

Menzies, R. J., 1956. A study of the microscopic structure of isopod setae. Ann. Magniez Nat. Hist. 12, 9: 698–700.

Redell, J. R., 1981. A review of the cavernicole fauna of Mexico, Guatemala and Belize. Bulletin 27. Texas Memorial Museum, Austin: 327 pp.

Rodríguez-Almaraz, G. A. & T. E. Bowman, 1995. *Sphaerolana karenae*, a new species of hypogean isopod crustacean from Nuevo León, México. Proc. Biol. Soc. Washington 108: 207–211.

Scoble, M. J., 1995. The Lepidoptera: Form, Function and Diversity. Oxford University Press, New York.

Spano, N. & A. Rapisarda, 1995. SEM carapace micro-morphology as a diagnostic character for some Decapoda (Fam. Leucosiidae). Crustaceana 68: 489–492.

Spano, N. & A. Rapisarda, 1997. Determination of subfossil *Ebalia* specimens by scanning electron microscopy (Decapoda, Brachyura, Oxystomata). Crustaceana 70: 758–761.

Townsend, V. R. & B. E. Felgenhauer, 1998. The cuticular scales of lynx spiders (Araneae, Oxyopidae). J. Morphol. 236: 223–231.

Wägele, J. W., 1993. Isopoda. In Harrison, F. W. & A. G. Hames (eds), Microscopic Anatomy of Invertebrates. Vol. 9. Crustacea. Wiley-Liss, New York: 529–617.

Watling, L., 1989. A classification system for crustacean setae based on the homology concept. In Felgenhauer, B. E., L. Watling & A. B. Thistle (eds), Functional Morphology of Feeding and Grooming in Crustacea. Crustacean issues 6, A.A. Balkema: Rotterdam, Netherlands: 15–26.

Hydrobiologia **467**: 63–69, 2002.
J. Alcocer & S.S.S. Sarma (eds), Advances in Mexican Limnology: Basic and Applied Aspects.
© 2002 *Kluwer Academic Publishers.*

Population growth of *Asplanchna sieboldi* fed two *Brachionus* spp. (Rotifera) raised on green alga and baker's yeast

S.S.S. Sarma*, P.S. Larios-Jurado & S. Nandini[1]
Carrera de Biología, Universidad Nacional Autonoma de Mexico, Campus Iztacala AP 314, CP 54090,
Los Reyes, Tlalnepantla, Estado de México, Mexico
E-mail: sarma@servidor.unam.mx
[1]*CyMA Project*

Key words: predator, prey, rotifer, food type, alga, yeast

Abstract

We conducted population growth experiments of *A. sieboldi* using *Brachionus calyciflorus* and *Brachionus patulus* as prey. The prey rotifers were mass cultured separately on *Chlorella vulgaris*, *Saccharomyces cerevisiae* or on their mixture. Data on population growth of *A. sieboldi* showed prey type and food density-related differences. At any given prey concentration, both *B. calyciflorus* and *B. patulus* raised on a mixture of alga and yeast, resulted in higher abundance of the predator than those raised solely on alga or yeast. The rate of population increase per day (*r*) of *A. sieboldi* increased with increasing prey density for both prey species. However, predators grown on *B. patulus* showed higher *r* values compared to those grown on *B. calyciflorus*.

Introduction

In rotifers, the nutritional quality of prey (e.g., *Brachionus plicatilis*) is important for the survival and growth of larval fishes in aquaculture (Watanabe et et al., 1983). For example, yeast-fed rotifers are inadequate to support the growth of many marine fish larvae (Watanabe, 1989). On the other hand, alga-fed rotifers are considered to be nutritionally excellent for proper growth and survival of fish larvae in both freshwater pisciculture and mariculture (Fernandez-Reiriz & Labarta, 1996). Larval fish are not the only natural predators for planktonic rotifers in nature. A variety of invertebrates such as insects and predatory genera of rotifers such as *Asplanchna*, *Asplanchnopus*, *Dicranophorus*, *Ploesoma*, *Proales*, *Cupelopagis*, and *Encentrum* also exert a great pressure on the abundance of herbivorous zooplankton (Stemberger & Gilbert, 1987; Bevington et al., 1995).

Most members of the rotifer genus *Asplanchna* are predatory. They engulf a variety of prey species which include ciliates, rotifers, cladocerans and copepods among others (Williamson, 1983; Amdt, 1993; Sarma, 1993). A good set of data is available on the food and feeding habits including feeding behaviour and hunger-related aspects of *Asplanchna* (Pourriot, 1977; Dumont, 1977; Nandini & Sarma, 1999). In general, many studies on *Asplanchna* have aimed at understanding the effect of density and diverstiy of prey species on the population growth and life table demography (Dumont & Sarma, 1995; Iyer & Rao, 1996).

Being a voracious predator, *Asplanchna* can be effectively used as a bioassay organism to test prey quality, particularly herbivorous rotifers which are also essential in the feeding of larval fishes. However, rarely the quality of prey grown on various diets has been tested using the population growth of *Asplanchna* (Sarma et al., 1998). In the population growth of zooplankton, density and nutritional quality of food item are the two important variables often considered. With reference to density of the food items used for the growth of zooplankton particularly planktonic rotifers, considerable information exists, from both field and laboratory studies (Walz, 1995; Ooms-Wilms et al., 1999). On the other hand, the role of food quality on the population growth of zooplankton has received importance only recently (Gulati & DeMott, 1997). Most of these studies have been conducted on herbivorous zooplankton. For example, Kilham et al.

(1997) have shown that the population growth rate of *Daphnia pulicaria* is significantly lower on algae deficient in nitrogen and phosphorus. However, it is not well-known as to how the nutritional quality of prey affects the population growth of their predators. In a study conducted on the population growth of *Asplanchna sieboldi* on *Brachionus calyciflorus* raised on *Chlorella*, waste water (from a food processing industry) and on a mixture of both, Sarma et al. (1998) found that the population growth of *Asplanchna* was maximal on rotifers raised exclusively on waste water.

The aim of the present study was to study the population growth of *Asplanchna sieboldi* fed *B. calyciflorus* and *B. patulus* which, in turn were raised on *Chlorella*, baker's yeast exclusively or on a 1:1 mixture of both.

Materials and methods

The prey and the predatory rotifer species used in this study have been maintained in our laboratory for more than 3 years. All the species were isolated from local waterbodies and clonal populations were established with parthenogetic females individually. For routine maintenance of cultures as well as for carrying out experiments we used reconstituted moderate hard water (EPA medium). The EPA medium was prepared by dissolving 96 mg $NaHCO_3$, 60 mg $CaSO_4$, 60 mg $MgSO_4$ and 4 mg KCl in 1 l of distilled water (Anonymous, 1985). The stock cultures of herbivorous rotifer species *Brachionus calyciflorus* and *B. patulus* were maintained separately using the green alga *Chlorella vulgaris* as the exclusive food. Mass culture of the alga was done in Bold Basal medium (Borowitzka & Borowitzka, 1988).

Prey rotifer species were mass cultured separately in 20 l aquaria containing 2×10^6 cell ml^{-1} of the green alga *Chlorella, Saccharomyces cerevisiae* (baker's yeast) or their mixture in equal proportion (numerically 1×10^6 cells ml^{-1} of alga and 1×10^6 cells ml^{-1} of yeast). Mass cultures of prey rotifer species were changed every alternate day with fresh EPA medium containing the same food concentration and commbination. In these cultures we usually obtained *B. calyciflorus* at a density of about 25 ind ml^{-1} and *B. patulus* at 50 ind ml^{-1}.

The predatory rotifer *Asplanchna siebodi* was maintained in a 5 l aquarium. They were fed on both the types of prey (at a density of 10 ind ml^{-1}) raised on various food combinations as mentioned above. We

were able to obtain *A. sieboldi* at a density of 0.5 ind ml^{-1} in the aquarium. *A. sieboldi* individuals were fed every day but the medium was replaced every alternate day.

For population growth of *A. sieboldi*, we used prey rotifers harvested daily during the exponential phase of their growth from the mass culture tanks. Experiments (for each of the two prey species separately) were conducted in 50 ml capacity glass jars containing 20 ml EPA medium. Each prey rotifer species was grown on algae alone, yeast alone or on their mixture and hence varied in their nutritional quality. We used 4 prey concentrations: 2.5, 5, 10 and 20 ind ml^{-1} for each of the three nutrional types. We maintained 3 replicates for each concentration. Thus for population growth of *A. sieboliodi* on *B. calycaiflorus*, we used 36 test jars (3 prey nutritional types × 4 prey densities × 3 replicates). Into each of these test jars, we introduced one young (age: 8±4 h after birth) individual of *A. sieboldi* using a wide bore Pasteur pipette under a stereomicroscope at 3× magnification. The number of prey per experimental jar was individually counted and introduced using a finely drawn Pasteur pipette. All experiments were conducted at 25 °C under continuous and diffused illumination. The initial pH of the test jars was adjusted to 7.5.

Following the addition of predators, at 24 h intervals, we counted the total number of *A. sieboldi* in each jar and transferred to new containers containing appropriate prey type, density and nutritional quality. Uneaten prey and dead *Asplanchna* individuals, if any, were discarded. Males of *Asplanchna* were not encountered during the study. For estimating the density of *Asplanchna*, we counted the entire contents of each jar. The experiment was terminated after day 10 when predator populations of most test jars began to decline. The rate of population increase (*r*) per day was calculated using the following exponential growth equation:

$$r = (\ln N_t - \ln N_0)/t, \tag{1}$$

where N_0 is the initial density (ind ml^{-1}) of *A. sieboldi*; N_t is final population density of the predator and *t* is the time in days.

We used different data points along the growth curve of *A. sieboldi* to calculate the mean per replicate. We took 4–6 data points on the growth curve during the exponential phase of the *Asplanchna* population as documented in Dumont et al. (1995).

The experimental design, data collection and analysis for the population growth of *A. sieboldi* fed *B. patulus* were the same as for *B. calyciflorus*.

Results

The population growth of *A. sieboldi* showed differences related to prey type, food density and the diet on which brachionids were raised. Regardless of prey species and thier rearing conditions, *Asplanchna* showed increased population growth with increasing food abundance. At any given prey concentration, both *B. calyciflorus* and *B. patulus* raised on a mixture of alga and yeast, resulted in higher abundance of the predator than those raised solely on algae or yeast. Within the categories of yeast-fed and alga-fed *B. calyciflorus* or *B. patulus*, prey raised on yeast alone supported higher population abundance of *A. sieboldi* (Figs 1–4).

There was a significant effect of prey type ($p < 0.05$, 2-way ANOVA) and prey density ($p < 0.01$) on the maximal population abundance of *A. sieboldi*, regardless of the brachionid species used. However, interaction of prey type × prey density was statistically insignificant for both *B. calyciflorus* and *B. patulus* (Table 1). The day at which maximal abundance values were reached varied from 6–8 days for *B. patulus* and 7–9 days for *B. calyciflorus*. There was no significant difference in the day at which maximal abundance values was reached either for *B. patulus* or *B. calyciflorus* ($p > 0.05$).

A. sieboldi reached peak abundance values (mean±standard error) of 0.43±0.10, 0.47±0.07, 0.65±0.20 and 1.28±0.20 ind ml^{-1} at prey (*B. calyciflorus* raised on algae) concentrations of 2.5, 5, 10 and 20 ind ml^{-1}, respectively. Comparable values of *A. sieboldi* grown on algae-fed *B. patulus* were 0.42±0.11, 1.07±0.15, 1.17±0.30 and 2.45±0.19 ind ml^{-1}. The highest peak population abundance (3.00±0.26 ind ml^{-1}) of *A. sieboldi* was obtained when *B. patulus*, raised on a mixture of alga and yeast, was used as prey. The lowest peak population density (0.43±0.10) of the predators were on alga-fed *B. calyciflorus* at 2.5 ind ml^{-1} prey density.

The rate of population increase per day (r) of *A. sieboldi* increased with increasing prey density, regardless of nutritional type. Between the two prey speices, *A. sieboldi* fed *B. patulus* showed higher r values compared to *B. calyciflorus*-fed predators (Fig. 5). Regardless of the prey species, the r values for *A.*

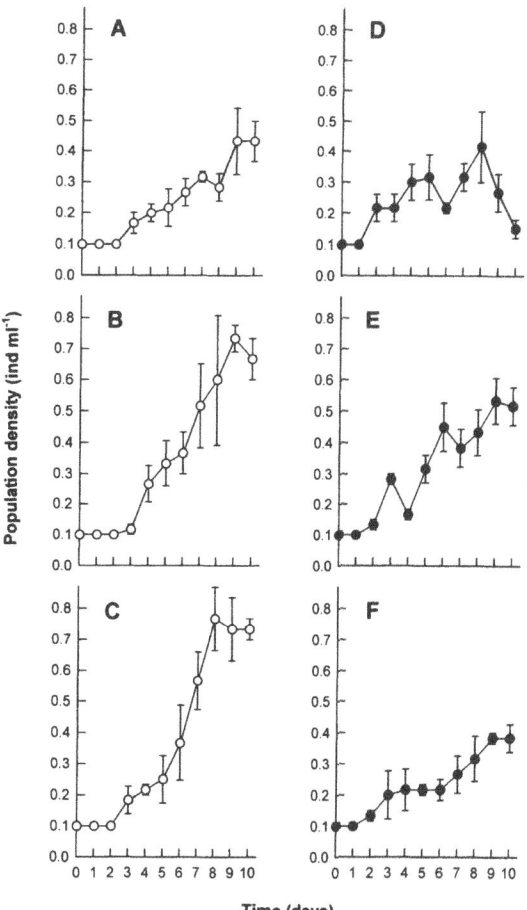

Figure 1. Population growth of *Asplanchna sieboldi* fed *B. calyciflorus* (open circles) and *B. patulus* (closed circles) at the prey density of 2.5 ind ml^{-1}. Prey species were cultured on only *Chlorella* (A, D), a mixture of alga and yeast (B, E) or only on yeast (C, F). Shown are the mean ± standard error based on three replicate recordings.

sieboldi were significantly influenced by both prey feeding conditions and prey density ($p < 0.01$, 2-way ANOVA; Table 1), but their interaction was not significant ($p > 0.05$). The range of r values for *A. sieboldi* in this study varied from 0.14±0.02 to 0.37±0.02.

Discussion

The prey brachionids used in the study for growing *Asplanchna* can be expected to have differences in their nutritional quality due to *Chlorella* or yeast (or their mixture). From a vast majority of publications concerning the nutrational quality of *Brachionus plicatilis*, it is evident that yeast-fed rotifers are inadequate for proper growth and survival of marine

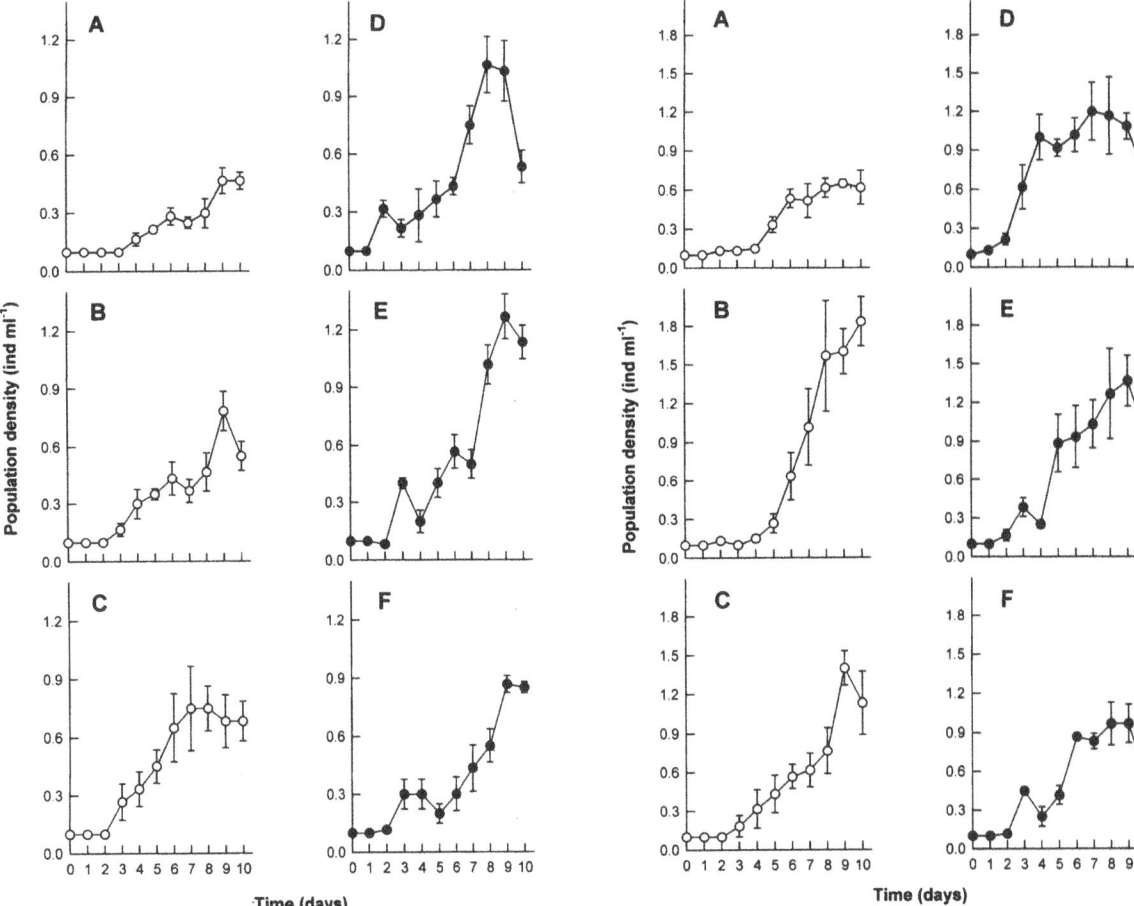

Figure 2. Population growth of *A. sieboldi* fed *B. calyciflorus* (open circles) and *B. patulus* (closed circles) at the prey density of 5 ind ml^{-1}. Prey species were cultured on only *Chlorella* (A, D), a mixture of alga and yeast (B, E) or only on yeast (C, F). Other details as in Figure 1.

Figure 3. Population growth of *A. sieboldi* fed *B. calyciflorus* (open circles) and *B. patulus* (closed circles) at the prey density of 10 ind ml^{-1}. Prey species were cultured on only *Chlorella* (A, D), a mixture of alga and yeast (B, E) or only on yeast (C, F). Other details as in Figure 1.

fish larvae (Watanabe, 1989). It is not known how invertebrate predators are affected by yeast-fed rotifers. From the present study, it is evident that the predatory *Asplanchna sieboldi* was not affected by their prey brachionids, raised on yeast. Sarma et al. (1998) have earlier shown that when *B. calyciflorus* raised on waste water (containing no phytoplankton) from a food processing industry were used as prey for *A. sieboldi*, the peak abundance values were higher than those when rotifers raised on green alga were used.

At present, we could not determine the biochemical composition of the prey *Brachionus* raised on different food types. Published information on the this genus indicates that the main difference in rotifers raised on alga and yeast lies in the content of highly unsaturated fatty acids (HUFA) rather than those of

total proteins and amino acid profiles (Frolov et al., 1991; Tamaru et al., 1993). Although HUFA is absent in yeast-fed brachionids, it is not known how HUFA affects the population growth of predatory rotifers. From the data, it is obvious that in terms of population growth, *A. sieboldi* was not affected by the absence of HUFA in *B. calyciflorus* or *B. patulus*.

Apart from nutritional quality, body size of prey rotifers grown in different diets may affect the rate of population growth of *Asplanchna*. Such a possibility has been shown for *B. calyciflorus* grown on waste water (Sarma et al., 1998). Although we did not measure the body size of brachionids grown on various diets in this study, Guevara et al., (1996) noted no differences in the body size of *B. plicatilis* grown on various diets including microalgae and baker's yeast. Nagata &

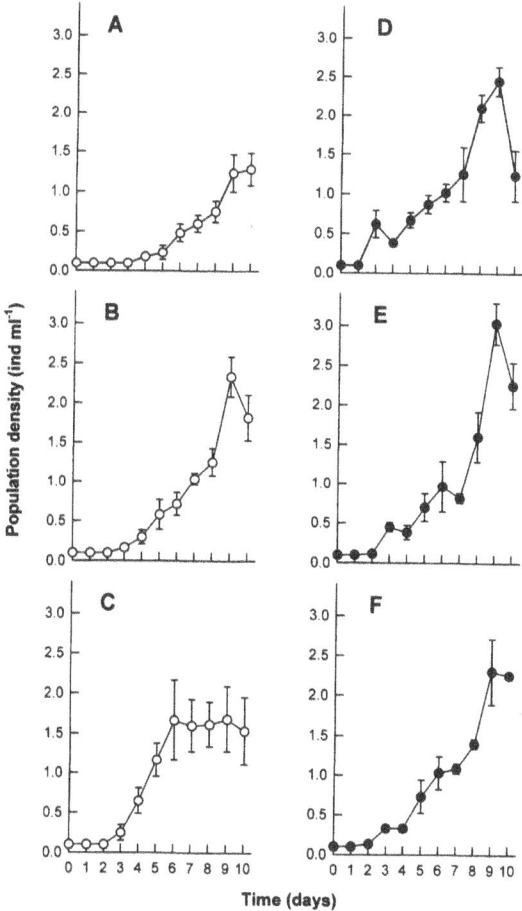

Figure 4. Population growth of *A. sieboldi* fed *B. calyciflorus* (open circles) and *B. patulus* (closed circles) at the prey density of 20 ind ml^{-1}. Prey species were cultured on only *Chlorella* (A, D), a mixture of alga and yeast (B, E) or only on yeast (C, F). Other details as in Figure 1.

Table 1. Analysis of variance (ANOVA) on the selected population variables of *Asplanchna sieboldi* grown on the prey *B. calyciflorus* and *B. patulus*. df = degrees of freedom, SS = sum of square, MS = mean square, F = f-ratio, *** = $p < 0.001$, ** = $p < 0.01$, * = $p < 0.05$, ns = not significant

Variable	df	SS	MS	F
Rate of population increase (r)				
Quality of *B. calyciflorus* (A)	2	0.028	0.01	11.52***
Prey density (B)	3	0.075	0.02	20.38***
A×B interaction	6	0.007	0.00	0.79 ns
Error	24	0.029	0.00	–
Quality of *B. patulus* (A)	2	0.011	0.01	6.52**
Prey density (B)	3	0.141	0.05	55.76***
A×B interaction	6	0.001	0.00	0.19 ns
Error	24	0.020	0.00	–
Day of maximal abundance				
Quality of *B. calyciflorus* (A)	2	0.89	0.44	0.27 ns
Prey density (B)	3	1.11	0.37	0.23 ns
A×B interaction	6	6.89	1.15	0.70 ns
Error	24			
Quality of *B. patulus* (A)	2	6.50	3.25	2.92 ns
Prey density (B)	3	6.33	2.11	1.90 ns
A×B interaction	6	5.50	0.92	0.8 ns
Error	24	26.67	1.11	–
Peak population abundance				
Quality of *B. calyciflorus* (A)	2	3.267	1.63	12.97***
Prey density (B)	3	8.714	2.90	23.06***
A×B interaction	6	1.158	0.19	1.53 ns
Error	24	3.023	0.13	–
Quality of *B. patulus* (A)	2	0.657	0.33	4.53*
Prey density (B)	3	23.27	7.76	106.99***
A×B interaction	6	0.463	0.08	1.07 ns
Error	24	1.74	0.07	–

Whyte (1992) have shown that the total energy content of rotifers fed on yeast cells was significantly higher than those fed on *Chlorella saccharophila*. This perhaps explains in our study the better population growth of *A. sieboldi* on the prey raised on yeast rather than *Chlorella vulgaris*.

The population growth curves of *A. sieboldi* recorded here were typical to those documented for other species of *Asplanchna* (*A. girodi*: Dumont & Sarma, 1995; *A. intermedia*: Iyer & Rao, 1996; *A. brightwelli*: Sarma et al., 1998). In general, species of *Asplanchna* reach asymptotic or retardation phase of population growth more rapidly (usually less than 7 days) as compared to brachionids (Sarma et al., 1999). Thus, the duration of the experiment for the growth of *A. sieboldi* in this study was within the period used for many other studies. Similarly, the range of prey densities used here were within the values reported from literature (Nandini & Sarma, 1999).

The maximum population growth rates are variously termed in rotifer literature. Some of them are: instantaneous population growth (Nandini & Sarma, 1999), maximal rate of population increase (Stemberger & Gilbert, 1985) or simply growth rates (Caswell, 1989). However, most workers agree on the mode of calculation of this value. In general, when the growth pattern is smooth, any two data points on the growth curve can be considered for fitting the logistic equation. Since this is extremely rare, several workers have

68

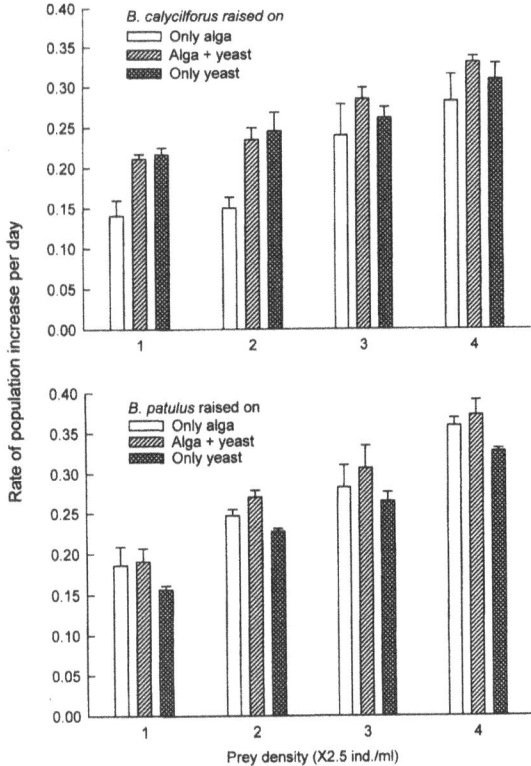

Figure 5. Rate of population growth (r) of *A. sieboldi* fed *B. calyciflorus* and *B. patulus* under various prey densities. Prey species were cultured on only *Chlorella*, a mixture of alga and yeast, or only on yeast. Shown are the mean and standard error based on three replicate recordings.

used more than one data point along the growth curve to obtain a mean value which is most representative of the trends (Dumont et al., 1995). We followed this approach here. For *Asplanchna*, the rates of population growth on various food items is known to vary (Dumont et al., 1995) greatly. Most values lie between 0.1 and 2.0 per day (Dumont & Sarma, 1995). In the present study, these values were within this broad range observed for other *Asplanchna* species.

In conclusion, our study shows that the nutritional quality of the prey possibly caused by alga and yeast, had no significant impact on the population growth of *A. sieboldi* unlike larval fish which are sensitive to the absence of unsaturated fatty acids.

Acknowledgments

We thank the authorities of ENEP-Iztacala for the facilities. SSSS and SN thank National System of Investigators for support (SNI-18723 and 20520).

References

Anonymous, 1985. Methods of measuring the acute toxicity of effluents to freshwater and marine organisms. US Environment Protection Agency EPA/600/4-85/0 13.

Arndt, H., 1993. Rotifers as predators on components of the microbial web (bacteria, heterotrophic flagellates, ciliates) – a review. Hydrobiologia 255/256: 231–246.

Bevington, D. J., C. White & R. L. Wallace, 1995. Predatory behaviour of *Cupelopagis vorax* (Rotifera; Collothecea; Atrochidae) on protozoan prey. Hydrobiologia 313/314: 213–217.

Borowitzka, M. A. & L. J. Borowitzka, 1988. Micro-algal Biotechnology. Cambridge University Press, London.

Caswell, H., 1989. Analysis of life table response experiments. 1. Decomposition of effects on population growth rate. Ecol. Modell. 46: 221–237.

Dumont, H. J, 1977. Biotic factors in the population dynamics of rotifers. Arch. Hydrobiol. Beih. 8: 98–112.

Dumont, H. J. & S. S. S. Sarma, 1995. Demography and population growth of *Asplanchna girodi* (Rotifera) as a function of prey (*Anuraeopsis fissa*) density. Hydrobiologia 306: 97–107.

Dumont, H. J., S. S. S. Sarma & A. J. Ali, 1995. Laboratory studies on the population dynamics of *Anuraeopsis fissa* (Rotifera) in relation to food density. Freshwat. Biol. 33: 39–46.

Fernandez-Reiriz, M. J. & U. Labarta, 1996. Lipid classes and fatty acid composition of rotifers (*Brachionus plicatilis*) fed two algal diets. Hydrobiologia 330: 73–79.

Frolov, A. V., S. L. Pankov, K. N. Geradze, S. A. Pankova & L. V. Spektorova, 1991. Influence of the biochemical composition of food on the biochemical composition of the rotifer *Brachionus plicatilis*. Aquaculture 97: 181–202.

Guevara, M., A. G. Gaspar & N. Mann, 1996. The use of microalgae and baker's yeast in the culture of *Brachionus plicatilis* from the Araya's saline. Acta Cientifica Venezolana 47: 255–261.

Gulati, R. D. & W. R. DeMott, 1997. The role of food quality for zooplankton: remarks on the state-of-the-art, perspectives and priorities. Freshwat. Biol. 38: 753–768.

Iyer, N. & T. R. Rao, 1996. Responses of the predatory rotifer *Asplanchna intermedia* to prey species differing in vulnerability: Laboratory and field studies. Freshwat. Biol. 36: 521–533.

Kilham, S. S., D. A. Kreeger, C. E. Goulden & S. G. Lynn, 1997. Effects of algal food quality on fecundity and population growth rates of *Dapnia*. Freshwat. Biol. 38: 639–647.

Nagata, W. D. & J. N. C. Whyte, 1992. Effects of yeast and algal diets on the growth and biochemical composition of the rotifer *Brachionus plicatilis* Muller in culture. Aquacult. Fish. Manag. 23: 13–21.

Nandini, S. & S. S. S. Sarma, 1999. Effect of hunger level on the prey capture behaviour, functional response and population growth of *Asplanchna sieboldi* (Rotifera). Freshwat. Biol. 42: 121–130.

Ooms–Wilms, A., G. Postema & R. D. Gulati, 1999. Population dynamics of planktonic rotifers in Lake Loosdrecht, the Netherlands, in relation to their potential food and predators. Freshwat. Biol. 42: 77–97.

Pourriot, R., 1977. Food and feeding habits of the Rotifera. Arch. Hydrobiol. Beih. 8: 243–260.

Sarma, S. S. S., 1993. Feeding responses of *Asplanchna brightwelli* (Rotifera): laboratory and field studies. Hydrobiologia 255/256: 275–282.

Sarma, S. S. S., S. Nandini & R. A. A. Stevenson, 1998. Nutritional quality of prey *Brachionus calyciflorus* affects the population growth of predatory rotifers (*Asplanchna sieboldi*) (Rotifera). Hidrobiologica 8: 73–80.

Sarma, S. S. S., M. A. Fernández–Araiza & S. Nandini, 1999. Competition between *Brachionus calyciflorus* Pallas and *Brachionus patulus* (Muller) (Rotifera) in relation to algal food concentration and initial population density. Aquatic Ecol. 33: 339–345.

Stemberger, R. S. & J. J. Gilbert, 1985. Body size, food concentration and population growth in planktonic rotifers. Ecology 66: 1151–1159.

Stemberger, R. S. & J. J. Gilbert, 1987. Defense of planktonic rotifers against predators. In Kerfoot, W. C. & A. Sih (eds), Predation: Direct and Indirect Effects on Aquatic Communities. The University Press of New England, Hanover, N.H. U.S.A.: 227–239.

Tamaru, C.S., R. Murashige, C. S. Lee, H. Ako & V. Sato, 1993. Rotifers fed various diets of baker's yeast and/or *Nanno-chloropsis oculata* and their effect on the growth and survival of striped mullet *Mugil cephalus* and milkfish *Chanos chanos* larvae. Aquacuhure 110: 361–372.

Walz, N., 1995. Rotifer populations in plankton communities: Energetics and life history strategies. Experientia 51: 437–453.

Watanabe T., 1989. Nutrition and growth. In Shephard, C. J. & N. R. Bromage (eds), Intensive Fish Farming. Blackwell Science Publ., London: 154–197.

Watanabe, T., C. Kitajima & S. Fujita, 1983. Nutritional values of live organisms used in Japan for mass propagation of fish: a review. Aquaculture 34: 115–197.

Williamson, C. E., 1983. Invertebrate predation on planktonic rotifers. Hydrobiologia 104: 385–396.

Hydrobiologia **467**: 71–78, 2002.
J. Alcocer & S.S.S. Sarma (eds), Advances in Mexican Limnology: Basic and Applied Aspects.
© 2002 *Kluwer Academic Publishers.*

Nitrogen fixation patterns displayed by cyanobacterial consortia in Alchichica crater-lake, Mexico

Luisa I. Falcón[1], Elva Escobar-Briones[2,*] & David Romero[3]
[1]*Marine Science Research Center, Stony Brook University, Stony Brook, NY 11794-5000, U.S.A.*
[2]*Unidad Académica Sistemas Oceanográficos y Costeros, Instituto de Ciencias del Mar y Limnología,*
Universidad Nacional Autónoma de México, A.P. 70-305, 04510 D.F. Mexico
[3]*Centro de Investigación sobre Fijación de Nitrógeno, Universidad Nacional Autónoma de México,*
Programa de Genética Molecular de Plásmidos Bacterianos, A.P. 565-A, 62210 Cuernavaca,
*Morelos, Mexico (*Author for correspondence)*

Key words: UV radiation, benthos, plankton, tufa, *Nodularia* cf. *spumigena*

Abstract

Alchichica is a saline crater-lake located in Mexico. Tufa grows on its periphery and a wind-driven *Nodularia* cf. *spumigena* bloom occurs annually. Fixation rates were assayed by the acetylene reduction method. Here, we describe the patterns of nitrogen fixation on two tufa forms before, during and after the bloom, as well as those from the planktonic cyanobacteria. We also analyzed the effect of ultraviolet radiation (under 390 nm) on the nitrogen fixation rates. Tufa showed light-stimulated nitrogen fixation, while *N.* cf. *spumigena* peaks in early morning and midnight. Both tufa forms diminished their nitrogen fixation rates after the planktonic bloom. UV radiation affected negatively nitrogen fixation rates in all forms.

Introduction

Nitrogen cycling in aquatic ecosystems includes the uptake of inorganic forms by phytoplankton, its transfer as organic nitrogen to levels in the food web and its remineralization back to the inorganic nitrogen pool by bacteria. Aquatic ecosystems are limited in reduced forms of nitrogen (Howarth et al., 1988; Herbst, 1998; Paerl, 1999). The hydrodynamics and trophic relations of the system determine the sources, pathways and availability of the limiting nitrogen forms such as nitrate and ammonia (Rastettler et al., 1992).

Few organisms have the capacity to break the triple bond of the dinitrogen molecule and reduce it to ammonia; these organisms, called diazotrophs, include diverse consortia of bacteria and cyanobacteria (Sprent & Sprent, 1990). The enzyme catalyzing this reaction, called nitrogenase, has been extensively characterized (Kirshtein et al., 1991; Dean & Jacobson, 1992). Diverse cyanobacterial species have specialized nitrogen-fixing cells, called heterocysts, in which photosystem II is absent and anaerobic conditions prevail (Haselkorn, 1986). Other species have mech-

anisms to balance the diazotroph and photoautotroph functions (Khamees et al., 1987). Nitrogen fixing organisms lacking heterocysts show different seasonal and diurnal patterns (Sprent & Sprent, 1990; Stal, 1999). These patterns include fixing nitrogen at night and photosynthesizing at day, or forming colonies where intercellular oxygen concentration diminishes towards the center, favoring nitrogen fixation and photosynthesis in the periphery (Khamees et al., 1987; Paerl et al., 1995).

Our knowledge on nitrogen-fixing bacteria in tropical aquatic ecosystems is limited (Coyer et al., 1996). Cyanobacteria occur both as part of the phytoplankton and the littoral benthos (Carpenter & Capone, 1992). In the benthos, cyanobacteria form symbiotic and multiple species assemblages within microbial mats and the tufa forms (spongy and columnar, Tavera & Komarek, 1996), the latter constitute some of the most ancient and abundant evidences of early life forms on Earth (Walter et al., 1992).

Previous studies have shown that cyanobacterial consortia may be exposed to ultraviolet radiation in most of the environments inhabited (Whitton, 1987).

Cyanobacteria have evolved a series of mechanisms for UV protection that include: photoreparation, reparation of DNA in the dark, production of photoprotective carotenoids and UV filtering pigments such as scytonemin and mycosporine-like amino acids (Garcia-Pichel et al., 1993; Holm-Hansen et al., 1993; Mitchell & Karentz, 1993; Brenowitz & Castenholz, 1997; Castenholz & Garcia-Pichel, 1997).

Area of study

Alchichica is a saline (T.D.S.$=7.2\pm0.1$ g l^{-1}) crater-lake located at moderate altitude (2200 m) in central Mexico (19° 24′ N–97° 24′ W). Its morphology, chemistry and biology have been described by Alcocer et al. (1993), Lugo et al. (1993) and Escobar et al. (1999).

Spongy and columnar tufa forms have been described in the littoral zone of the lake by Tavera & Komárek (1996). The former type includes cyanobacteria with and without heterocysts, while in the latter type heterocysts are absent. An heterocystous cyanobacterial species, *Nodularia* cf. *spumigena*, occurs as a free-living form in the water column; large densities of this species coincide with the end of the wind-driven winter upwelling (Alcocer et al., 2001).

The pelagic and littoral systems are comprised by different primary producers; however, *N.* cf. *spumigena* is exported as biogenic carbon to the deep benthos and the littoral environment in the form of detritus. It is presumed that there are other organisms from the phytoplankton and zooplankton that contribute to the carbon lost to the bottom of the lake (Escobar et al., 1999).

The major objective of this study was to describe the patterns of nitrogen fixation by three cyanobacterial consortia in the saline crater-lake Alchichica in central Mexico based in the hypothesis that heterocyst planktonic and tufa forms (with and without heterocysts cyanobacteria and with non-heterocyst cyanobacteria) may have different rates. An experimental study was conducted to describe the effect of UV radiation on nitrogen fixation rates.

Materials and methods

Tufa samples were hand collected from the sublittoral zone and transported to the laboratory with water in plastic containers. Experiments were set up to evaluate the nitrogen fixation rates on a 24-h basis, measured as ethylene formation assayed with the acetylene reduction method (Bebout, 1992). Experiments on benthic cyanobacteria consortia were done prior to the *N.* cf. *spumigena* bloom (March), during (May) and after the bloom (July). Experiments on *N.* cf. *spumigena* were carried out only once, during the bloom. Nitrogen fixation was measured to define the patterns on the benthic consortia and the planktonic form. Due to the biological differences on the three cyanobacterial consortia each was analyzed separately. Results found during the experiments run in March allowed us to identify the time of nitrogen fixation peaks on the benthic forms. Tufa fixation peaks were tested for differences before, during and after the planktonic bloom. Differences for day/night nitrogen fixation in all three forms were tested considering the presence of heterocystic and non-heterocystic cyanobacteria in the crater-lake. All three forms were tested to define differences in fixation rates for both UV treatments. Data were tested with *t*-Student for independent samples and ANOVA statistics ($p<0.05$).

Four replicates of one square centimeter and one cubic centimeter were sampled from each tufa and planktonic form to evaluate the nitrogen fixation rates on two light treatments: absence and presence of UV radiation. A Lee filter #226 (absorbing 90% of radiation under 390 nm) was used to evaluate the absence of UV radiation. Three blank assays (containing only water) were included to measure spontaneous acetylene reduction; additionally, three sample replicates per tufa type and planktonic form (without acetylene) were used to evaluate if they naturally form ethylene. Each replicated sample and blank was incubated for 6 h in 58 ml flasks with 20 ml of water and 5 ml of acetylene (12% of the flask volume), which was bubbled in the water using a 5 ml syringe. The 6 h incubation periods corresponded to 24–6, 6–12, 12–18 and 18–24 h, samples were taken at the end of each period. After the incubation, a 10 ml gas sample was removed by syringe and analyzed in a Varian 3300 gas chromatograph fitted with a hydrogen flame ionization detector. The stainless steel column used was 0.32 cm (outer diameter) and 200 cm in length, packed with Porapak N (80–100 mesh). Data were recorded using a Varian 4290 integrator. The ethylene production was determined as the integration of the curve area and the values were transformed to μM m^{-2} h^{-1} (according to Bebout, 1992). Acetylene was produced by reaction of CaC_2 with water. Other factors such as temperature and salinity were kept constant and similar to the environmental measurements in order to quantify the effects caused by UV radiation only. After quantifying the nitrogen fixation during 24-h cycles for each tufa form in March, peaks were identified. Additional experiments carried out in both tufa forms considered peak-hours only (6–12 h and 12–18 h).

Results

March. a. Tufa

The nitrogen fixation values were in the range of 2.15–55.71 μM C_2H_4 m^{-2} h^{-1} in the spongy tufa when UV radiation was absent and in the range of 0.63–25.72 μM C_2H_4 m^{-2} h^{-1} when UV radiation was present (Fig. 1a). The columnar form presented values of 0.52–47.19 μM C_2H_4 m^{-2} h^{-1} when UV radiation was absent and of 0.24–20.26 μM C_2H_4 m^{-2} h^{-1} when UV radiation was present (Fig. 1b). Both tufa forms had a nitrogen fixation maximum for the 6–12 and 12–18 h intervals, under both UV radiation treatments (Table 1). Both peaks were significantly different ($p < 0.05$) than the rest of the values. The 6–12 h peak for the spongy form in absence of UV radiation was different than the non-peak values ($p < 0.0063$); differences were also shown for this peak in presence of UV radiation ($p < 0.0347$). The 12–18 h peak for the spongy form showed differences in both UV treatments with the non-peak values, in absence ($p < 0.0249$) and in presence ($p < 0.0021$). The columnar form also showed differences between the peak and non-peak values: at 6–12 h ($p < 0.0012$) in absence and ($p < 0.0016$) in presence of UV radiation and at 12–18 h ($p < 0.0001$) in absence and ($p < 0.0002$) in presence of UV radiation.

May. a. Planktonic form

The 24-h nitrogen fixation cycle for *N*. cf. *spumigena* showed a continuous increase in the treatment with absence of UV radiation with a minimum value of 12.47 μM C_2H_4 m^{-2} h^{-1} at 24–6 h and a maximum of 92.91 μM C_2H_4 m^{-2} h^{-1} recorded at 18–24 h. In presence of UV radiation (Table 1), nitrogen fixation rates showed a minimum of 4.72 μM C_2H_4 m^{-2} h^{-1} at 24–6 h and a maximum of 45.95 μM C_2H_4 m^{-2} h^{-1} at 18–24 h (Fig. 1c).

b. Tufa

The spongy tufa showed nitrogen fixation values between 16.95 and 24.55 μM C_2H_4 m^{-2} h^{-1} in absence of UV radiation and of 4.92–21.43 μM C_2H_4 m^{-2} h^{-1} during the treatment in presence of UV radiation (Table 1). The columnar form had nitrogen fixation values from 15.18 to 29.56 μM C_2H_4 m^{-2} h^{-1} in absence of UV radiation and ranged from 10.87 to 19.65 μM C_2H_4 m^{-2} h^{-1} in presence of UV radiation (Fig. 2a and b).

July. a. Tufa

The nitrogen fixation values were in the range of 3.66–21.77 μM C_2H_4 m^{-2} h^{-1} in the spongy tufa when UV radiation was absent and in the range of 1.72–19.16 μM C_2H_4 m^{-2} h^{-1} when UV radiation was present. The columnar form showed values of 1.62–5.08 μM C_2H_4 m^{-2} h^{-1} when UV radiation was absent and from 3.13 to 13.68 μM C_2H_4 m^{-2} h^{-1} when UV radiation was present (Fig. 2a and b).

Day/night differences ($p < 0.05$) in the nitrogen fixation rates of both tufa forms were observed during the 24 h cycles in March (tufa) and May (*N*. cf. *spumigena*). These occurred both in presence of UV ($p < 0.0000$) and absence of UV ($p < 0.0029$) for the spongy form and in presence of UV ($p < 0.0003$) and absence of UV ($p < 0.0105$) for the columnar form. *N*. cf. *spumigena* showed no clear day/night differences in the nitrogen fixation rates ($p < 0.2332$) neither in presence nor in absence of UV ($p < 0.4567$).

All nitrogen fixation values obtained during the three sets of experiments for both tufa (March, May and July) and planktonic form (May) were tested to observe differences ($p < 0.05$) between both UV treatments. Both tufa forms did not show differences for UV treatments, ($p < 0.0523$) for the spongy form and ($p < 0.2806$) for the columnar form. *N*. cf. *spumigena* showed differences for both UV treatments ($p < 0.0019$).

Nitrogen fixation rates of both tufa forms before, during and after the planktonic bloom were tested ($p < 0.05$). The spongy form showed differences between nitrogen fixation rates (Table 1) before and after the bloom ($p < 0.0018$) as well as during and after the bloom ($p < 0.0405$). The columnar form showed as well differences (Table 1) before and after the bloom ($p < 0.0388$) and during and after the bloom ($p < 0.0028$).

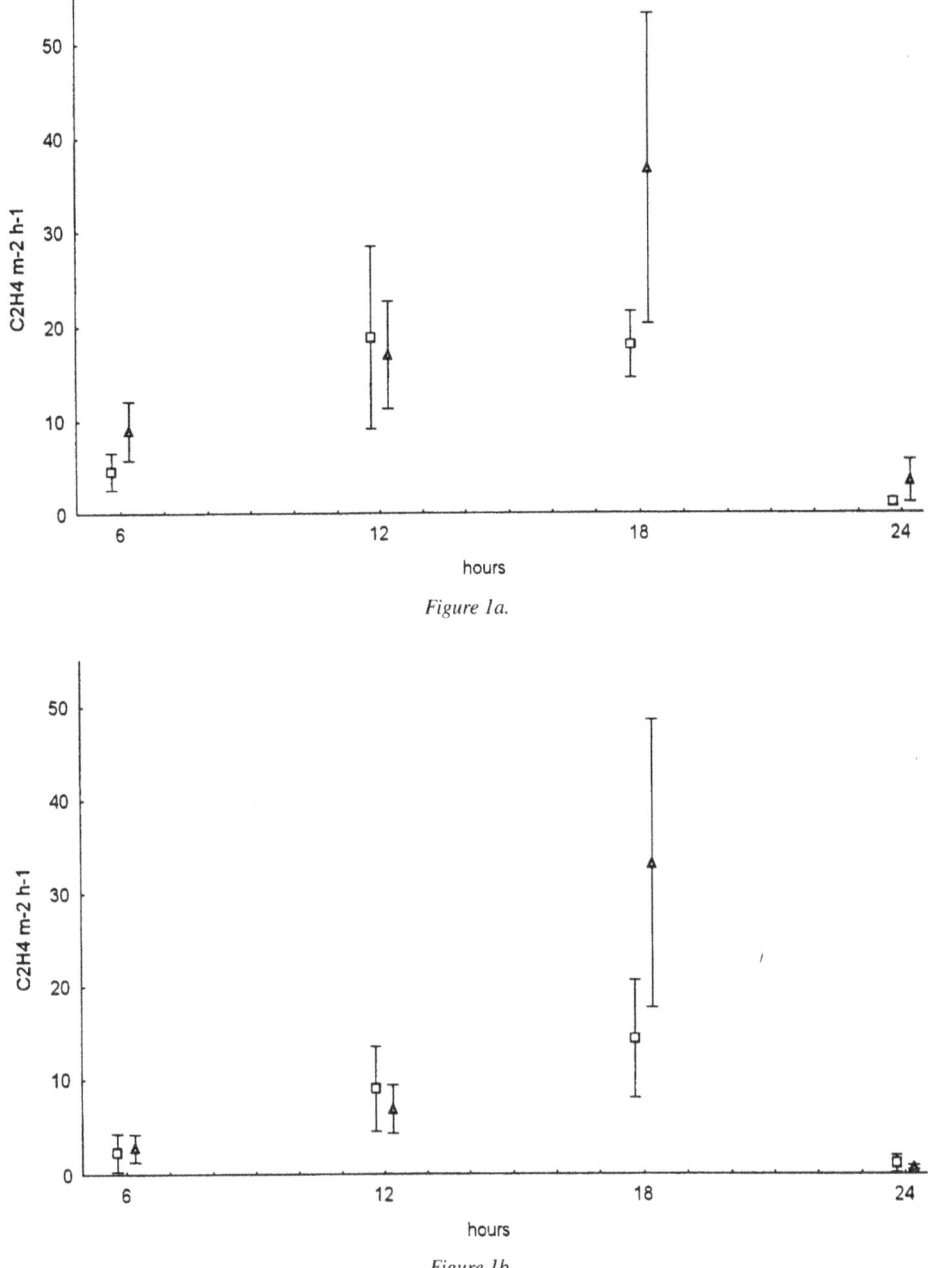

Figure 1a.

Figure 1b.

Figure 1. Nitrogen fixation rates in nmoles C_2H_4 (mean and standard deviation) during 24 hour cycle in presence (□) and absence (▲) of UV radiation: (a) for spongy tufa, (b) for columnar tufa and (c) for *N.* cf. *spumigena.*

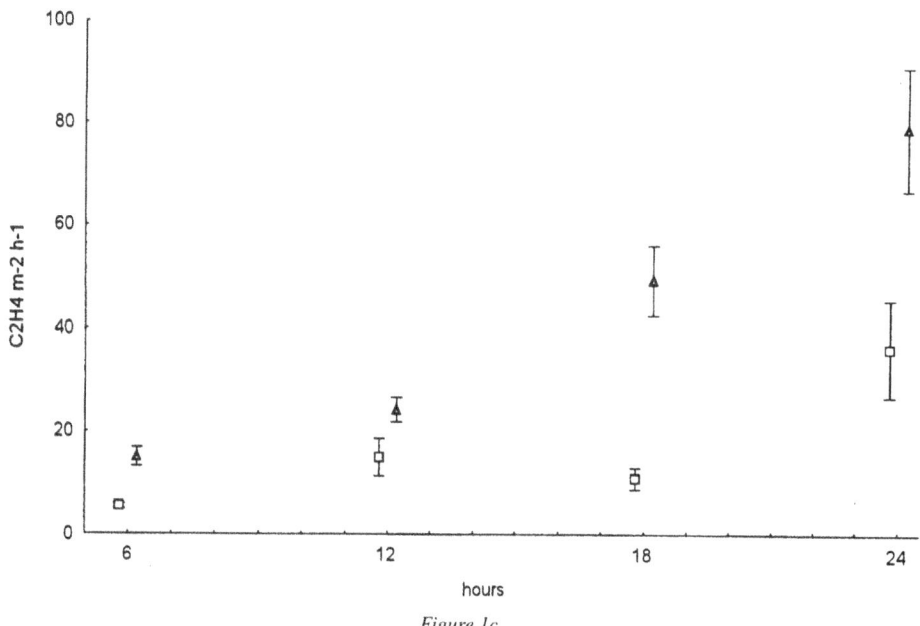

Figure 1c.

Table 1. Nitrogen fixation values in $\mu M\ C_2H_4\ m^{-2}\ h^{-1}$ in Alchichica crater lake

Period	Experiment	Cyanobacterial consortium	Time of measurement			
			T1 0.00–6.00	T2 6.00–12.00	T3 12.00–18.00	T4 18.00–24.00
Pre-bloom	UV	Spongy	4.57±2.02	18.80±9.67	18.0±3.51	1.11±0.42
		Columnar	3.07±1.46	8.95±4.48	14.27±6.31	1.07±0.91
	No UV	Spongy	8.95±3.16	16.97±5.68	36.77±16.50	3.43±2.30
		Columnar	2.87±1.44	6.82±2.53	33.12±15.45	0.71±0.24
Bloom	UV	Spongy		8.43±3.18	16.01±4.04	
		Columnar		13.41±1.99	13.84±4.04	
		Nodularia cf. *spumigena*	5.43±0.49	14.96±3.62	10.88±2.14	36.17±9.36
	No UV	Spongy		19.58±4.31	20.84±2.11	
		Columnar		19.89±4.03	22.19±5.71	
		Nodularia cf. *spumigena*	15.03±1.79	24.22±2.39	49.24±6.72	78.91±12.11
Post-bloom	UV	Spongy		7.60±7.90	4.36±2.93	
		Columnar		9.88±3.49	5.80±2.05	
	No UV	Spongy		7.74±3.94	10.19±7.95	
		Columnar		2.46±0.99	3.02±1.43	

Discussion

Rates of nitrogen fixation are variable over the daily cycle. Stal (1999) proposes three main nitrogen fixation patterns present in microbial mats during 24 h periods, depending on the presence or absence of heterocystous and non-heterocystous cyanobacteria. During this research, we found that both tufa forms fix nitrogen at day, although only the spongy form has heterocysts. Stal (1999) proposes a diurnal nitrogen fixation pattern for heterocystous forms, but he also proposes a similar pattern for non-heterocystous anaerobic cyanobacteria, which could be the case in this study. Diazotrophic activity in non-heterocystous forms is poorly understood. It is possible that in microbial mats and tufa forms, sulfate-reduction in-

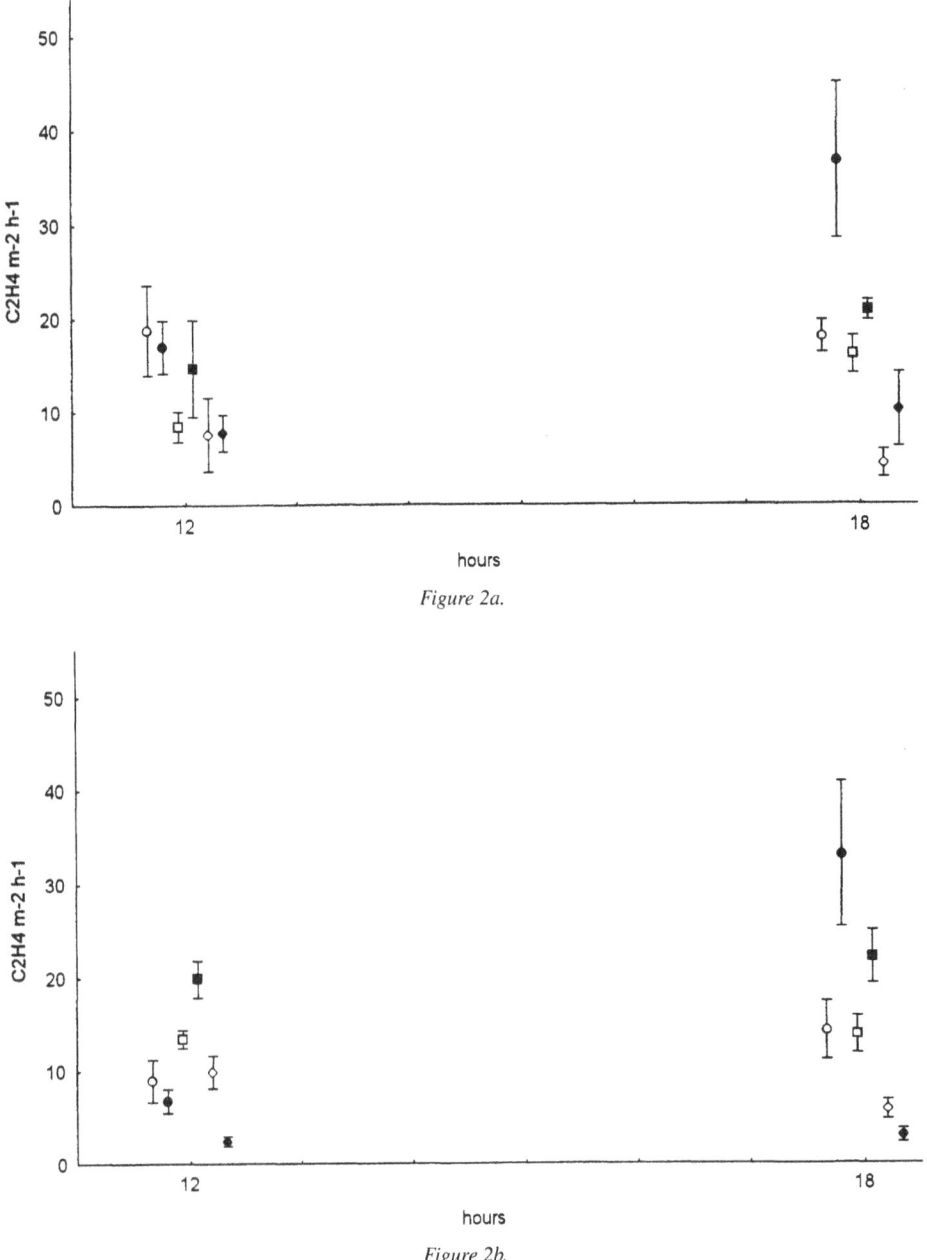

Figure 2a.

Figure 2b.

Figure 2. (a) Nitrogen fixation rates (mean and standard deviation) at 12.00 and 18.00 h for spongy tufa in presence (UV) and absence (n/UV) of UV radiation in March (UV □) (n/UV ■), May (UV ◊) (n/UV ♦) and July (UV ○) (n/UV ●). (b) Nitrogen fixation rates (mean and standard deviation) at 12.00 and 18.00 h for columnar tufa in presence (UV) and absence (n/UV) of UV radiation during March (UV □) (n/UV ■), May (UV ◊) (n/UV ♦) and July (UV ○) (n/UV ●).

hibits photosynthesis due to the production of sulfide (Stal, 1995; Stal, 1999) were non-heterocystous forms could be less affected by sulfide than heterocystous forms (Castenholz, 1976). Many factors affect nitrogen fixation by cyanobacteria in freshwater, causing the different patterns observed during diurnal and seasonal cycles (Sprent & Sprent, 1990). The planktonic form showed nitrogen fixation activity both at day (6–12 h) and night (18–24 h). Vanderhoef et al. (1975) proposed that planktonic cyanobacteria could be affected in their diurnal diazotrophic activity by high irradiance, showing maximum rates at night. At night, nitrogen fixation could be favored since diurnal photosynthetic activity causes the accumulation of carbohydrates (Stal, 1995). Nitrogen fixation in Alchichica crater-lake tufa and planktonic cyanobacteria occurred in presence of light, which indicates that this process is coupled to photosynthesis (Verschuren et al., 1999).

Cyanobacteria exposed to UV radiation show a decrease on their biomass production, photosynthetic and nitrogen fixation rates; the depletion of the ozone layer caused by human activities has increased the damage to limnetic and pelagic communities exposed to UV radiation (Babu et al., 1998). *N.* cf. *spumigena* nitrogen fixation was reduced 2.5 times in the presence of ultraviolet radiation, showing significant differences in presence and absence of UV. Tufa nitrogen fixation was reduced in 0.4–1.7 times in the presence of UV radiation, although no significant differences between treatments was found. Because tufa comprises a whole microbial community, dynamic relationships could be part of the fixation patterns found (Des Marais, 1995) hiding the UV effect over the nitrogen fixation response.

Nitrate tends to inhibit nitrogen fixation by cyanobacteria while on the other hand phosphate influx to phosphate deficient lakes triggers nitrogen-fixing cyanobacterial blooms (Sprent & Sprent, 1990). Both tufa forms continued fixing nitrogen at the same rate in presence of the planktonic bloom, indicating benthic nitrogen fixation is not affected by planktonic oxygen generation. At the end of the *N.* cf. *spumigena* bloom, the tufa fixation rates diminished, which could be caused by tissue saturation by reduced nitrogen forms that were generated during the bloom, although presently there are no measurements to support this.

A tropical warm-monomictic crater-lake like Alchichica is characterized by uniformly low productivity caused by permanent nutrient limitation. A strongly stratified column favors the transport of nutrients out of the euphotic zone by sinking of detritus, returning via upwellings. Mixing could be moving the remineralized nitrogen back to the surface waters annually. An energy-expensive process as nitrogen fixation could be favored in these tropical systems due to the low concentration of nutrients in the surface (Libes, 1992). It can be argued that all of the reduced forms of nitrogen should be incorporated directly to the food chain of the crater-lake; Escobar et al. (1999) showed that biogenic carbon originated from the *N.* cf. *spumigena* bloom is not transferred within the pelagic system. Most of this material accumulates and remains in the anoxic sediments at the bottom of the lake. It may as well be wind driven to the littoral zone where it may contribute to benthic littoral production. Diverse lake species of chironomids and cladocerans (Verschuren et al., 1999) may be functionally adapted to a diet of filamentous cyanobacteria. A third route would be to recycle the carbon and nitrogen in the euphotic zone through viral lysis (Thingstad et al., 1993). Historical records in African lakes have related changes in water column cyanobacteria consortia to progressive reduction in nutrient availability and supply in the catchment area due to loss by sequestration.

Conclusion

Nitrogen fixation rates in both tufa forms diminished after the bloom period of planktonic *N.* cf. *spumigena*. This is suggested to be a seasonal response to nitrogen availability. The cyanobacterial consortia of both tufa and planktonic forms fixed nitrogen in presence of light. The peak in *N.* cf. *spumigena* occurred at midnight. The tufa had a fixation pattern with peaks at noon and again in the evening, in contrast with *N.* cf. *spumigena* that showed a small peak at noon and a second peak at midnight. Nitrogen fixation rates were significantly lower in *N.* cf. *spumigena* when exposed to UV, diminishing 2.5 times. Both tufa forms diminished their nitrogen fixation rates when exposed to UV (0.4–1.7 times) although differences between UV treatments were not significant.

Acknowledgements

This study was financed by projects CONACyT 25430-T, DGAPA IN-204597 and IN-217298, additional support is due to the Institutes of CML and FN at UNAM. We are grateful to J. Alcocer, L. Peralta and L. Oseguera for support in the field and C. Ezcurdia and E. Valencia–Morales in the laboratory. We thank the reviewers to this paper for their valuable comments and contributions.

References

Alcocer, J., A. Lugo, S. Estrada, M. Ubeda & E. Escobar, 1993. Littoral chironomids of a Mexican Plateau athalassohaline lake. Verh. int. Ver. Limnol. 25: 444–447.

78

Alcocer, J., A. Lugo, E. Escobar, M. R. Sánchez & G. Vilaclara, 2001. Water column stratification and its implications in the tropical warm monomictic lake Alchichica, Puebla, Mexico. Verh. int. Ver. Limnol. 27. In press.

Babu, G. S., P. C., Joshi & P. N. Viswanathan, 1998. UVB-induced reduction in biomass and overall productivity of cyanobacteria. Biochem. Biophys. Res. Commun. 244: 138–142.

Bebout, B. M., 1992. Interactions of Carbon and Nitrogen Cycling in Microbial Mats and Stromatolites, PhD Dissertation, University of North Carolina at Chapel Hill: 134 pp.

Brenowitz, S. & R. W. Castenholz, 1997. Long-term effects of UV and visible irradiance on natural populations of a scytonemin-containing cyanobacterium (*Calothrix* sp.). FEMS Microbiol. Ecol. 24: 343–352.

Carpenter, E. J. & D. G. Capone, 1992. Nitrogen fixation in *Trichodesmium* blooms. In Carpenter, E. J., D. G. Capone & J. G. Reuter (eds), Marine Pelagic Cyanobacteria: *Trichodesmium* and other Diazotrophs. Kluwer Academic Publishers, Dordrecht: 211–218.

Castenholz, R. W., 1976. The effect of sulfide on the blue-green algae of hot springs. I. New Zealand and Iceland. J. Phycol. 12: 54–68.

Castenholz, R. W. & F. Garcia-Pichel, 1997. Cyanobacterial responses to UV-radiation. In Whitton, B. A. & M. Potts (eds), Ecology of Cyanobacteria: Their Diversity in Time and Space. Kluwer Academic Publishers, Dordrecht: 78–89.

Coyer, J. A., A. Cabello-Pasini, H. Swift & R. Alberte, 1996. N$_2$ fixation in marine heterotrophic bacteria: Dynamics of environmental and molecular regulation. Proc. Natn. Acad. Sci. U.S.A. 93: 3575–3580.

Dean, D. & M. R. Jacobson, 1992. Biochemical genetics of the Nitrogenase. In Stacey, G., R. H. Burris & H. J. Evans (eds), Biochemical Genetics of the Nitrogenase in Biological Nitrogen Fixation. Chapman & Hall, N.Y.: 763–833.

Des Marais, D. J., 1995. The biogeochemistry of hypersaline microbial mats. Adv. Microb. Ecol. 14: 251–274.

Escobar, E., J. Alcocer, E. Cienfuegos & P. Morales, 1999. Carbon stable isotopes ratios of pelagic and littoral communities in Alchichica crater-lake, Mexico. Int. J. Salt Lake Res. 7: 345–355.

Garcia-Pichel, F., C. W. Wingard & R. W. Castenholtz, 1993. Evidence regarding the UV sunscreen role of a mycosporine-like compound in the cyanobacterium *Gloeocapsa* sp. Apl. Envir. Microbiol. 59: 170–176.

Haselkorn, R., 1986. Organization of the genes for nitrogen fixation in photosynthetic bacteria and cyanobacteria. Ann. Rev. Microbiol. 40: 525–547.

Herbst, D., 1998. Potential salinity limitations in nitrogen fixation in sediments from Mono Lake, California. Int. J. Salt Lake Res. 7: 261–274.

Holm-Hansen, O., D. Lubin & E. W. Heibling, 1993. Ultraviolet radiation and its effect on organisms in aquatic environments. In Young, A. R., L. O. Björn, J. Moan & W. Nultsch (eds), Environmental UV Photobiology. Plenum Press, N.Y.: 379–425.

Howarth, R. W., R. Marino, J. Lane & J. J. Cole, 1988. Nitrogen fixation in freshwater, estuarine, and marine ecosystems. 1. Rates and importance. Limnol. Oceanogr. 33: 669–687.

Khamees, H. S., J. R. Gallon & A. E. Chaplin, 1987. The pattern of acetylene reduction by cyanobacteria grown under alternating light and darkness. Brit. Phycol. J. 22: 55–60.

Kirshtein, J. D., H. W. Paerl & J. Zehr, 1991. Amplification, cloning and sequencing of a *nifH* segment from aquatic microorganisms and natural communities. Appl. Envir. Microbiol. 57: 2645–2650.

Libes, S. M., 1992. An Introduction to Marine Biogeochemistry. John Wiley & Sons, Inc: 734 pp.

Lugo, A., J. Alcocer, M. Sanchez & E. Escobar, 1993. Trophic status of tropical lakes indicated by littoral protozoan assemblages. Verh. int. Ver. Limnol. 25: 441–443.

Mitchell, D. L. & D. Karentz, 1993. The induction and repair of DNA photodamage in the environment. In Young, A. R., L. O. Björn, J. Moan W. Nultsch (eds), Environmental UV Photobiology. Plenum Press, N.Y.: 345–377.

Paerl, H. W., J. L. Pickney & S. A. Kucera, 1995. Clarification of the structural and functional roles of heterocysts and anoxic microzones in the control of pelagic nitrogen fixation. Limnol. Oceanogr. 40: 634–638.

Paerl, H. W., 1999. Physical-chemical constraints on cyanobacterial growth in the Oceans. In Charpy, L. & A. W. D. Larkum (eds), Marine Cyanobacteria. Bull. Inst. Oceanol. Monaco. 19: 319–349.

Rastettler, E. B., R. B. McKane, G. R. Shaver & J. B. Melillo, 1992. Changes in C storage by terrestrial ecosystems: how C-N interactions restrict responses to CO_2 and temperature. In Wisniewski, J. & A. Lugo (eds), Natural Sinks of CO_2. Kluwer Academic Publishers, Dordrecht: 45–61.

Sprent, J. I. & P. Sprent, 1990. Nitrogen-fixing Organisms: Pure and Applied Aspects. Chapman & Hall, London: 256 pp.

Stal, L. J., 1995. Physiological ecology of cyanobacteria in microbial mats and other communities. Trans. Rev. 84: 1–32.

Stal, L. J., 1999. Nitrogen fixation in microbial mats and stromatolites. In Charpy, L. & A. W. D. Larkum (eds), Marine Cyanobacteria. Bull. Inst. Oceano. Monaco. 19: 357–363.

Tavera, R. & J. Komárek, 1996. Cyanoprokaryotes in the volcanic lake of Alchichica, Puebla State, Mexico. Algal Stud. 83: 511–538.

Thingstad, T. F., M. Heldal, G. Bratback & I. Dundas, 1993. Are viruses important partners in pelagic food webs? Trends. Ecol. Evol.: 209–213.

Vanderhoef, L. N., P. J. Leibson, R. J. Musil, C. Y. Huang, R. E. Fiehweg, J. W. Wiliams, D. L. Waxkwitz & K. T. Mason, 1975. Diurnal variation in algal acetylene reduction (nitrogen fixation) *in situ*. Plant Physiol. 55: 273–276.

Verschuren, D., C. Cocquyt, J. Tibby, C. N. Roberts & P. R. Leavitt, 1999. Long-term dynamics of algal and invertebrate communities in a small, fluctuating tropical soda lake. Limnol. Oceanogr. 44: 1216–1231.

Walter, M. R., J. P. Grotzinger & J. W. Schopf, 1992. Proterozoic stromatolites. In Schopf, J. W. & C. Klein (eds), The Proterozoic Biosphere: A Multidisciplinary Study. Cambridge University Press, N.Y.: 253–260.

Whitton, B. A., 1987. Survival and dormancy of blue-green algae. In Henis, Y. (ed.), Survival and Dormancy of Microorganisms. Wiley, N.Y.: 109–167.

Hydrobiologia **467**: 79–89, 2002.
J. Alcocer & S.S.S. Sarma (eds), Advances in Mexican Limnology: Basic and Applied Aspects.
© 2002 *Kluwer Academic Publishers.*

Phytoplankton of cenotes and anchialine caves along a distance gradient from the northeastern coast of Quintana Roo, Yucatan Peninsula

Malinali Sánchez[1], Javier Alcocer[1,*], Elva Escobar[2] & Alfonso Lugo[1]

[1]*Limnology Laboratory, Environmental Conservation & Improvement Project, UIICSE, FES Iztacala, UNAM, Av. de los Barrios s/n, Los Reyes Iztacala, 54090 Tlalnepantla, Estado de Mexico, Mexico*
Fax: +52-5277-1829; E-mail: jalcocer@servidor.unam.mx
[2]*Instituto de Ciencias del Mar y Limnologia, UNAM, Apdo, Postal 70-305, Coyoacan 04510, Mexico, D.F., Mexico*
(*Author for correspondence)

Key words: karst, tropical limnology, diatoms, species richness, Mexico, Cenote, Yucatan, suspended algae, phycoflora, sinkholes, tycoplankton

Abstract

This work details the taxonomic composition of suspended algae (phytoplankton and tycoplankton) communities in five cenotes (sinkholes) and two anchialine caves in northeastern Quintana Roo, Mexico. The sample set of cenotes are Casa, Nohoch Nah Chich, Maya Blue, Cristal, and Carwash, as well as the two associated caves leading from the cenotes of Maya Blue and Cristal. The site distribution represents a distance gradient with respect to the coastline with which we observe the effects of tidal movement and the mixing of waters (e.g. saline water and freshwater) on the composition of the suspended algae communities. Two sample sets were taken, one at the end of the dry season (March–April 1995) and the second at the end of the rainy season (September–October 1995) with the goal of comparing the contrasting climatic conditions of the region. A total of 79 species were identified, of which, diatoms were the most important with respect to species richness with a total of 75% of species. The floristic composition is very similar between the freshwater cenotes. The distance of a cenote site with respect to the coastline was a determining factor in the species composition. Casa Cenote is the most distinct of the sample set for the presence of marine species due to its proximity to the coastline. The tides are a large determining factor of the floristic composition of Casa Cenote with 24% all species identified in this study found exclusively in this system. The anchialine system species are transported from the cenotes and the adjacent cave systems. The largest percentage or species (95%) are freshwater, and only 5% of the total number of identified species are of marine origin. It is recognized that the most distant cenotes from the coast, Carwash and Cristal, as well as Maya Blue and Nohoch Nah Chich, are the most similar, despite being part of different cave systems. In these inland systems the marine species decreased drastically (2.4% in Nohoch Nah Chich and no marine species in the remaining cenotes). Marine species are found at the halocline of the caves.

Introduction

The Yucatan Peninsula is a typical karst environment where highly permeable limestone promotes the formation of complex submerged cave systems and sinkholes, known locally as cenotes (Redell, 1981). The orientation of many of these cave systems gives rise to extensive interconnected passage systems which are parallel to each other and run perpendicular to the coastline due to the fracture and fault orientation of the region (Iliffe, 1993).

In the littoral karst, tidal loading pushes the underlying marine water inland; this displaces from below the shallow fresh water which flows towards the coast under the force of gravity. As a result, anchialine systems form where dense saline water underlies low-density light fresh water. These layers are separated by a marked halocline (Stock et al., 1986). The salinity of the shallow fresh water decreases with distance from the coast, so that fresh water is found at a distance inland (Iliffe, 1993).

The submerged passages of the Yucatan Peninsula cave systems are inhabited by endemic biota, which have been well documented (Illiffe, 1993).

A transition zone exists between open sunlit cenotes and the total darkness of caves. Possibly the most important interaction between the open cenote and the closed cave is the transfer of loose and particulate photosynthetic organic matter (including organisms) from both terrestrial and aquatic sources from the illuminated cenotes to the complete darkness of the cave. Apparently of lesser magnitude are the chemosynthetic processes, which produce organic matter directly within the caves, which in turn may be transported to the downstream cenote in the system.

Phytoplankton are the dominant organisms amongst the primary producers within aquatic ecosystems. However, the phycoflora communities of karstic systems are poorly studied despite the fundamental importance of this knowledge with respect to the biogeochemistry of the submerged systems and the microbial ecology, which in turn may shed light on the ecology of the troglobite macrofauna and nutrient cycling (Martin et al., 1995).

The aim of the present study was to determine the suspended algae (planktonic and benthic algae) community composition in five cenotes and two anchialine caves in northeast Quintana Roo, Mexico. For comparative purposes, five cenotes distributed between the three karstic systems of Nohoch Nah Chich, Naranjal (which includes cenotes Maya Blue and Cristal), and Aktun Ha (which includes Car Wash cenote) were selected to form a perpendicular transect with increasing distance to the coast. This spatial distribution of sampling points allows for evaluation of tidal influence on the suspended algae composition of these systems.

Study site

The study region is located in the northeast part of the state of Quintana Roo and is bordered by 20° 11′ and 20° 17′ latitude north and 87° 23′ and 87° 29′ longitude west (Fig. 1). The sampling sites are located as follows: Casa Cenote (Cenote Manati, Cenote Tankah, in the Nohoch Nah Chich System) (20° 15.97′ N, 87° 23.41′ W), Nohoch Nah Chich System Main Entrance (20° 17.93′ N, 87° 24.20′ W), Maya Blue in the Naranjal System (20° 11.61′ N, 87° 29.74′ W), Cristal also in the Naranjal System (20° 12.50′ N, 87° 28.98′ W), and, Carwash in the Aktun Ha System (20° 16.48′ N, 87° 29.20′ W).

Casa Cenote is located on the coast and is characterized by active exchange with marine water. However, Casa Cenote, like all anchialine caves, is vertically stratified with respect to salinity (Table 1). The upper stratum, the epicline, is fresh to brackish water and slightly acidic. The lower stratum, the hypocline, is marine water and slightly basic. The intermediate layer, the halocline, is the transition zone between the two layers. The other cenotes are fresh water and slightly acidic (Table 1). Physical and chemical variables displayed minor differences among fresh water masses as well as between salt water ones. Further information on physical, chemical and microbiological characteristics are given in Alcocer et al. (1998, 1999).

The study area is characterized by limestone of Miocene and Pliocene age (Back & Hanshaw, 1970) with a maximum altitude of less than 200 m above sea level (Lesser & Weidie, 1988). The environmental temperature is generally stable over the year, fluctuating from 23 °C in January to a maximum of 28 °C in May. Seasonal variation on Yucatan Peninsula is defined not by temperature variation but by precipitation. The rainy season is from May to October with a precipitation between 500 and 1500 mm per year, and average 1000 mm (Back & Hanshaw, 1970), and a 'dry' season from November to April, when 35% of the annual precipitation falls. Infiltration is nearly matched by evaporation, while overland runoff is negligible (Lesser & Weidie, 1988).

Materials and methods

Two sample sets were taken to compare the contrasting seasons of the region. One set was taken at the end of the dry season (March–April of 1995) and another at the end of the rainy season (September–October of 1995). The number of samples from each of the cenotes and anchialine cave water columns varied from one to three depending on environmental heterogeneity which was established through vertical temperature, salinity and dissolved oxygen profiles using a multiparameter water quality monitoring probe (Hydrolab DS3/SVR3). Where stratification was observed (thermocline, halocline, and/or oxycline) three levels of samples were taken: surface, bottom, and cline depth. This sampling strategy occurred at Casa Cenote and in the caves. The remaining cenotes were sampled at mid-water column since homogeneous water columns were observed. A Niskin water-sampling bottle with 6-l capacity was used in the cenotes. From

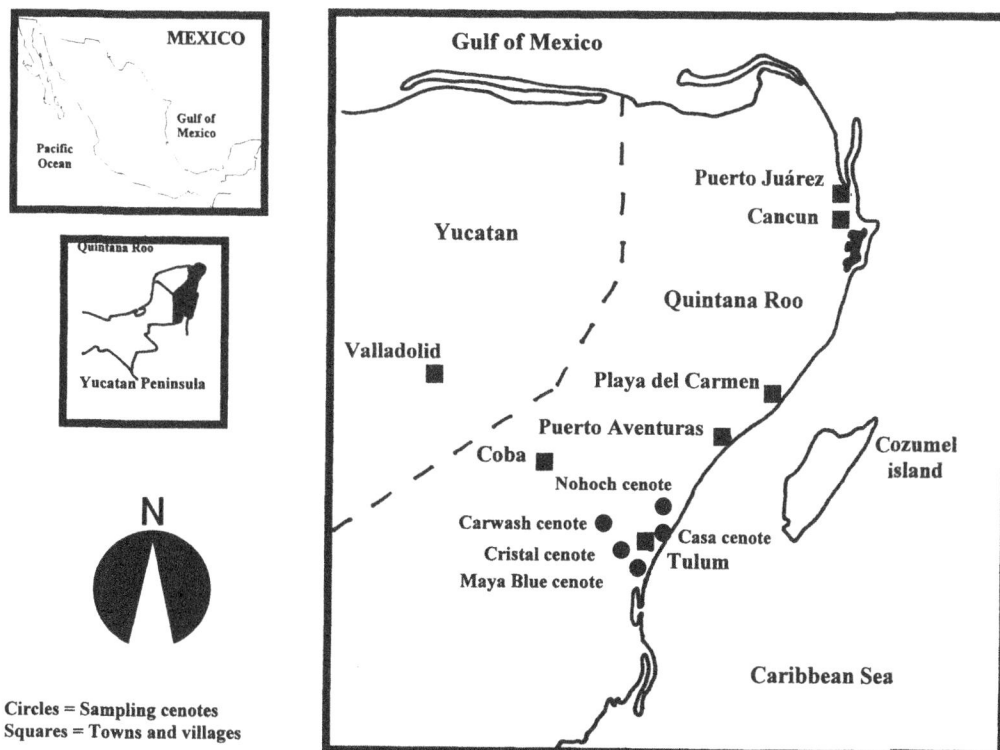

Figure 1. Location map of sampling sites. (Modified from Iliffe, 1992).

within the caves, water samples were taken directly into 0.5 l Nalgene sample bottles. Additionally, vertical and horizontal suspended algae samples were taken with 25 μm mesh net to capture dispersed, less abundant organisms. Finally, artificial substrates (glass slides) were located with the same purpose. All biological samples were preserved with 1% Lugol's iodine and stored in darkness until the final laboratory analysis was undertaken (Vollenweider, 1974).

Taxonomic species identification was undertaken using the keys of Hustedt (1930a, 1930b, 1959), Hustedt & Jensen (1985), Patrick & Reimer (1966, 1975), Bourrely (1966, 1968, 1970), Krammer & Lange-Bertalot (1986, 1988, 1991a, 1991b), Geitler (1932) and Komárek & Fott (1983). Contrast-phase microscope observations were undertaken. Diatom frustules were cleaned using Hasle & Fryxell (1970) techniques and were observed using a scanning electron microscope (SEM). The taxonomic classification followed Krammer & Lange-Bertalot's criteria (1986, 1988, 1991a, 1991b).

The resulting data set was used to develop a presence/absence matrix for the wet and dry sampling seasons. Subsequently, a cluster analysis (agglomerat-ive classification, simple union) was undertaken based on the Jaccard Index that measures the degree of affinity between two or more data sets. By this means, the cenotes and the caves were grouped based on the number of common species, using the statistics package ANACOM (Version 3.0, 1991).

Results

General composition

Seventy-nine suspended algae species were identified (Table 2), belonging to four divisions. The Chromo-phyta division constituted 75% of the total, while Cyanophyta were 14%, Chlorophyta 10%, and the Euglenophyta 1%. Seventy-eight species were found in the cenotes, while only 40 species were identified in the cave samples.

Sixty-three species of algae (80%) are benthic and could be considered as suspended algae or tycoplank-ton, meanwhile sixteen species (20%) are truly plank-tonic. Cyanophyceae (64% planktonic, 36% benthic), Euglenophyceae (100% planktonic), Chlorophyceae (86% planktonic, 14% benthic) and Zygophyceae

Table 1. Minimum-maximum concentrations of physical and chemical parameters of the studied cenotes and anchialine caves. (Temp – temperature, TDS – total dissolved solids, DO – dissolved oxygen). (Modified from Alcocer et al., 1998)

	Depth (m)	Temp (°C)	pH (pH units)	TDS (g/l)	DO (mg/l)
Cenotes:					
Casa Cenote	7	25.6–28.3	6.76–7.92	12.5–34.6	2.17–7.02
Nohoch Nah Chich	7	24.7–26.4	6.88–6.95	1.7–2.0	1.88–2.34
Maya Blue	5	25.2–26.6	6.77–7.31	1.7–2.9	0.64–4.07
Cristal	5	25.2–26.2	6.76–6.83	2.0–2.5	0.93–2.62
Carwash	6	25.1–28.1	6.73–7.47	0.5–1.7	2.98–4.68
Caves:					
Maya Blue	variable	25.2–25.9	6.91–7.32	2.9–34.3	0.7–1.7
Cristal	variable	25.22–25.5	6.83–7.42	2.5–32.8	0.9–2.36

(100% planktonic), with a total of 20 species, were mostly planktonic. On the other hand, Bacillariophyceae (2% planktonic, 98% benthic) with 59 species was almost exclusively benthic.

The group most important with respect to richness were the diatoms, with 59 species (75%). The pennate diatoms represented the largest contribution with 71% (56 species), while the central diatoms were only 4% (3 species) of the total diatoms identified. The most important diatoms in the cenotes and the caves which were identified in all sampling sites were *Achnanthes minutissima* var. *minutissima*, *Cocconeis placentula* var. *lineata*, *Cyclotella meneghiniana*, *Cymbella gracilis*, *C. microcephala*, *C. minuta*, *C. muelleri*, *Gomphonema gracile*, *Navicula cryptotenella*, *Navicula minuscula*, *Nitzchia amphibioides*, *Synedra acus*, *S. ulna* and *Terpsinoe musica*.

Cenotes

Similar floristic composition was found in Maya Blue, Cristal and Car Wash cenotes (70% common species). Nohoch Nah Chich was found to have the greatest number of species from the Cyanophyta group with 10 species (24%) in comparison with an average of four (10%) in the remaining sampled cenotes. Casa Cenote has 50 species of which 12 are marine and exclusive to this system amongst the sampling sites. Diatoms were generally found to make the greatest contribution to the suspended algae distribution in all the systems. All the species found in the caves were also identified in the cenotes; the species richness of the cenotes was greater than that of the caves.

Sixty-eight species were registered in the dry season in the cenotes and 38 from within the caves. During the rainy season the species total was 71 in the cenotes and 25 from within the caves. The temporal variation of the specific richness in the cenotes was lower (≈5%), while within the caves, the number of species was lower by a third between the dry and the rainy season (Table 3).

Casa Cenote was found to have the greatest global species richness with 50 species, seconded by Maya Blue with 46, Nohoch Nah Chich with 41, Cristal with 33, and finally Carwash with 32 species. During the dry season, the cenote with the greatest species richness was again Casa Cenote with 42 species, followed by in decreasing order, Nohoch Nah Chich with 40, Maya Blue with 35, Cristal with 29, and Carwash with 26 species. The number and composition of species was similar across the three water strata identified in Casa Cenote: 32 in the surface, 29 in the halocline, and 30 in the bottom layer (Table 3). No differences were found in the number and the abundance of marine suspended algae in Casa Cenote between the dry and the rainy season (12–24% of the total during the dry season and 11–22% of the total during the rainy season).

During the rainy season, the highest specific richness was found in Casa Cenote with 43 species, followed by Maya Blue with 37 species. Cenotes Cristal and Carwash maintained similar species richness to the dry season with 24 and 27 species, respectively. Nohoch Nah Chich Main Entrance had the lowest species richness with 19 species. The vertical profiles of Casa Cenote did not differ between the rainy and the dry season with respect to the number and the com-

Table 2. Taxonomic list of the phytoplankton species encountered in the sampled cenotes and caves. P – planktonic species, B – benthic species, FW – freshwater species, SW – sea water species

Class	Species	P/B	FW/SW	CA	NO	MB	CR	CW	MBC	CRC
Cyanophyceae	*Chroococcus* sp.	P	FW	X	X		X	X		
	Chroococcus minimus	P	FW					X		
	Aphanocapsa sp.	P	FW		X	X			X	
	Merismopedia elegans	P	FW		X	X				
	Merismopedia minima	P	FW		X			X		
	Limnothrix amphigranulata	B	FW		X	X				
	Phormidium tenue	P	FW		X	X	X	X	X	
	Spirulina sp.	P	FW	X	X	X			X	
	Lyngbya sp.	B	FW		X		X		X	X
	Oscillatoria agardhii	B	FW		X	X			X	
	Oscillatoria limosa	B	FW	X	X					
Bacillariophyceae	*Cyclotella meneghiniana*	P	FW	X	X	X	X	X	X	X
	Melosira nummuloides	B	SW	X						
	Terpsinoe musica	B	FW	X	X	X	X	X	X	
	Fragilaria fasciculata	B	FW	X	X	X	X	X	X	X
	Gramatophora sp.	B	SW	X						
	Licmophora sp.	B	SW	X						
	Podocystis sp.	B	SW	X						
	Synedra acus	B	FW	X	X	X	X	X	X	X
	Synedra aff. *amphicephala*	B	FW	X	X	X			X	X
	Synedra ulna	B	FW		X	X	X	X	X	X
	Eunotia minor	B	FW		X	X	X	X	X	X
	Eunotia monodon	B	FW		X		X		X	X
	Achnanthes amoena	B	SW	X						
	Achnanthes brevipes var. *intermedia*	B	SW	X					X	X
	Achnanthes delicatula sp. *hauckiana*	B	FW	X		X		X	X	
	Achnanthes minutissima var.	B	FW				X	X	X	X
	Achnanthes minutissima var. *minutissima*	B	FW	X	X	X	X	X	X	
	Achnanthes temperei	B	SW	X						
	Cocconeis pediculus	B	FW	X						
	Cocconeis placentula var. *lineata*	B	FW	X	X	X	X	X	X	X
	Amphora sp.	B	FW	X						
	Amphora angusta	B	SW	X	X		X			
	Amphora coffeafformis	B	FW	X		X			X	X
	Amphora coffeafformis var. *acutiuscula*	B	FW	X						
	Amphora lineolata	B	FW	X		X	X			
	Cymbella gracilis	B	FW			X	X	X	X	X
	Cymbella microcephala	B	FW	X	X	X	X	X	X	X
	Cymbella minuta	B	FW		X	X	X	X		
	Cymbella muelleri	B	FW	X	X	X	X	X	X	X
	Diploneis elliptica	B	FW	X	X	X				
	Entomoneis (Amphiprora) gigantea	B	SW	X						X
	Gomphonema gracile	B	FW	X	X	X	X	X	X	X
	Gomphonema aff. *insigne*	B	FW			X		X		X
	Gomphonema intricatum var. *vibrio*	B	FW					X		X
	Gomphonema parvulum	B	FW	X	X	X	X	X	X	
	Gyrosigma obscurum	B	FW			X	X	X	X	X
	Gyrosigma scalproides	B	FW	X						
	Navicula bicephala	B	FW		X	X		X	X	
	Navicula cryptotenella	B	FW	X	X	X	X	X	X	X
	Navicula digito-radiata	B	FW	X						
	Navicula halophila	B	FW	X		X				
	Navicula minuscula	B	FW		X	X	X	X	X	X

Continued on p. 84

Table 2. Continued

Class	Species	P/B	FW/SW	CA	NO	MB	CR	CW	MBC	CRC
	Navicula phyllepta	B	FW	X						
	Navicula platyventris	B	SW	X						
	Navicula radiosa	B	FW		X	X	X	X	X	X
	Navicula zeta	B	SW	X						
	Neidium sp.	B	FW		X	X		X	X	x
	Pleurosigma elongatum	B	FW	X						
	Pleurosigma salinarum	B	FW	X						
	Stauroneis kriegeri	B	FW		X	X	X			
	Rhopalodia gibba	B	FW	X						
	Bacillaria paradoxa	B	FW	X	X	X	X		X	
	Nitzchia amphibia	B	FW	X	X	X			X	
	Nitzchia amphibioides	B	FW	X	X	X	X	X	X	X
	Nitzchia constricta	B	SW	X						
	Nitzchia frustulum	B	FW	X	X	X		X	X	
	Nitzchia linearis	B	FW			X	X			
	Nitzchia scalaris	B	FW			X				
	Nitzchia sigmoidea	B	FW	X	X	X	X		X	
Euglemophyceae	*Phacus acuminatus*	P	FW		X	X				
Chlorophyceae	*Schroedera* sp.	P	FW	X		X				
	Pediastrum duplex	P	FW	X						
	Chlorella vulgaris	P	FW			X	X	X	X	
	Monoraphidium minutum	P	FW				X	X		X
	Scenedesmus quadricauda	P	FW	X	X	X				
	Chlamydomonas sp.	P	FW			X				
	Ulothrix sp.	B	FW	X	X			X		
Zygophyceae	*Spyrogira* sp.	P	FW							X

position of species present (29 at the surface, 30 in the halocline, and 31 at the bottom) (Table 3).

The species composition in Casa Cenote, Maya Blue, Cristal and Carwash did not vary between the two seasons. Nohoch Nah Chich varied in number of species present, as mentioned with 40 species identified during the dry season and a reduced set of 19 during the rainy season. However, all species identified during the rainy season, were also identified in the dry season samples; no new species were identified during the rainy season. The lower number of species was principally in the diatom group, with the following species absent in the rainy season sampling: *Achnanthes minutissima* var. *minutissima*, *Cymbella muelleri*, *Eunotia minor*, *E. monodon*, *Gomphonema gracile*, *G. parvulum*, *Navicula bicephala*, *N. radiosa*, *Nitzchia amphibia*, *N. frustulum*, *Synedra acus*, *S.* aff. *amphicephala* and *S. ulna*.

Anchialine caves

Within the caves, Maya Blue registered the greatest number of species with 36, while Cristal registered 27. The size of the species in each of the cave levels was not a determining factor in their location within the water column despite the presence of the halocline (pycnocline). All the species found were large (i.e. >5 μm). However, different sizes were identified at the three levels: from 5 to 20 μm (e.g. *Acnanthes minutissima* var. *minutissima*, *Cymbella microcephala* and *Gomphonema gracile*), from 20 to 30 μm (e.g., *Fragilaria fasciculata*, *Synedra acus*, *Amphora coffeafformis*, *Cocconeis placentula* var. *lineata* and *Cymbella muelleri*) and >30 μm (e.g. *Lyngbya* and *Synedra ulna Cyclotella meneghiniana*). The majority of the identified species were located within the epicline and/or the halocline. The species of marine origin were only present within the hypocline.

Eighty-six percent (31) of the identified species in Maya Blue cave were present in cenote Maya

Blue. The species identified in the cave, but that were not sampled in cenote Maya Blue (14%, 5 species) were *Achnanthes minutissima, Gomphonema intricatum* var. *vibrio, Eunotia monodon, Lyngbya* sp. and *Achnanthes brevipes* var. *intermedia*. The first four of these species were found in other sampled systems, for example in cenote Cristal, which is part of the same explored cave system and which is located in close proximity to Maya Blue. *Achnanthes brevipes* var. *intermedia* is the unique marine origin species sampled at this site. This species was also identified in Casa Cenote.

Seventy-percent (19) of the species identified in Cristal cave were also sampled in cenote Cristal. The remaining 30% (8 species) were among the species identified in the cenote and the cave of Maya Blue, which is a reflection of the proximity and communication – Naranjal system – between the two cenotes (Illiffe, 1993). This is the case of *Amphora coffeaeformis, Gomphonema* aff. *insigne, G. intricatum* var. *vibrio* and *Gyrosigma obscurum*. Two marine origin species, *Achnanthes brevipes* var. *intermedia* and *Entomoneis (Amphiprora) gigantea*, were also identified, and these were similarly found in Casa Cenote. *Spyrogira* sp. is a species that was only found in Cristal cave and which was not registered in any of the cenotes, or in the cave associated with Maya Blue. *Spyrogira* sp. is characterized as inhabiting lakes and ponds (Ortega, 1984) and thus may have been washed into the cave via percolation through fissures in the limestone.

The species composition in the caves showed seasonal variation. A greater number of species were identified during the dry season, especially in the Maya Blue cave. For example, the following diatoms were absent during the rainy season: *Bacillaria paradoxa, Cymbella gracilis, C. muelleri, Gyrosigma obscurum, Navicula radiosa, Nitzchia amphibia, N. amphibioides, N. sigmoidea, Synedra acus, S. ulna* and *Terpsinoe musica*.

Certain species were found exclusively in the dry season sampling or in the rainy season sampling. For example, during the dry season *Achnanthes minutissima* was identified in Cristal and Carwash cenotes as well as in the two caves associated with these sites, and *Lyngbya* sp. in cenote Cristal and in both caves. During the rainy season *Chlamydomonas* sp. was found in Maya Blue and *Chlorella vulgaris* in Maya Blue, Cristal, Carwash and in the Maya Blue cave.

During the dry season, Maya Blue cave showed notable differences in the epicline, where 32 species were recorded. This richness is comparable to that found in the cenotes. In the halocline and the hypocline, only seven species were identified at each level. In the cave associated with Cristal cenote, the greatest species richness was found at the halocline (20 species), while in the epicline (15) and the hypocline (15) fewer species were identified. During the rainy season, both caves registered lower totals. Few species were found in the cave associated with Maya Blue cenote at all three strata of the water column (seven in the epicline and four in each of the other two strata). The cave associated with Cristal cenote had the greatest specific species richness in the halocline (nine species) and fewer in the epicline (6) and the hypocline (8).

Cluster analysis

The cluster analysis shows two principal groups: the first is composed of Nohoch Nah Chich, Maya Blue, Cristal and Carwash cenotes, including the associated caves, and the second group is represented by Casa Cenote which is characterized by brackish and marine waters. The similar composition between these four cenotes is due to their freshwater and that the water bodies experience limited physicochemical variation (Alcocer et al., 1998).

The Jaccard coefficient shows a great similarity in the species composition between the two sampling seasons for the same cenote or cave, such as is the case with Casa Cenote, Cristal and Carwash cenotes, as well as the cave associated with Cristal cenote. A high degree of similarity was found between the two most inland sites of Cristal and Carwash (Fig. 2), which was even greater than the degree of similarity between Cristal cenote and Maya Blue despite the two cenotes belong to the Naranjal system with a direct conduit passage between them (Coke & Young, 1989). These cave systems have very complex structures of, which very little is understood in relation to the water transport through them.

Elevated water velocities in the fractures have been measured (e.g. 1–3 cm/s in Carwash and Maya Blue) due to the volume of water that must flow through reduced spaces (Moore et al., 1992; Beddows, 1999).

Discussion

The karstic systems of the present study, including the cenotes and the anchialine caves, were found to

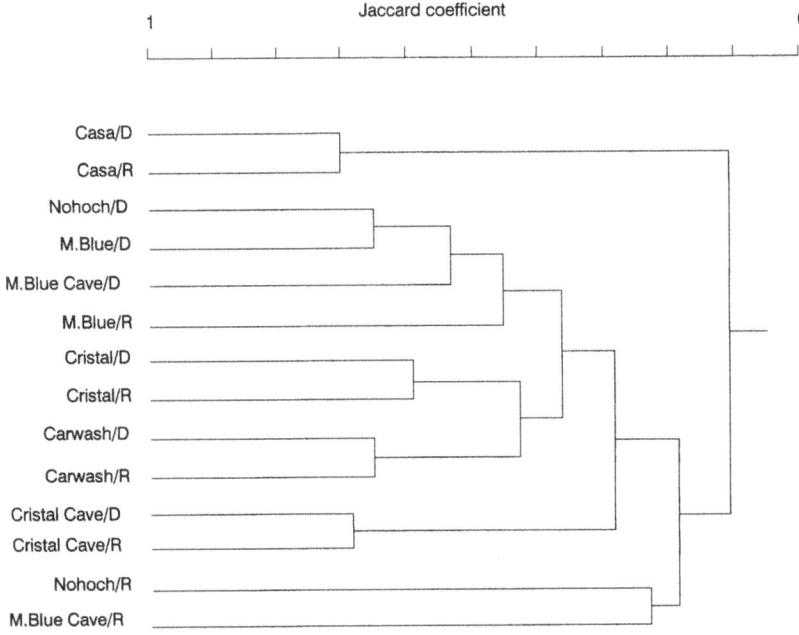

Figure 2. Dendrogram of the biological similarity – taxonomic composition – (using Jaccard Index) of the cenotes and the caves of the northeast Quintana Roo. (D – dry season, R – rainy season, M. Blue – Maya Blue).

have species in common (*Chlamydomonas, Scenedesmus, Ulothrix, Phacus, Lyngbya, Oscillatoria, Phormidium, Merismopedia, Aphanocapsa, Chrooccocus, Achnanthes brevipes, A. temperei, Cocconeis placentula, Gomphonema gracile, Synedra acus, S. ulna, Terpsinoe musica, Eunotia monodon* and *Nitzchia scalaris*) with other cenotes studied by López-Adrián et al. (1993), López-Adrián & Herrera-Silveira (1994) and Herrera-Silveira et al. (1997). The listed diatom species, along with *Cyclotella meneghiniana* and *Nitzchia amphibioides*, have been reported as characteristic species of the diatom flora found in open cenotes such as those of the Yucatan in general and those of this study.

Based on their trophic characteristics, the systems studied in the northeast of Quintana Roo are found to be similar to those described by López-Adrián et al. (1993) and Herrera-Silveira et al. (1997) in the northern region of the Yucatan Peninsula. They share similar alga assemblages and species richness dominated by diatoms (especially pennate diatoms).

The marine water characteristics of Casa Cenote influences the species composition. Of the total number of species recorded for this cenote (50), 24% of these, all diatoms, are characteristic of marine environments, while a further 16% of the total species present, also diatoms, have been identified in hy-

Table 3. Species richness in the studied cenotes and caves. (Dry – dry season, Rainy – rainy season)

	Global	Dry	Rainy
Cenotes:	**78**	**68**	**71**
Casa Cenote	50	42	43
Epicline	42	32	29
Halocline	30	29	30
Hypocline	35	30	31
Nohoch Nah Chich	41	40	19
Maya Blue	46	35	37
Cristal	33	29	24
Carwash	32	26	27
Caves:	**40**	**38**	**25**
Maya Blue	36	31	13
Epicline	33	32	7
Halocline	9	7	4
Hypocline	10	7	4
Cristal	27	26	19
Epicline	16	15	6
Halocline	21	20	9
Hypocline	16	15	8

persaline water in continental settings (Giffen, 1963; Navarro, 1983; Krammer & Lange-Bertalot 1986, 1988, 1991a, 1991b; Moreno et al., 1996).

The influence of tides as a means of dispersal of marine phytoplankton to the studied cenotes is the greatest in Casa Cenote, given that this is the site where the greatest percentage representation of marine species was observed (24% of the total number of species). Marine species decrease drastically with relative distance from the coast of the sampling site; Nohoch Nah Chich Main Entrance has only 2.4% of marine representation (the second closest site to the coast), and the remaining sites have no marine species. However, in the hypocline of both anchialine caves marine species were also identified although in low representation (5% on average). Given that the distance to the ocean of Maya Blue and Cristal caves is similar to that of the associated cenotes, the absence of marine phytoplankton in the cenotes is because these cenotes do not have saline water in them (with phytoplankton present) such as is present in the hypocline of the caves. This information suggests that distance to the coastline is a determining factor for the presence of marine species such that fewer marine species are present at greater distances inland, and that marine water at the site is required for the presence of marine species. The dense marine water that is propelled inland by the tides forms pockets of saline water in depressions along the length of these karstic systems. It is very probable that only in those cenotes deep enough to reach the marine water layer are marine species present.

The species richness of the cenotes and the caves is similar to that identified in other cave systems of this region. For example, in four cenotes in the state of Yucatan 37 genera were identified (López-Adrián et al., 1993), while in the cenotes of Quintana Roo we found a slightly higher figure (42 genera). In the cenote of Dzitya or Chen-ha located in the state of Yucatan, 30 species were identified (Sánchez-Molina et al., 1994), a number comparable to that found in the cenotes of this study which present lower species richness, e.g. Cristal (34 species) and Carwash (32).

The diatoms identified in this study are characteristic species of the microphytobenthos (Krammer & Lange-Bertalot, 1986, 1988, 1991a, 1991b), with the exception of *Cyclotella meneghiniana* that is a planktonic species found in both fresh and saline environments. The presence of benthic diatoms in the water column is certainly due to the action of current which propel these organisms into the shallow water

column. This effect has been observed by Herrera-Silveira et al. (1997) in various water bodies of the Yucatan Peninsula. These authors observed, like us, a dominance of pennate in comparison to the central diatoms. It should also be considered divers disturbing the benthos multiple times a day during the busy tourist season, or even us on the sampling dives. The higher species richness during the dry season may then be related to more divers during the dry season. We think this result would warrant further study, because it could be a major disturbance to the groundwater communities connected.

The observed suspended algae species in the caves and dark caverns are species that have been propelled from the adjacent cenotes. In the case of the caverns (twilight zone between the cave and the cenote), this is since some species may live in twilight zones, such as cyanobacteria (Gounot, 1994). In the studied caves, an elevated number of diatoms were identified which surely come from the walls and the floor of the well lit cenotes. The percolation or the displacement of soil and pond organisms from the forest is an additional source of suspended algae, not only to the caves but also to the cenotes. This is the case for *Spyrogira* sp. which, as previously mentioned, is characterized as a lake and pond species (Ortega, 1984).

The species richness of the caves was always found to be greater in the epicline and the halocline, with a lower number of organisms in the hypocline (which were of marine origin); notwithstanding, species of different sizes were found in the three sampled strata. This suggests that the density stratification in the profile (pycnocline or halocline) of the caves constitutes a physical barrier to the downward displacement of organisms, conversely, it may be argued that the salinity gradient limits the presence of freshwater species in the high salinity strata. This is suggested by the fact that dead cells are encountered in the hypocline of the caves, in contrast to the other levels and the cenotes, where no dead organisms were registered.

The greatest number of organisms were found in Maya Blue cave within the epicline. These species were exclusively freshwater, which suggests that the inflow of cenote water was most important due to the proximity to the cave. The marine species *Achnanthes brevipes* var. *intermedia* (Krammer & Lange-Bertalot, 1991b) was identified in the hypocline, however, of all the species encountered in this strata (7) it was the only marine species. Its presence in the cave suggests its displacement by tidal flows, which apparently have sufficient force to displace marine species to the

88

interior of the peninsula. However, a minor concern is the possibility that its presence is from contamination off of the diving gear. In the case of Cristal cave, the greatest species richness was identified in the halocline, where they were retained in the density interface. The majority of the identified species in this cave were found also in the cenote since they are freshwater species which arrived there from the inflow of cenote water into the cave. Only two species, *Achnanthes brevipes* var. *intermedia* and *Entomoneis* (*Amphiprora*) *gigantea*, are of marine origin (Krammer & Lange-Bertalot, 1991b) and were found in the hypocline. This discussion indicates that within the studied caves the associated cenotes are the most important source of organisms beyond those related to marine sources.

Reduced species richness was found in the caves during the rainy season, which is probably related to the increased discharge of fresh water to the ocean and the elevated flow rates, which have been recorded for these systems during this season. The flow rates increase with decreasing distance from the coastline (Moore et al., 1992) which impedes the accumulation of organisms within the caves.

Acknowledgements

The authors greatly appreciated the economic support provided in part by the General Administration of Affairs of the Academic Personnel of the UNAM through the PAPIIT-IN203894 project. Similarly, we acknowledge M. Madden and the CEDAM Cave Diving Team for the logistic support and other facilities extended to us, without which this study would have been impossible. A special thanks to C. Stevens for sharing his knowledge and experiences in the karst environment and, similarly to V. Urbieta (F. Medicina, UNAM), for her help in undertaking the sampling in such difficult sampling sites; L. Peralta and L.A. Oseguera (FES Iztacala, UNAM) for support with the field research. The electron micrographs were under the supervision of J. Sepúlveda (Laboratorio de Microscopía Electrónica, I. Fisiología Celular, UNAM). Finally, we thank Patricia A. Beddows (University of Bristol) for useful comments and linguistic advice on this manuscript.

References

Alcocer, J., A. Lugo, L. E. Marín & E. Escobar, 1998. Hydrochemistry of waters from five cenotes and evaluation of their suitability for potential drinking-water supplies, northeastern Yucatan, Mexico. Hydrogeol. J. 6: 293–301.

Alcocer, J., A. Lugo, M. R. Sánchez, E. Escobar & M. Sánchez, 1999. Bacterioplankton from cenotes and anchialine caves of Quintana Roo, Yucatan Peninsula, Mexico. Rev. Biol. Trop. 47: 73–80.

Back, W. & B. B. Hanshaw, 1970. Comparison of chemical hydrogeology of the carbonate Peninsulas of Florida and Yucatan. J. Hydrol. 10: 330–368.

Beddows, P. A., 1999. Conduit hydrogeology of a tropical coastal carbonate aquifer: Caribbean coast of the Yucatan Peninsula. Master of Science Thesis, McMaster University. Hamilton, Canada. xiii: 162 pp.

Bourrely, P., 1966. Les algues dþeau douce. Initiation á la systématique. Tome I: Les algues vertes. Boubée, Paris: 572 pp.

Bourrely, P., 1968. Les algues d'eau douce. Initiation á la systématique. Tome II: Les algues jaunes et brunes. Chrysophycées, pheophycées, xantophycées et diatomées. Boubée, Paris: 517 pp.

Bourrely, P., 1970. Les algues dþeau douce. Initiation á la systématique. Tome III: Les algues bleues et rouges. Les eugléniens, peridiniens et cryptomonadines. Boubée, Paris: 512 pp.

Coke, J. & T. M. Young, 1989. Cenote Naharon, Tulum, Q. R. Mexico. (map).

Geitler, L., 1932. Cyanophyceae. Akademische verlagsgesellschalt. m.b.H, Leipzig: 1196 pp.

Giffen, M. H., 1963. Contributions to the diatom flora of South Africa. Hydrobiologia 21: 201–265.

Gounot, A. M., 1994. Microbial ecology of groundwaters. In Gibert, J., D. L. Danielopol & J. A. Stanford (eds), Groundwater Ecology. London Academic Press, London: 189–215.

Hasle, G. R. & G. A. Fryxell, 1970. Diatoms: cleaning and mounting for light and electron microscopy. Trans. am. microsc. Soc. 89: 469–474.

Herrera-Silveira, J. A., F. A. Comín, S. López, & I. Sánchez, 1997. Limnological characterization of aquatic ecosystems in Yucatan Peninsula (SE Mexico). Verh. int. Ver. Limnol. 26.

Hustedt, F., 1930a. Rabenhorst's kryptogamen-flora. Band VIII. Die kieselalgen. 1 teil: Einleitung und centricae. Otto Koeltz Science Publishers, Koenigstein: 920 pp.

Hustedt, F., 1930b. Bacillariophyta (Diatomeae). In Pascher, A. (ed.), Die Suesswasser Flora Mitteleuropas 10. Gustav Fischer Verlag, Jena: 1–466.

Hustedt, F., 1959. Rabenhorst's kryptogamen-flora. Band VII. Die kieselalgen. 2 teil: Pennatae. Otto Koeltz Science Publishers, Koenigstein: 845 pp.

Hustedt, F. & N. G. Jensen, 1985. The Pennate Diatoms. Koeltz Scientific Books, Koenigstein: 918 pp.

Iliffe, T. M., 1993. Fauna troglobia acuática de la Península de Yucatán. In Salazar-Vallejo, S. I. & N. Emilia Gonzáles (eds), Biodiversidad Marina y Costera de México. Comisión Nacional para la Biodiversidad & CIQRO, Mexico: 673–686.

Komárek, J. & B. Fott, 1983. Das phytoplankton des süsswassers Chlorophyceae (Grünalgen). Ordrung: Chlorococcales. Schweizerbartsch, Stuttgart: 1044 pp.

Krammer, K. & H. Lange-Bertalot, 1986. Bacillariophyceae. 1 teil: Naviculaceae. Fischer, Stuttgart: 876 pp.

Krammer, K. & H. Lange-Bertalot, 1988. Bacillariophyceae. 2 teil: Bacillariaceae, Epithemiaceae, Surirellaceae. Fischer, Stuttgart: 596 pp.

Krammer, K. & H. Lange-Bertalot, 1991a. Bacillariophyceae. 3 teil: Centrales, Ffragilariaceae, Eunotiaceae. Fischer, Stuttgart: 575 pp.

Krammer, K. & H. Lange-Bertalot, 1991b. Bacillariophyceae. 4 teil: Acnanthaceae, kritische ergänzungen su Navicula (lineolatae) und Gomphonema. Fischer, Stuttgart: 437 pp.

Lesser, J. A. & A. E. Weidie, 1988. Region 25, Yucatan Peninsula. In Back, W., J. S. Rosenshein & P. R. Seaber (eds), The Geology of North America. Geological Society of America, Hydrogeology Boulder, Colorado: 237–241.

López-Adrián, S., I. Sánchez, R. Tavera, J. Komárek, J. Komarkova & M. Villasuso, 1993. Estudio ecológico de los cuerpos de agua continentales de la Península de Yucatán. Aspectos ficológicos. Programa de Ecología Terrestre. SEP, UAY, FMVZ, Mérida.

López-Adrián, S. & J. A. Herrera-Silveira, 1994. Plankton composition in a cenote, Yucatan, Mexico. Verh. int. Ver. Limnol. 25: 1402–1405.

Martin, H. W., R. L. Brigmon & T. L. Morris, 1995. Diving protocol for sterile sampling of aquifer bacteria in underwater caves. NSS Bulletin 57: 24–30.

Moore, Y. H., R. K. Stoessell & D. H. Easley, 1992. Freshwater/sea-water relationship within a ground-water flow system, northeastern coast of the Yucatan Peninsula. Ground water 30: 343–350.

Moreno, J. L., S. Licea & H. Santoyo, 1996. Diatomeas del Golfo de California. SEP-FOMES, PROMARCO, UABCS, Baja California: 273 pp.

Navarro, J. N., 1983. A survey of the marine diatoms of Puerto Rico. Botánica Marina 26: 393–408.

Ortega, M. M., 1984. Catálogo de algas continentales recientes de México. Instituto de Biología UNAM, México: 566 pp.

Patrick, R. & C. W. Reimer, 1966. The diatoms of the United States. Exclusive of Alaska and Hawaii. Vol. I. Monographs of the Academy of Natural Sciences of Philadelphia. NAS, Philadelphia: 688 pp.

Patrick, R. & C. W. Reimer, 1975. The diatoms of the United States. Vol. II. Monographs of the Academy of Natural Sciences of Philadelphia. NAS, Philadelphia: 213 pp.

Redell, J. R., 1981. A review of the cavernicole fauna of Mexico. Guatemala and Belize. Bull. 27. Texas Memorial Museum, Austin: 327 pp.

Sánchez-Molina, I., C. Zetina-Moguel, R. Medina-González & L. Pérez-Aranda, 1994. Phytoplankton composition in cenote Dzitya, Yucatan, Mexico. Lake Reserv. Mgmt. 9: 111.

Stock, J. H., T. M. Iliffe & D. Williams, 1986. The concept of 'anchialine' reconsidered. Stygologia 2: 90–92.

Vollenweider, R. A., 1974. A manual on methods for measuring primary production in aquatic environments. Blackwell Scientific Publications, IBP Handbook No. 12: 225 pp.

Hydrobiologia **467**: 91–98, 2002.
J. Alcocer & S.S.S. Sarma (eds), Advances in Mexican Limnology: Basic and Applied Aspects.
© 2002 *Kluwer Academic Publishers.*

Phytoplankton composition and biomass in a shallow monomictic tropical lake

Judith García-Rodríguez[1] & Rosaluz Tavera[2,*]
[1]*Hydrobiology Laboratory, Center of Biological Research, Universidad Autónoma del Estado de Morelos in Cuernavaca City, Morelos, Mexico*
[2]*Phycology Laboratory, Biology Department, School of Sciences, Universidad Nacional Autónoma de México, Apdo. Postal 70-620, C.U. Coyoacán, 04510, Mexico City, Mexico*
(*Author for correspondence)

Key words: tropical limnology, monomictic lakes, phytoplankton, Zempoala, Mexico, clear-water phase

Abstract

Lake Zempoala was studied throughout 16 months in 1996–1997. It is a shallow monomictic lake situated at 2800 masl at the Neovolcanic Belt, well within the Mexican tropical zone. Most of the phytoplankton species in this lake may be characterized as temperate, according to their geographical distribution. A break down in phytoplankton biomass was observed before the lake's circulation, and open to question if a clear-water phase could be present in a tropical lake.

Introduction

Tropical limnology is commonly associated with floodplain, usually polymictic lakes (Melack, 1995); however, in tropical mountainous regions lakes frequently display thermal stratification. Some well-known examples are Lakes Macubaji and Valencia in Venezuela (Lewis & Weibezahn, 1976), Bunyoni and Mulehe in Africa (Talling, 1957), and Dabagri, Dorotiri and Sacabico in Costa-Rica (Jones et al., 1993).

Warm monomictic lakes have been already documented in Mexico (De Buen, 1944; Macek et al., 1994). According to Alcocer et al. (1999) the majority of such lakes are small and deep. Lake Zempoala described here seems to belong to this author's monomictic lakes category; however, it has a maximum depth of 5 m. The annual stagnation of its water column may be explained by the particular climate regime at Lake Zempoala area, as it has been observed also in shallow lakes from Brazil (Payne, 1986; Sommer, 1989).

Regarding to the biology of Lake Zempoala, the combination of high irradiance due to its latitude (19°) and cold sub-humid weather due to its elevation (2800 masl) lead to the selection of some algal species that are usually found in temperate and subtropical lakes. Probably this combination of light and temperature, together with scarce populations of zooplankton and planktivorous fish, favor (*v. gr.*) the dominancy of *Asterionella formosa* in Zempoala, which otherwise would be considered as typical from temperate zones (García-Rodríguez & Tavera, 1998).

Study area

Lake Zempoala is located in the central region of Mexico, in Morelos State, over the Neovolcanic Belt in the Northern edge of the Mexican tropic (Fig. 1A). It belongs to the 'Lagunas de Zempoala' National Park at 65 km SE from Mexico City (19° 03′ N and 99° 18′ 42″ W).

The basin of Lake Zempoala is endorreic. Precipitation and running off from the watershed are the only source of water income whose budget is not balanced: annual precipitation and evaporation are 1200–1500 mm and 1600 mm, respectively (Hernández, 1990), with the rainy season in June–September, whereas there is a lack of precipitation during October–May. The Lake Zempoala surface decreases yearly from 123 400 to 105 640 m^2.

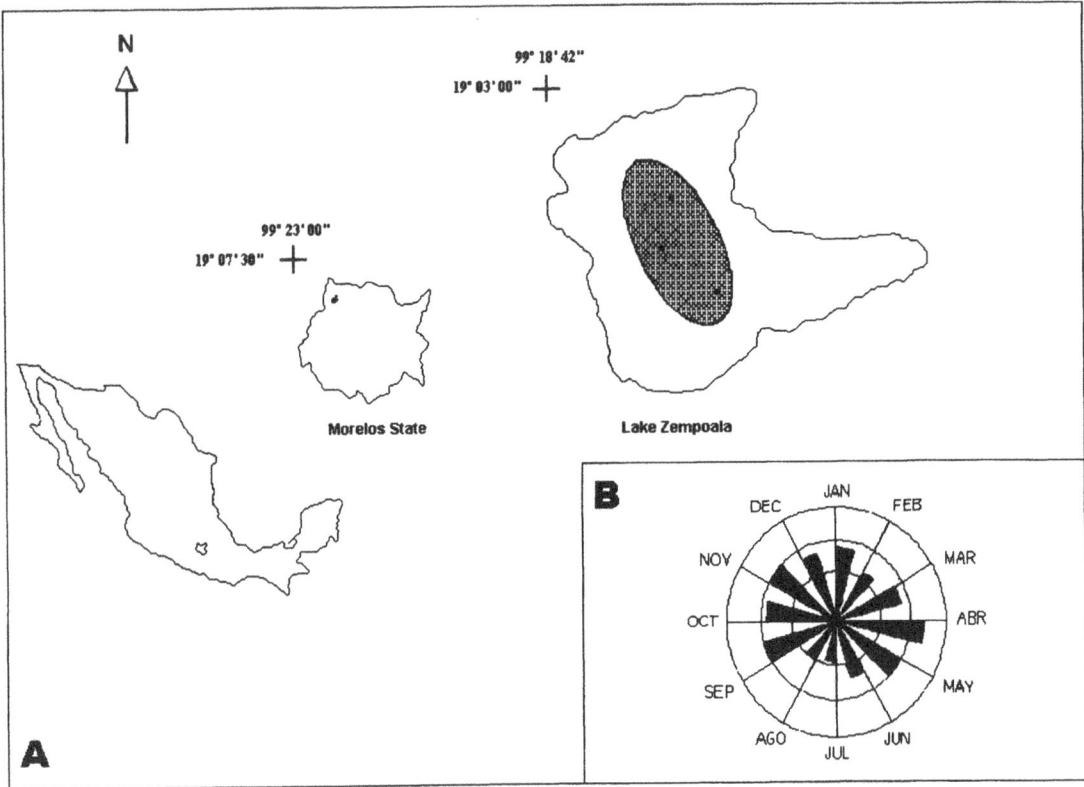

Figure 1. (A) Location of Lake Zempoala. (B) Diagram of monthly average of wind intensity (m s^{-1}), according to Beaufort's scale. Internal circle represents wind of 2 m s^{-1}; external circle represents wind of 4 m s^{-1}.

The lake is of volcanic origin (damming). The natural landscape in the lake's area is Quaternary volcanic and tectonic mountains. Annual temperature fluctuates from −3 to 22 °C with a mean value of 12 ° C (García & Vidal-Zepeda, 1990), and mixed and coniferous forests are the plant coverage; however lately, clearing of the forest for agricultural use has modified them (D'Luna et al., 1990), avoiding wind protection to the lake.

According to Beaufort's scale the prevailing wind energy during spring fall and winter, reaches 10–20 watts m^{-2} (Pérez-Villegas, 1990) in the area where the lake is located. The speed of wind during the year is generally 4 m s^{-1} and above (Fig. 1B), but wind blowing in summer is less than 2 m s^{-1}.

Methods

Sampling program considered monthly samples (April 1996–July 1997), which were collected throughout the water column in few sites (3) in the central part of the lake (Fig. 1A). The water column was sampled at surface, 1, 2.5 and 5 m using a 3-l Van Dorn water sampler. Temperature, pH and dissolved oxygen concentrations were probe-measured with conventional portable equipment. The oxygen saturation percentages were calculated following Mortimer's nomogram (Wetzel, 1975).

Chlorophyll *a* concentrations were determined by the Boyd's method (Vollenweider, 1974) with acidic correction. Phytoplankton samples were preserved in one percent lugol for the Kollkwitz–Utermöhl counting in 3 ml chambers, with an inverted Wild M-40 microscope. Biomass of multicellular organisms was estimated on an average of living cells. Biovolume (μm^3) was transformed to wet weight (μg l^{-1}) following Javornický's method (1978). A Tukey statistical analysis applied to counting from all the collected samples (192) did not reveal significant differences between phytoplankton biomass from the collected sites.

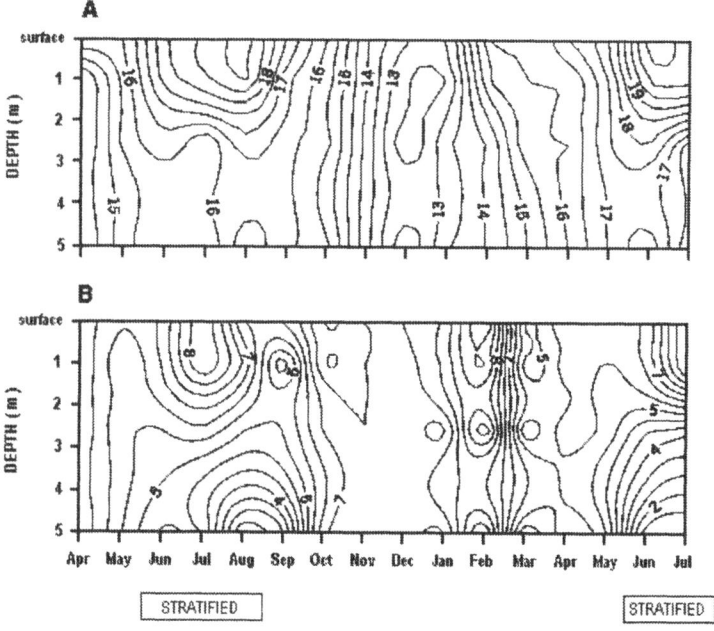

Figure 2. Depth–time diagrams of isopleths. (A) Temperature (°C); (B) Dissolved oxygen (mg l^{-1}).

Results and discussion

A well-developed thermocline existed at a depth of 2.5 m (Fig. 2A). It was strongly established in the warm rainy wet season. In the middle of spring and during summer (from May to August), solar irradiance produced surface water heating. In summer (June to August), when the water column remained stable because of weak wind, differences in temperature reached the mean lake's depth (about 2 m). Those differences ranging from 1 to 4 ° C were enough to establish durable stratified conditions (Fig. 2A). In contrast, favored by the wind action, mixing of the water column started in September and thermocline became eroded. Henceforth the lake seemed to remain mixed until the end of next spring.

Dissolved oxygen (D.O.) profiles followed temperature distribution in the water column (Fig. 2B) and also during summer there were important differences between the surface (7.6–8.6 mg l^{-1}) and the lake's bottom D.O. concentrations (1–2.2 mg l^{-1}). Thus, enhanced by the thermal stratification, biological activity seemed to drive the hipolimnetic D.O. to those very low levels, meanwhile the higher photosynthesis in the epilimnion significantly contributed to the surface D.O. oversaturation: 121% in July and 110% in August of 1996, and 124% in July of 1997.

Values of pH never dropped under 6.3 at the bottom of the lake and at the surface they ranged from 7 to 8.4. In the periods when the lake was stratified, the higher water surface pH values were apparently associated with the more intense activity of photosynthetic organisms at the epilimnion.

Larger peaks of algal biomass appeared concentrated at the epilimnetic zone during summer; probably the water column's stratification was the main factor preventing their even vertical distribution, which was observed during the mixing period (Fig. 3). The increase of water temperature (16–21 °C) brought about by the high summer irradiance promoted an increase in algal biomass. Afterwards, this higher biomass probably prevented underwater light penetration reducing depth of the photic zone (self-shading). In spite of the fact that most organisms remained inside the epilimnion, high densities of organisms led to increase competition and biomass tended to decrease towards the end of summer (Figs 3, 4A). Moreover, this situation happens often in tropical polymictic lakes (Tavera & Castillo, 1998) and explains why only algae debris (mainly pieces of diatoms and euglenophytes) were observed in Lake Zempoala bottom during the summer.

It was particularly noteworthy that at the beginning of September the observed decrease in biomass was so steep that a characteristic low phytoplankton

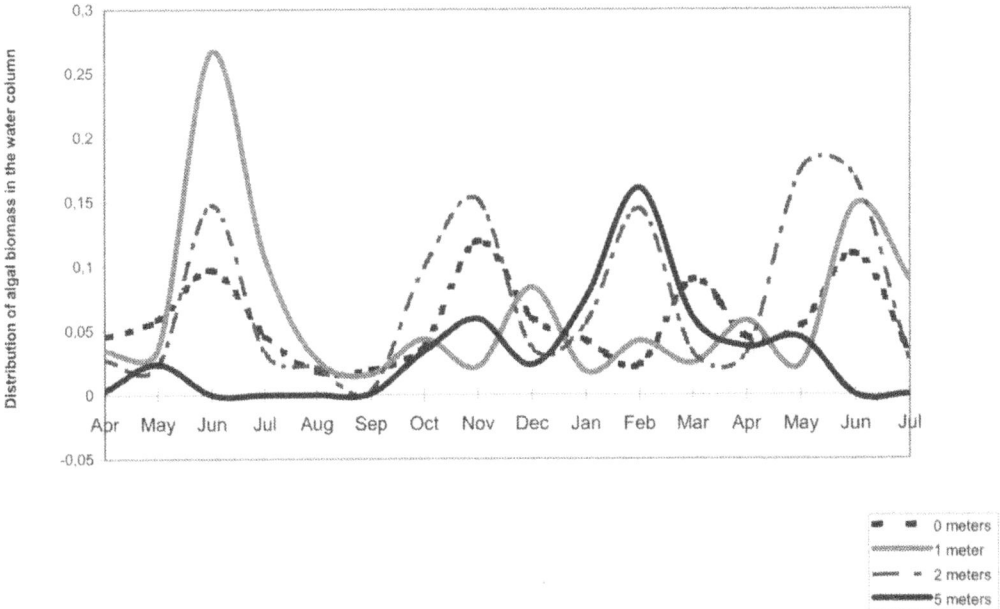

Figure 3. Percentage of algal biomass as it is distributed with depth. During the summer's stratification of the water column phytoplankton were absent at 5 m.

level called the clear-water phase appeared, which was evident from chlorophyll *a* and Secchi depth water-transparency values as well (Fig. 4A). However, such low level usually occurs between spring and summer (Sommer et al., 1986; Lampert & Sommer, 1997). On the other hand, as long as dominancy of *A. formosa* was evident (maximum *Asterionella*'s biomass was eight times the maximum biomass of the next more-abundant alga), it is reasonable to debate if the biomass decrease could be misjudged as a clear-water phase. In the graph of Figure 4B, a comparison between total and partial biomass is presented. The partial biomass consists of all phytoplanktonic species excluding *A. formosa*. The comparison shows that, independently of biomass fluctuations of any particular species, the whole community was notably affected from the end of August to the beginning of September.

As it has been defined for the PEG-model of seasonal succession (Sommer et al., 1986; Lampert & Sommer, 1997), the clear water phase is decided by herbivory, and it will be relevant to conduct an analysis that compares zooplankton and phytoplankton fluctuations, mainly because as has been observed in other lakes, when the phytoplankton peaks do not correspond well with PEG-model steps of succession (Vyhnálek et al., 1991), nutrients could be playing an important role too. This may be the case in Zempoala; from the few phosphorus determinations made in this

lake (García-Rodriguez & Tavera, 1998), the lowest value (60 μg l^{-1}) corresponded to June of 1996. It could be possible that the summer's crash of algal biomass could also be due to a gradually exhaustion of nutrients, not only to herbivorous overexploitation of algae.

In this paper, only an analysis of the phytoplankton variations is presented. It seems to suggest that this community exhibited similar changes as those that have been described during the phytoplankton succession in temperate lakes. Lake Zempoala water transparency in September was due to decay of the whole biomass (not only *A. formosa*), thus it was relevant to search for the possible role of the species associated with *Asterionella* in the phytoplankton community.

In the spring and beginning of the summer of 1996 and 1997 (Fig. 5), phytoplankton were composed by an ensemble of species <20 μm (*Monoraphidium griffithii, Achnanthes minutissima, Nitzschia frustulum, Trachelomonas stokesiana, Trachelomonas hispida* and *Oocystis marssonii*), and some by >20–<90 μm Chlorophyta (*Scenedesmus opoliensis, Staurastrum sebaldi, Gregiochloris lacustris*), and the diatom *Gomphonema acuminatum*. With some differences between years, other larger algae (>90 μm) were present: *Fragilaria crotonensis, Epithemia turgida* and *Dinobryon sociale* in 1996 while *Ceratium*

A

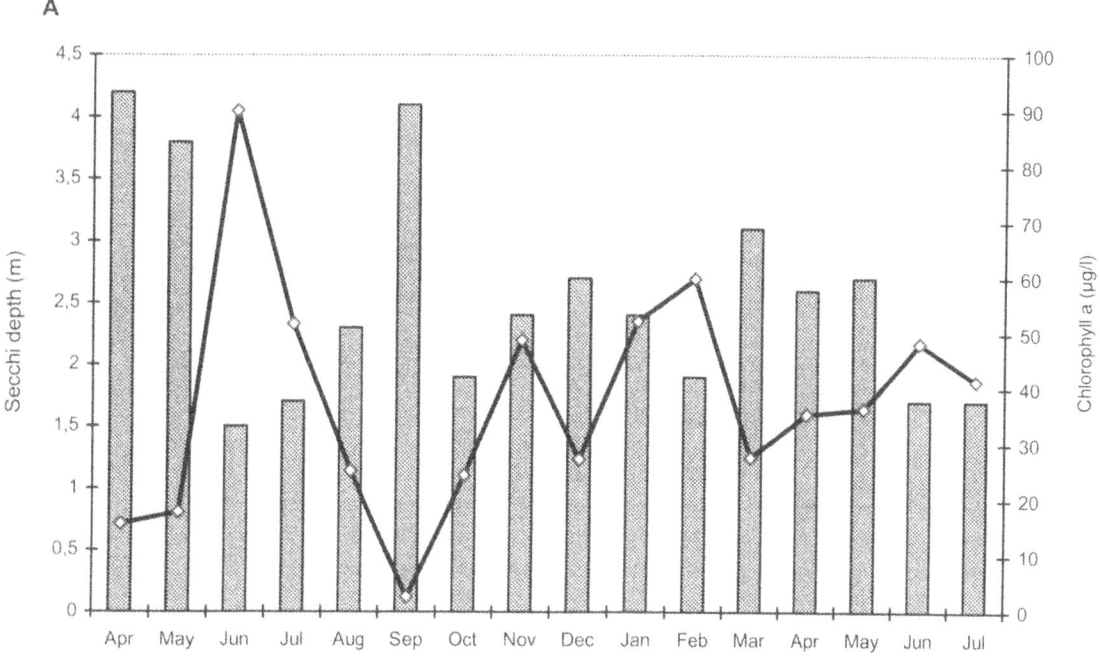

B

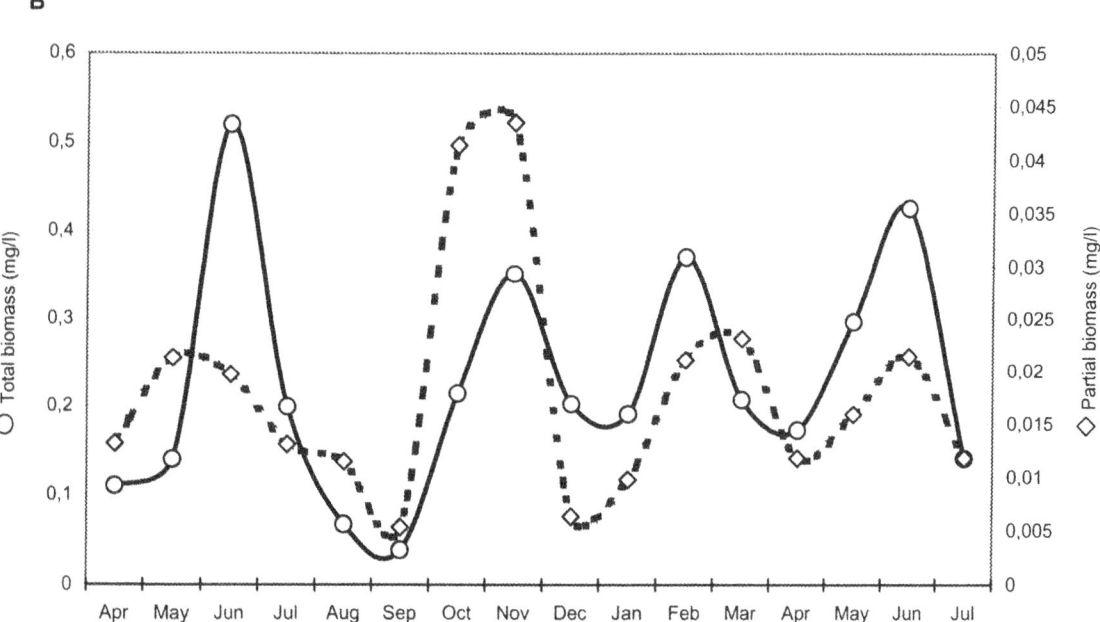

Figure 4. (A) Monthly epilimnetic values of Chlorophyll *a* (continual black line). Monthly values of water column transparency (gray-filled bars) closely follow Chlorophyll concentration. (B) Monthly values of total (continual black line) and partial biomass of algae (dashed gray line). Differences between total and partial biomass (μg/l) are almost one order magnitude.

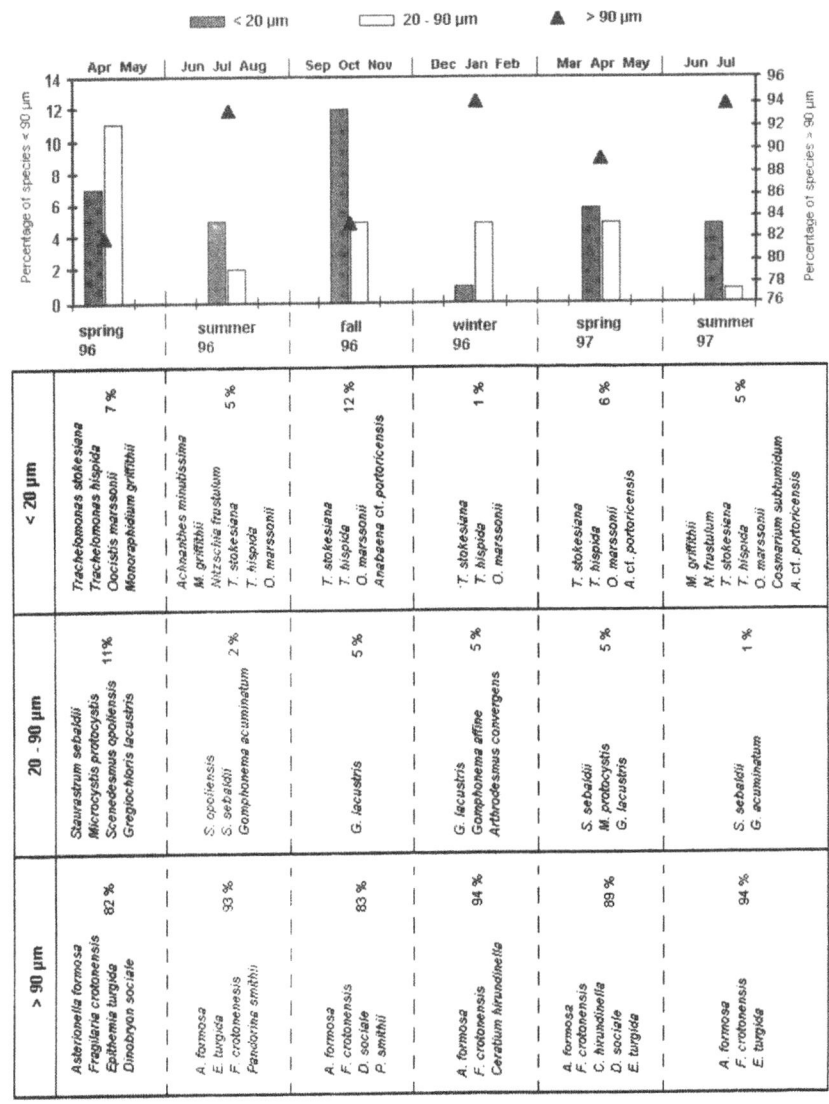

Figure 5. The graph on the top shows the percentage of phytoplankton separated by their size. The table below shows each season size-range composition of species, and gives an idea about fluctuations between edible and inedible species along the study.

hirundinella was present only in 1997. *Pandorina smithii* on the contrary, appeared only in 1996, and *Asterionella formosa* was recurrent all the time. None of all these algae (with exception of *A. formosa*) was particularly abundant. Toward the end of the summer, when the nutrients were lower, and probably exhausted (García-Rodríguez & Tavera, 1998), biomass of each phytoplanktonic species could be seriously growth-limited and drastically decreased. Lake's mixing started in fall (September) and could trigger nutrients replenishment throughout the water column, especially phosphorus (higher value of 650 μg l^{-1} corresponded to September of 1996, (García-Rodríguez

& Tavera, *op. cit*). If algae took advantage of a no-nutrient limitation, populations would restore their biomass at that time.

Considering the size-classes of algae <20–90 μm, small species (<20 μm) with short generation times proliferated early in the fall (Fig. 5). A bloom of *Anabaena* cf. *portoricensis*, (a nitrogen's fixer filamentous cyanoprokaryote) emerged in October of 1996, followed by small Euglenophyta and Chlorophyta during November–December. Immediately afterwards, in winter of 1996 (December–February), still during mixing and when small species already declined, populations of large diatoms and green colonial algae

(>90 μm) strikingly increased their biomass in January to February of 1997. Henceforth to the end of the mixing period (May of 1997), more or less the same algal species of all sizes fluctuated in biomass without further increase. The next stagnant period in summer of 1997 found out the more or less same fluctuation in the previous spring–summer of 1996: a recovering of inedible species (>90 μm) and a decrease of the edible large species (20–90 μm).

It cannot be ignored that beside the phytoplankton composition changes in this lake, the largest species (over 90 μm in size Fig. 5) had biomass fluctuations apparently not related to changes in the edible phytoplankton component. The large-species assemblage included *A. formosa*, *Fragilaria crotonensis*, *Pandorina smithii*, *Dinobryon sociale* and *Ceratium hirundinella*: algae that are usually protected due to their large size from herbivory.

Although the special combination of climatic conditions (light and temperature mainly) is probably supporting the success of *Asterionella formosa*, its rather high biomass may be the outcome of some other factors as well, such as low density of suitable predators (R. Trejo, pers. comm.) and this particular specie's capacity to exploit resources better and/or faster than other diatoms. Some experiments have shown that below 20 °C under culture conditions, the silicate requirements of *A. formosa* being lower than those for *Synedra ulna*, could competitively displace the latter (Tilman et al., 1981). If the phytoplankton decline occurs every year, it seems that *A. formosa* is able to grow faster than other diatoms, being less nutrient-limited, independent of the initial population densities. Perhaps through time and combined with its protection to grazing, this advantage has resulted in the actual dominance of *Asterionella*. In 1986, in sediments collected in this lake (Caballero, 1995) a rich flora of diatoms was found with *Fragilaria crotonensis* as the most abundant species. By the time *F. crotonensis* was abundant (Caballero, pers. comm.) *Asterionella formosa* was not found.

Conclusions

Lake Zempoala is a monomictic, clinograde lake, supersaturated with oxygen at the surface during summer stagnation. The geographical distribution of its phytoplankton species is mainly temperate (García-Rodríguez & Tavera, 1998). Reaching the surface of Lake Zempoala, there are little variations in high

irradiances near the time of the solstice, (Lewis, 1987). Thus, combined with temperature and circulation events in the water column, irradiances must influence the species composition. In the frame of the presented assortment of physical and biological characteristics of Lake Zempoala, it is acceptable to propose that some succession of phytoplankton may be occurring. It is worthy to find in a tropical lake a significant annual decline of phytoplankton; however, according to Sommer et al. (1986), the clear-water phase takes place in lakes between the spring and summer's phytoplankton peaks. In Lake Zempoala, the 'clear-water phase' appears to be shifted to the end of the summer, and it will be necessary to incorporate a study of the zooplankton community and additional data of nutrients to elucidate the possible succession pattern in Zempoala.

The phytoplankton were always dominated by *Asterionella formosa*, a species described as typical from temperate regions and which, as well as the most of other algae registered in this lake, have a geographical distribution almost restricted to temperate zones (García-Rodríguez & Tavera, 1998). This feature is clearly not typical for a tropical lake, but probably is common to find it, as it seems to be repeated in several lakes from the elevated regions in this country (Tavera & Komárek, 1996), and especially from the Mexican plateau (Tavera et al., 2000).

Acknowledgements

Authors thank very much the critical review of the manuscript and English improvement to Dr J. Alcocer and Dr S.S.S. Sarma from Limnology Laboratory, Environmental Conservation & Improvement Project. UIICSE. *Campus* Iztacala, UNAM. We also are grateful to Prof. J. Sánchez from Faculty of Sciences, UNAM for the final English correction of the manuscript.

References

Alcocer, J., A. Lugo, E. Escobar, M. R. Sánchez & G. Vilaclara, 1999. Water column stratification and its implications in the tropical warm monomictic Lake Alchichica, Puebla, Mexico. Verh. int. Ver. Limnol. 27 (in press).

Caballero, M., 1995. Late Quaternary paleolimnology of Lake Chalco, the Basin of Mexico. PhD thesis. Hull University, U.K.: 286 pp.

De Buen, F., 1944. Los lagos Michoacanos. II. Pátzcuaro. Rev. Soc. Mex. Hist. Nat. 5: 99–125.

98

D'Luna, A., C. Meza, J. Moncada, O. Oropeza & A. Palacio, 1990. Carta influencia del hombre en el medio ambiente. V.1.1. Atlas Nacional de México. Instituto de Geografía, UNAM. Vol II Naturaleza.

García, E. & R. Vidal-Zepeda, 1990. Carta temperaturas extremas. IV.4.5. Atlas Nacional de México. Instituto de Geografía, UNAM. Vol II Naturaleza.

García-Rodríguez, J. & R. Tavera, 1998. Fitoplancton del lago Zempoala. Bol. Soc. Bot. Mex. 63: 85–100.

Hernández, M. E., 1990. Carta Clima. Medidas de aridez. IV.4.9 (A.B.) Atlas Nacional de México. Instituto de Geografía, UNAM. Vol II Naturaleza.

Javornický, P., 1978. Ekológia sladkovodných rias. In Hindák, F. (ed.), Sladkovodné Riasy. Bratislava, Slovenské pedagogické nakaladatelstvo: 9–61.

Jones, J. R., K. Lohmann & U. G. Umana, 1993. Water chemistry and trophic state of eight lakes in Costa Rica. Verh. int. Ver. Limnol. 25: 899–905.

Lampert, W. & U. Sommer, 1997. Limnoecology: the Ecology of Lakes and Streams. Oxford Univ. Press: 382 pp.

Lewis, W. M. & F. H. Weibezahn, 1976. Chemistry, energy flow, and community structure in some Venezuelan fresh waters. Arch. Hydrobiol./Suppl. 50(2/3): 145–207.

Lewis, W. M., 1987. Tropical Limnology. Ann. Rev. Ecol. Syst. 18: 159–184.

Macek, M., G. Vilaclara & A. Lugo, 1994. Changes in protozoan assemblage structure and activity in a stratified tropical lake. Marine Microbial Food Webs, 8: 235–249.

Melack, J. M., 1995. Recent developments in tropical limnology. Verh. int. Ver. Limnol. 26: 211–218.

Payne, A. I., 1986. The Ecology of Tropical Lakes and Rivers. J. Wiley & Sons, Chichester: 301 pp.

Pérez-Villegas, G., 1990. Carta Clima. Energía del viento dominante. IV.4.2, IV.4.3. Atlas Nacional de México, 1990. Instituto de Geografía, UNAM. Vol II Naturaleza.

Sommer, U., Z. M. Glewicz, W. Lampert & A. Duncan, 1986. The PEG-model of seasonal succession of planktonic events in fresh waters. Arch. Hydrobiol. 106: 433–471.

Sommer, U. (ed.), 1989. Plankton Ecology. Succession in Plankton Communities. Springer-Verlag: 369 pp.

Talling, J. F., 1957. Diurnal changes of stratification and photosynthesis in some tropical African waters. Proc. Royal Soc. B. 147: 57–83.

Tavera, R. & J. Komárek, 1996. Cyanoprokaryotes in the volcanic Lake Alchichica, Puebla State, Mexico. Archive für Hydrobiol/Algological Studies 83: 511–538.

Tavera, R. & S. Castillo, 2000. An eutrophication-induced shift in the composition, frequency and abundance of the phytoplankton in Lake Catemaco, Veracruz, México. In Munawar, M., S. Lawrence, I. F. Munawar & D. Malley (eds), Aquatic Ecosystems of Mexico: Status and Scope. Ecovision World Monograph Series. Backhuys Pub., Leiden, The Netherlands: 103–117.

Tavera, R., E. Novelo & A. Comas, 2000. Chlorococcalean algae (s.l.) from the Ecological Park of Xochimilco, Mexico. Archive für Hydrobiol/Algological Studies 100 (in press).

Tilman, D., M. Mattson & S. Langer, 1981. Competition and nutrient kinetics along a temperature gradient: An experimental test of a mechanistic approach to niche theory. Limnol. Oceanogr. 26: 1020–1033.

Vollenweider, R. A. (ed.), 1974. A manual on Methods for Measuring Primary Production in Aquatic Environments. IBP Handbook No. 12, Oxford, Blackwell: 225 pp.

Vyhnálek, V., J. Komárková, J. Sed'a, Z. Brandl, K. Šimek & N. Johanisová, 1991. Clear-water phase in the Římov Reservoir (South Bohemia): Controlling factors. Verh. int. Ver. Limnol. 24: 1336–1339.

Wetzel, R. G., 1975. Limnology. Saunders College Publishing: 743 pp.

Hydrobiologia **467**: 99–108, 2002.
J. Alcocer & S.S.S. Sarma (eds), Advances in Mexican Limnology: Basic and Applied Aspects.
© 2002 *Kluwer Academic Publishers.*

Seasonal variations of zooplankton abundance in the freshwater reservoir Valle de Bravo (Mexico)

P. Ramírez García[1], S. Nandini[1,*], S.S.S. Sarma[2], E. Robles Valderrama[1], I. Cuesta[3] &
Maria Dolores Hurtado[1]

[1]*Unidad de Investigación Interdisciplinaria en Ciencias de la Salud y la Educación, Cyma Project, UNAM,
Campus Iztacala, CP 54090 Tlalnepantla, Los Reyes AP 314, Edo. de México, Mexico*
E-mail: nandini@servidor.unam.mx
[2]*Carrera de Biología, UNAM, Campus Iztacala, CP 54090 Tlalnepantla, Los Reyes AP 314,
Edo. de México, Mexico*
E-mail: sarma@servidor.unam.mx
[3]*CNA/Subdir. Gral. Téc/Gerencia de Saneamiento y Calidad del Agua/Subgerencia de Estudios
de Calidad del Agua e Impacto Ambiental. Av. San Barnabé No. 549 San Jerónimo Lidice, Mexico*
(*Author for correspondence)

Key words: Mexico, man-made reservoir, zooplankton, Rotifera, Cladocera, Copepoda

Abstract

Information on the density and diversity of zooplankton from drinking water reservoirs in Mexico is meagre. This is important not only from the point of view of lake management but also for providing clean drinking water for human populations. In the present work, we provide quantitative information on the seasonal variations of zooplankton and selected physico-chemical variables from Valle de Bravo, a large man-made reservoir in the State of Mexico. Based on the nutrient data, this reservoir can be regarded as mesotrophic. However, we found a high density of phytoplankton. Among Cyanophyceae, *Anabaena, Microcystis, Nostoc* and *Oscillatoria* were encountered, particularly during the warmer months. *Microcystis* blooms were observed from June to September. Diatoms dominated the phytoplankton during the remaining months of the year. Among zooplankton, Rotifera comprised the highest number of species. The most common species occurring throughout the year were *Keratella chochlearis, Polyarthra vulgaris, Trichocerca capucina, Trichocerca similis, Asplanchna priodonta* and *Synchaeta pectinata*. The highest abundance of *K. cochlearis* was observed in May at site 5 (340 ind l^{-1}). *P. vulgaris* occurred in high numbers throughout the year. Both *Trichocerca capucina* and *T. similis* were encountered in low numbers (mean annual average: 10 ind l^{-1}). Among the crustacean zooplankton, *Bosmina longirostris* occurred in higher numbers (up to 105 ind l^{-1}) regardless of the time of the year. Copepods were predominantly from the genera *Mesocyclops* and *Megacyclops*; calanoids were rare. Naupliar stages were dominant during winter months (up to 670 ind l^{-1}). These results have been discussed from the point of view of *Microcystis* control in water bodies.

Introduction

Mexican limnological studies have been conducted predominantly in large and/or natural waterbodies (Torres-Orozco et al., 1996; Sarma & Elías-Gutiérrez, 1999; Alcocer et al., 2000). These are relatively stable and almost permanent. Zooplankton diversity and density of these waterbodies have been used in both, basic areas such as ecological interactions, and applied fields such as fisheries and water contamination

studies. Studies on the zooplankton diversity in Mexico began in the 1930s (Ahlstrom, 1932) and a good quantity of information exists not only from large waterbodies but also from ephemeral ponds. Quantitative analysis of zooplankton, on the other hand, is fairly recent (Suarez-Morales et al., 1986; Torres-Orozco & Estrada-Hernandez, 1997) from large reservoirs and so far indicated the dominance of one of the three major components of freshwater zooplankton (rotifers, cladocerans and copepods) during certain months of the year. However, similar information on the density and diversity of zooplankton from drinking water reservoirs is relatively meagre. This information is important not only from the point of view of lake management, but also for providing clean drinking water for human populations.

Most Mexican waterbodies are alkaline or nearly neutral (Suarez-Morales et al., 1993; Torres-Orozco & Estrada-Hernandez, 1997; Sarma & Elías-Gutiérrez, 1997) and are rarely acidic. Drinking water reservoirs so far studied indicate alkaline pH conditions. Being tropical (except those from high altitude locations), these waterbodies have a mean temperature range of 25±3 °C (Sarma et al., 1998; Torres-Orozco & Zanatta, 1998) while that of dissolved oxygen is 3–8 mg l^{-1} (Sarma & Elías-Gutiérrez, 1999). Valle de Bravo, a drinking water reservoir supporting watersupply for a human population of 6 000 000, is one of the imporant waterbodies in the State of Mexico. Studies conducted on this lake have aimed at characterizing the morphometry and physicochemical structure of the lake, qualitative analysis of bacteria and plankton and quantitative and qualitative analysis of the fish population (Anon., 1998).

Since an in-depth analysis of zooplankton helps in efficient management of a water body, in this study we provide quantitative information on the seasonal variations of zooplankton and selected physico-chemical variables from Valle de Bravo.

Materials and methods

Valle de Bravo located at an altitude of 1830 m at 19° 21′ 30″ N and 100° 11′ 00″ W is approximately 1730 ha area with a mean depth of 19.4 m. It is a perennial waterbody with watershed of 547 km^2 (Pedro-Ramírez et al., 1998). For the purpose of sampling, we selected a total of five stations (Fig. 1). The southern and eastern sites are close to the industrial sites and a township. Monthly zooplankton samples were obtained from each of these sites for the period May, 1998 to April, 1999. Concurrenty, water samples were taken for measuring selected physico-chemical variables. For zooplankton samples, we filtered 80 l of water using a plankton net of 50 μm mesh size. Samples were collected from the surface (~0.5 m) during the morning hours. Although we collected the samples for some months at fortnightly intervals, for presentation we pooled the data and expressed it on a monthly basis. Zooplankton samples were preserved in 10%

formalin at the site itself. At the time of sampling, we measured the surface water temperature and pH, conductivity and Secchi transparency. Analysis of other variables (dissolved oxygen, free ammonia, dissolved ammonia, nitrite, nitrate and phosphate) were conducted in the laboratory using standard procedures (APHA, 1985).

Identification of zooplankton species was done using standard literature (Koste, 1978; Dussart & Defaye, 1995; Korovchinsky & Smirnov, 1998). Only rotifers and cladocerans were identified to species level. For quantitative analysis, we counted the number of individuals for each species present in aliquot of 1 ml from the concentrate (to 100 ml) of field-collected zooplankton. The data were later converted to the actual quantity of water filtered from the lake. We used 3–4 aliquots for each sample. Density of zooplankton was expressed as number of individuals per litre. Quantitative analysis of copepods included adults and various copepodite stages as well as naupliar stages.

Results

Data on the selected physico-chemical variables through seasons from Valle de Bravo are presented in Figure 2. The range of temperature throughout the study period varied from 20 to 26 °C. Throughout the study, there were no significant differences in any of the physico-chemical variables measured between the sites. The range of DO values varied from 6 to 8 mg l^{-1}, the lowest being during the winter and highest during summer. Similarly pH of the waterbody indicated an alkaline condition. An elevated pH (8.9) was observed in summer while the values were near-neutral in winter (pH=6.9). Conductivity measurements revealed an annual average of 152 μS cm^{-1}. The highest values of over 200 μS cm^{-1} were recorded from sites 1 and 5. Highest free ammonia values (0.93 mg l^{-1}) were observed during the summer months, particularly at sites 1 and 5. Dissolved ammonia values ranged from 0.0 to 0.49 mg l^{-1} depending on the sampling period. Levels of nitrite nitrogen were much lower (maximum: 0.14 mg l^{-1}) and even absent during certain months in Valle de Bravo. The highest concentration of nitrate nitrogen was at site 1 (1.03 mg l^{-1}). Phosphate values were in the lower range (annual mean value=0.019 mg l^{-1}). The annual mean value of the N:P (nitrate nitrogen:phosphate ratio) ranged from 0.98 to 5.68, indicating nitrogen limitation. Only once in April a much higher N:P ratio of 28:32 was recor-

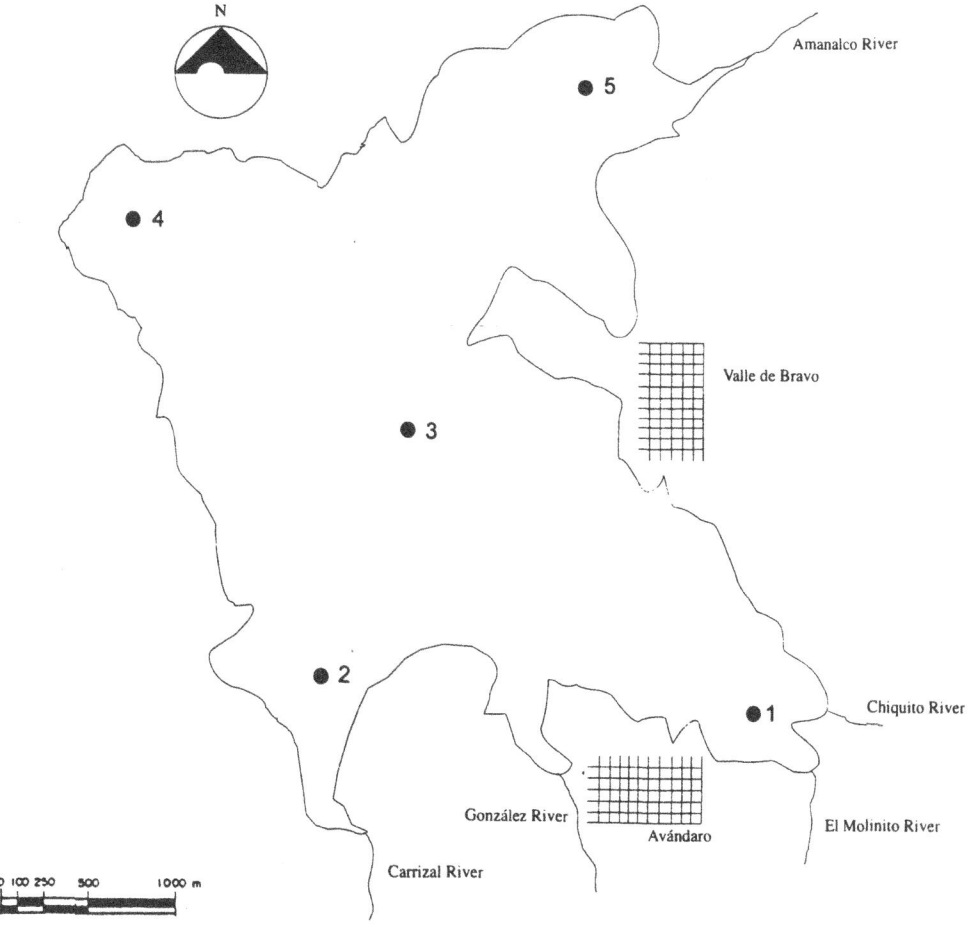

Figure 1. Collection sites from Valle de Bravo reservoir (figure not to scale).

ded in the reservoir. Secchi disc transparency values ranged from 0.73 to 2.01 m depending on the period of sampling (Fig. 2). In general, higher transperancy values were recorded during the winter months.

A list of phyto- and zooplankton encountered in Valle de Bravo during the study period is presented in Table 1. Four genera of Cyanophyceae, *Anabaena, Microcystis, Nostoc* and *Oscillatoria* were encountered, particularly during the warmer months. *Microcystis* blooms were observed from June to September. Diatoms dominated the phytoplankton during the remaining months of the year. Among zooplankton, Rotifera comprised the highest number of species. The most common species occurring throughout the year were *Keratella chochlearis, Polyarthra vulgaris, Trichocerca capucina, Trichocerca similis, Asplanchna priodonta* and *Synchaeta pectinata* (Fig. 3). High densities of *Keratella cochlearis* (245–345 ind l^{-1}) were recorded at two of the five sites studied (sites

1 and 5). The highest abundance of *K. cochlearis* was observed in May. Among the rotifers, *P. vulgaris* occurred in high numbers in most of the tested sites and throughout the year. The highest densities were found at site 5 which was as high as 1000 ind l^{-1} particularly during August–October. Both *Trichocerca capucina* and *T. similis* were encountered in low numbers, with maximum densities of 63 and 159 ind l^{-1}, respectively. *Asplanchna priodonta* occurred throughout the year but maximal abundance (48 ind l^{-1}) was recorded in the winter months. Highest numbers of *Synchaeta pectinata* were observed at sites 1 and 5 during October (about 700 ind l^{-1}).

Among the crustacean zooplankton, *Bosmina longirostris* occurred in higher numbers (up to 105 ind l^{-1}) regardless of the time of the year. In general, site 5 had a higher annual mean (about 30 ind l^{-1}) while the least abundance (about 10 ind l^{-1}) was recorded from site 2. *Daphnia laevis* densities were very low; gener-

Table 1. List of phytoplankton and zooplankton encountered in Valle de Bravo

Phytoplankton family	Taxa
Bacillariophyceae	*Anomoeneis*
	Fragilaria
	Mastogloea
	Melosira
	Phormidium
	Rhizosolenia
Chlorophyceae	*Ceratium*
	Closteriopsis
	Cosmocladium
	Dictyosphaerium
	Mougetia
	Spirogyra
	Staurastrum
Cyanophyceae	*Anabaena*
	Microcystis
	Nostoc
	Oscillatoria
Zooplankton	
Rotifera	
Philodinidae	*Rotaria* sp.
Brachionidae	*Brachionus patulus* (O. F. Müller, 1786)
	Keratella cochlearis (Gosse, 1851)
	Platyias quadricornis (Ehrenberg, 1832)
Euchlanidae	*Euchlanis dilatata* Ehrenberg 1832
	E. incisa Carlin (1939)
Mytilinidae	*Mytilina ventralis* (Ehrenberg, 1832)
Colurellidae	*Lepadella acuminata* (Ehrenberg, 1834)
Lecanidae	*Lecane aculeata* (Jakubski, 1912)
	L. bulla (Gosse, 1851)
	L. ludwigi (Eckstein, 1883)
	L. lunaris (Ehrenberg, 1832)
Notommatidae	*Cephalodella gibba* (Ehrenberg, 1838)
Trichocercidae	*Trichocerca capucina* (Wierzejski & Zacharias, 1893)
	T. similis (Wierzejski, 1893)
Gastropodidae	*Ascomorpha ovalis* (Bergendal, 1892)
Synchaetidae	*Polyarthra vulgaris* Carlin 1943
	Synchaeta oblonga Ehrenberg, 1831
	S. pectinata Ehrenberg, 1832
Asplanchnidae	*Asplanchna brightwelli* (Gosse, 1850)
	A. priodonta (Gosse, 1850)
Testudinellidae	*Pompholyx sulcata* (Hudson, 1885)

Continued on p. 103

Table 1. Continued

Phytoplankton family	Taxa
Conochilidae	*Conochilus unicornis* Rousselet 1892
Hexarthridae	*Hexarthra intermedia* Wiszniewski 1929
Filiniidae	*Filinia longiseta* (Ehrenberg, 1834)
Collothecidae	*Collotheca* sp.
Cladocera	
Bosminidae	*Bosmina longirostris* (O. F. Müller, 1785)
Daphnidae	*Daphnia laevis* Birge 1878
	Ceriodaphnia lacustris Birge 1893
	Simocephalus vetulus (O. F. Müller, 1776)
Chydoridae	*Chydorus sphaericus* O. F. Müller 1785
	Alona sp.
	Camptocercus sp.
Copepoda	*Mesocyclops* sp.
	Megacyclops sp.
	Calanoidea

ally less than 1 ind l^{-1}. Sites 3 and 4 supported higher densities of *D. laevis* as compared to the other stations. Densities of *Ceriodaphnia lacustris* were also low (<1 ind l^{-1}) except during the month of November at site 5 where 20 ind l^{-1} were recorded.

Copepods were predominantly from the genera *Mesocyclops* and *Megacylops*, calanoids were rare. Naupliar stages were dominant during winter months (up to 670 ind l^{-1} in site 4). Adults and copepodite stages were also observed in higher numbers during winter (annual mean about 30 ind l^{-1}) (Fig. 4).

Species diversity was calculated using Shannon–Wiener's formula and the values ranged from 2.17 to 2.60. The lowest value was obtained from site 2 and the highest from site 4. Data on species evenness ranged from 0.63 to 0.75.

Discussion

Recent reviews on Mexican limnology (Alcocer et al., 2000) have pointed out the lack of information from drinking water reservoirs in Mexico. A number of workers have measured various physico-chemical variables from various freshwater bodies in Mexico. These studies range from high altitude ponds to typical tropical systems (Sarma et al., 1996; Torres-Orozco & Zanatta, 1998). Data on the temperature, nutrient concentrations, pH range, dissoved oxygen values, and Secchi transparency were in general agreement with limnological characteristics of Mexican waterbodies. Data on plankton exists in the form of checklists for phytoplankton (Chavez, 1986) but there is no quantitative data on the seasonal abundance of zooplankton. Since high altitude tropical lakes have been compared to temperate ones (Dejoux & Iltis, 1992), studies on this reservoir are also interesting to compare whether patterns conform to tropical or temperate water bodies.

Based on the nutrient data, this reservoir can be regarded mesotrophic. However, we found a high density of phytoplankton, indicated by a low Secchi transparency particularly during the summer months. There are numerous studies indicating the importance of phosphates and nitrates in controlling the abundance of phytoplankton and thereby, zooplankton (Schindler, 1974; Barica, 1990). A better indicator of nutrient status of lakes appears to be the ratio of nitrate nitrogen to orthophosphates (N:P). It has been shown

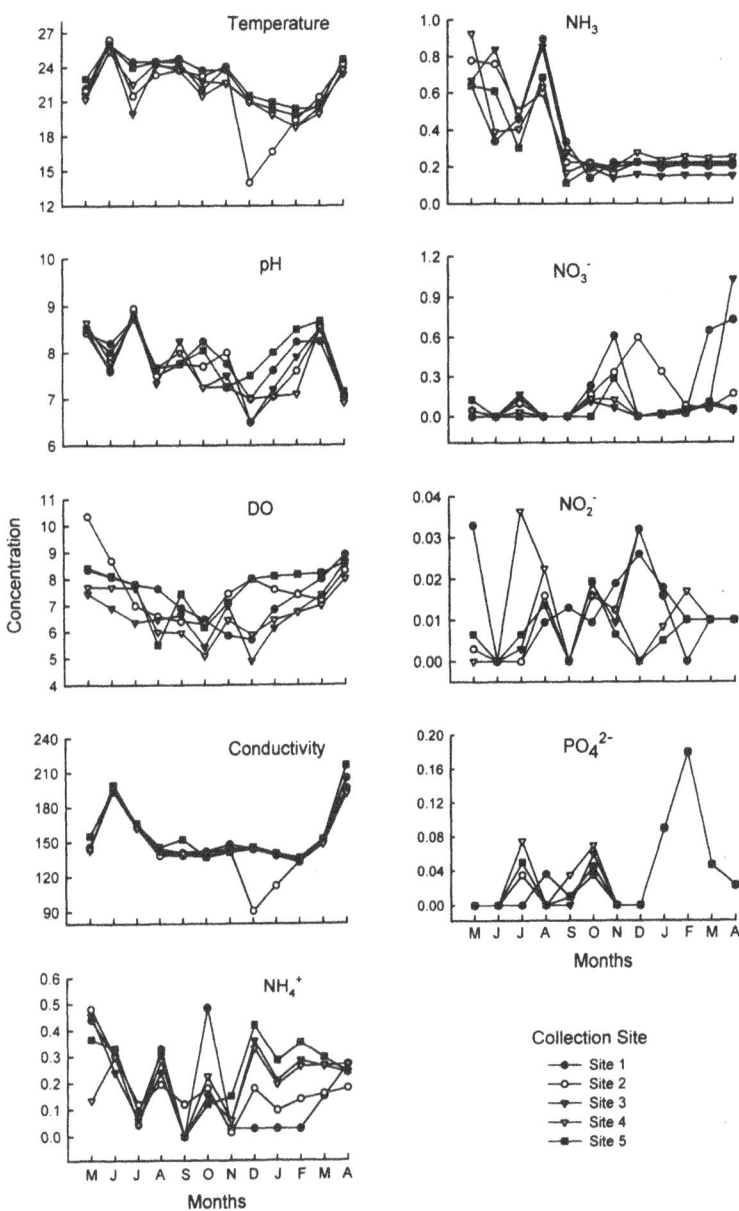

Figure 2. Seasonal changes in temperature (°C), pH, dissolved oxygen (mg l^{-1}), conductivity (μS cm^{-1}), dissolved ammonia (mg l^{-1}), free-ammonia (mg l^{-1}), nitrate (mg l^{-1}), nitrite (mg l^{-1}) and phosphate (mg l^{-1}) from Valle de Bravo reservoir.

that a ratio less than 10 results in nitrogen limitation which favours cyanobacterial blooms (Forsberg et al., 1978; Smith, 1983; Paerl & Tucker, 1995). In the present study, regardless of the site, these ratios ranged from 0.98 to 5.68. Blooms of cyanobacteria, particularly *Microcystis* were found from May to August which were replaced by diatoms during the winter. In tropical water bodies at sea level, on the other hand, nitrogen limitation often leads to perennial cy-

anobacterial blooms (Zafar, 1986). Lathrop (1988) and Canfield et al. (1989), however, suggest that complete reliance on N:P ratios may not always be sufficient to explain autotroph succession in water bodies.

Rotifers in the present study dominated both qualitatively and numerically. Sarma and Elias-Gutiérrez (1998) have shown that as many as 50 rotifers can be found in a single waterbody in Mexico. Among the rotifers, we did not encounter any predatory taxa.

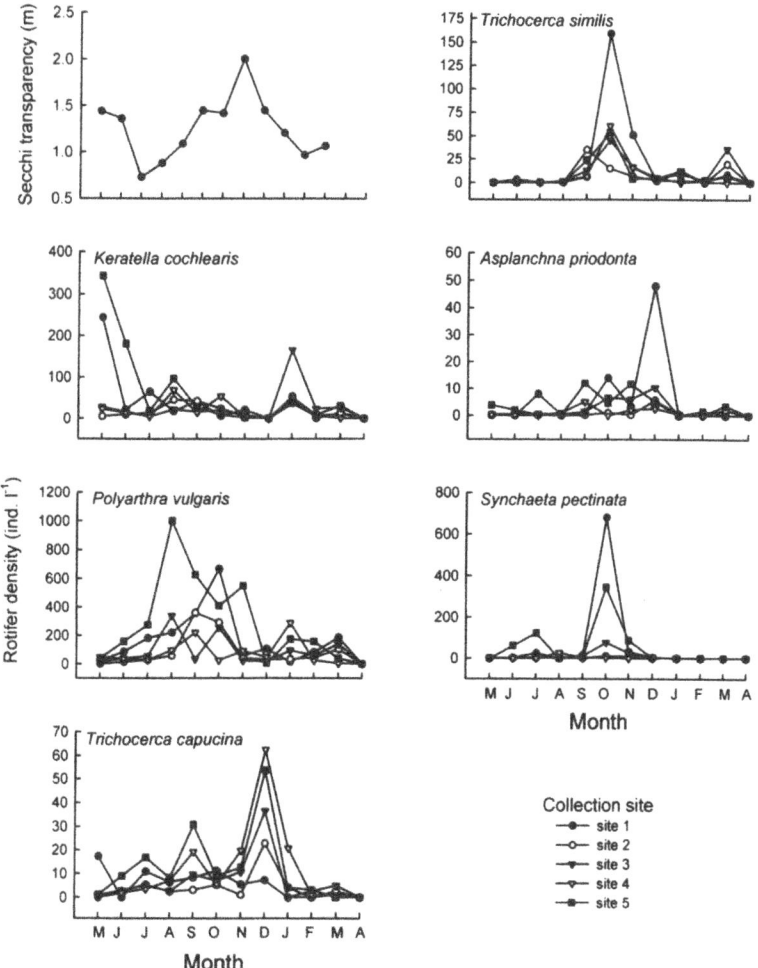

Figure 3. Seasonal variations in Secchi transparency (m) and the abundance of rotifers (ind l^{-1}) from Valle de Bravo reservoir.

Asplanchna priodonta may consume other smaller rotifers, but in general is considered to be predominantly phytoplanktophagous (Hoffmann, 1974). Since our sampling was restricted to the limnetic zone, only planktonic species of rotifers and crustaceans were encountered. However, occasionally (once or twice during the study period) non-planktonic and periphytic rotifers were found in our samples in low abundance (1–10 ind l^{-1}) (see Table 1). The seasonal abundance of rotifers in general showed a single peak during different months of the study period depending on the species. Unlike other tropical lakes, the genus *Brachionus* was conspicuously absent in Valle de Bravo throughout the study. *Polyarthra vulgaris* and *Trichocerca capucina* have been earlier recorded in the large neotropical lake, Lake Catemaco (area=7254 ha) (Torres & Zanatta, 1998), but were found in densit-ies less than 0.1 ind l^{-1}. In Valle de Bravo however, mean densities of *Polyarthra* ranged up to 1000 ind l^{-1} while those of *T. capucina* up to 54 ind l^{-1} at site 5. The highest densities of *Synchaeta pectinata* and *Asplanchna priodonta* were observed during the winter months.

Among cladocerans, *D. laevis* was encountered in Valle de Bravo in all the study sites. Higher annual mean abundance (2 ind l^{-1}) of *D. laevis* was observed in sites 3 and 4. This species was earlier observed from Mexico by Van de Velde et al. (1978). The ability of this cladoceran species to consume *Microcystis* is interesting. Nandini et al. (2000) have observed that *D. laevis* could consume toxic *Microcystis*. In their study, using life table demography and population growth, *D. laevis* offered *Microcystis* showed responses comparable to those fed the green alga *Chlorella vulgaris*.

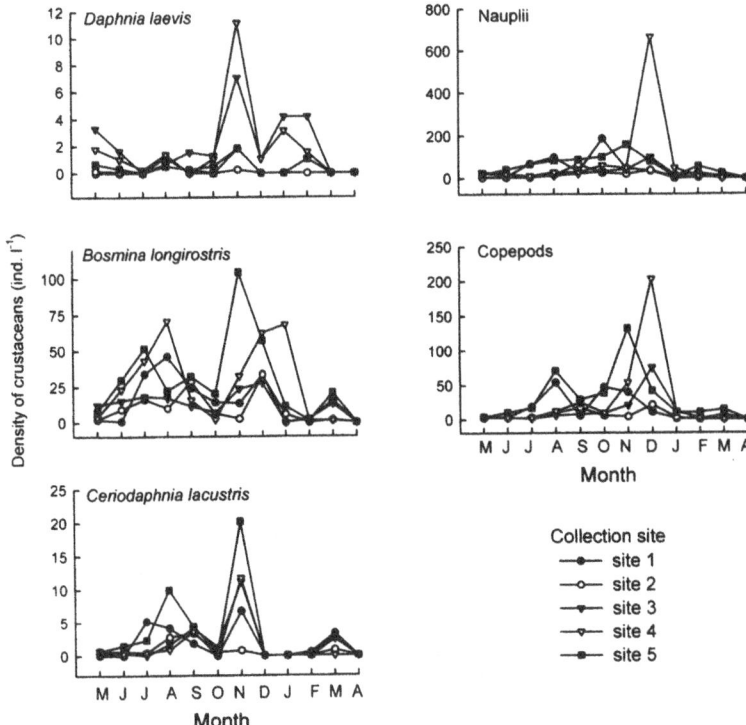

Figure 4. Seasonal variations in the abundance of crustacean zooplankton (ind l^{-1}) from Valle de Bravo reservoir.

However, the highest densities of *D. laevis* were recorded during the winter months when *Microcystis* blooms were absent. Laboratory observations also indicate that this species, though capable of growing on a diet of *Microcystis*, flourishes at temperatures of 20±2 °C. *B. longirostris* was observed in high densities (15–30 ind l^{-1}) throughout the year. DeMott & Kerfoot (1982) have shown that *Bosmina longirostris* is capable of coexisting with cyanobacterial blooms, using its fifth pair of limbs to screen and avoid feeding on cyanophyceae. *Ceriodaphnia lacustris* showed distinct preference for winter months; a trend also shown by adult, copepodite and naupliar stages of cyclopoid copepods.

Competition and predation are two factors strongly affecting the structure of zooplankton communities in freshwater ecosystems. Gilbert (1988) reviewed the negative relationship between the density of rotifers and cladocerans. In the present study, we found no singificant relationship between rotifer density and the abundance of cladocerans. There may be serveral reasons for this. Firstly, as shown recently, evasive rotifer species such as *Polyarthra*, the most abundant rotifer in our study, are unaffected by the cladoceran abundance (Kak & Rao, 1998). Secondly, so far, only

larger cladocerans have the ability to suppress the abundance of rotifers. In our study, the only large cladoceran species observed was *D. laevis*, the density of which was relatively low (usually <2 ind l^{-1}) throughout the year. Smaller cladocerans such as *Bosmina* and *Ceriodaphnia*, more abundant in our study, do not have the same adverse effect of interference competition with rotifers. Cyanobacterial blooms are also known to restructure the zooplankton population of a water body in favour of smaller species (Gilbert, 1996). This was evident during our study when the summer months with high cyanobacterial populations were dominated by the small *Polyarthra vulgaris* and *Bosmina longirostris* while the winter months had higher densities of the large *Daphnia laevis*.

Among the sites studied, on an average, we obtained higher peak abundances of *K. cochlearis*, *Polyarthra vulgaris*, *Bosmina longirostris* and *Ceriodaphnia lacustris* at site 5, which was close to an industrial township. This site also receives organic wastes which probably contributed to the higher abundance of zooplankton. Annual mean conductivity value, nitrite and ammonia levels were also higher at this site (Fig. 2).

In conclusion, our study showed that although nutrient levels were low, relatively high densities of phytoplankton and zooplankton were found in Valle de Bravo. Cyanobacterial blooms were reflected in low Secchi transparency values. However, the total zooplanktonic abundance showed no identifiable patterns with cyanobacteria. Crustacean density was inversely related to Secchi transparency data suggesting that the available food was unsuitable for consumption by zooplankton.

Acknowledgements

We thank UNAM-DGAPA Project (IN206198) for financial support. We are grateful for the help received from CNA (Mexico) Water Quality Department managed by Ing. Carlos Israel.

References

Ahlstrom, E. H., 1932. Plankton Rotatoria from Mexico. Trans. am. Microscop. Soc. 51: 242–251.

Alcocer, J., E. Escobar & L. E. Marin, 2000. Epicontinental aquatic systems of Mexico in the context of hydrology, climate, geography and geology. In Munawar, M., S. G. Lawrence, I. F. Munawar & D. F. Malley (eds), Aquatic Ecosystems of Mexico: Status and Scope. Ecovision World Monograph Series, Backhuys Publishers, Leiden, The Netherlands: 1–13.

Anonymous, 1998. Determinación de la población de peces en la presa Valle de Bravo, Estado de México, con fines de manejo del embalse. Comisión Nacional de Agua, Mexico DF: 70 pp.

Barica, J., 1990. Seasonal variability of N:P ratios in eutrophic lakes. Hydrobiologia 191: 97–103.

Canfield, D. E. Jr., E. Phillips & C. M. Duarte, 1989. Factors influencing the abundance of blue–green algae in Florida lakes. Can. J. Fish. Aquat. Sci. 46: 1232–1237.

Chavez, M. M., 1986. Contribución al conocimiento de la estructura y composición de las comunidades fitoplanctonicas de Valle de Bravo, Edo. de Mex. Tesis licenciatura, ENEP Iztacala, UNAM, Mexico.

Dejoux, C. & A. Iltis (eds), 1992. Lake Titicaca. A Synthesis of Limnological Knowledge. Monographiae Biologicae. Kluwer Acad. Publ. Dordrecht: 573 pp.

DeMott, W. R. & W. C. Kerfoot, 1982. Competition among cladocerans: nature of the interaction between *Bosmina* and *Daphnia*. Ecology 63: 1949–1966.

Dussart, B. H. & D. Defaye, 1995. Introduction to the Copepoda. Vol. 7. SPB Academic Publishing, Amsterdam: 277 pp.

Forsberg, C., S. O. Ryding, A. Forsberg & A. Claesson, 1978. Research on recovery of polluted lakes. 1. Improved water quality in Lake Boren and Lake Ekoln after nutrient reduction. Verh. int. Ver. Limnol. 20: 825–832.

Gilbert, J. J., 1996. Effect of temperature on the response of planktonic rotifers to a toxic cyanobacterium. Ecology 77: 1174–1180.

Gilbert, J. J., 1988. Supression of rotifer populations by *Daphnia*: a review of the evidence, the mechanisms and the effect on zooplankton community structure. Limnol. Oceanogr. 33: 1286–1303.

Hoffmann, W., 1985. Food selection of a predomonantly phytophagous population of the plankton rotifer *Asplanchna priodonta* (GrosserPoenitzer See) in Eastern Holstein (West Germany). Fauna. Oecol. Mitt. 5: 365–373.

Kak, A. & T. R. Rao, 1998. Does the evasive behaviour of *Hexarthra* influence its competition with cladocerans? Hydrobiologia 387/388: 409–419.

Korovchinsky, N. & N. N. Smirnov, 1998. Introduction to the 'Cladocera' (Ctenopoda, Anomopoda, Onychopoda and Haplopoda). Supplemented for America. Study Material, ENEP Iztacala, UNAM, Mexico: 143 pp.

Koste, W., 1978. Rotatoria. Borntraeger, Berlin, 2 vols.: 673 pp., 234 plates.

Lathrop, R. C., 1988. Evaluation of whole lake nitrogen fertilization for controlling blue–green algal blooms in a hypereutrophic lake. Can. J. Fish. Aquat. Sci. 45: 2061–2075.

Nandini, S., S. S. S. Sarma & G. Pedro-Ramírez, 2000. Life table demography and population growth of *Daphnia laevis* Birge, 1878 (Anomopoda: Crustacea) in relation to different densities of *Microcystis aeregunosa* and *Chlorella vulgaris*. Crustaceana 73: 1273–1286.

Paerl, H. W. & C. S. Tucker, 1995. Ecology of blue-green algae in aquaculture ponds. J. World Aquaculture Soc. 26: 109–131.

Pedro-Ramírez, G., V. Victor-Olvera, V. M. Pulido & A. Duran-D, 1998. Presence of *Vibrio cholerae* in a freshwater reservoir of Valle de Bravo (Mexico State, Mexico). Int. Rev. Hydrobiol. 83: 647–650.

Sarma, S. S. S. & M. Elías-Gutiérrez, 1999. Rotifers from four natural bodies of central Mexico. Limnologica 29: 475–483.

Sarma, S. S. S. & M. Elías-Gutiérrez, 1997. Taxonomic studies of freshwater rotifers (Rotifera) from Mexico. Pol. Arch. Hydrobiol. 44: 341–357.

Sarma, S. S. S. & M. Elías-Gutiérrez, 1998. Rotifer diversity in a central Mexican pond. Hydrobiologia 387/388: 47–54.

Sarma, S. S. S., A. A. Stevenson-Raymundo & S. Nandini, 1998. Influence of food (*Chlorella vulgaris*) concentration and temperature on the population dynamics of *Brachionus calyciflorus* Pallas (Rotifera). Ciencia ergo sum 5: 77–81.

Sarma, S. S. S., M. Elías-Gutiérrez & S. Carmen-Soto, 1996. Rotifers from high altitude crater lakes at Nevado de Toluca, State of Mexico (Mexico). Hidrobiologica 6: 33–38.

Schindler, D. W., 1974. Eutrophication and recovery in experimental lakes: implications for lake management. Science 184: 897–899.

Smith, V. H., 1983. Low nitrogen to phosphorus ratios favour dominance by blue–green algae in lake phytoplankton. Science 221: 669–671.

Suárez-Morales, A., A. Vázquez & E. Solís, 1993. Preliminary investigations on the zooplankton community of a mexican euthropic reservoir, a seasonal survey. Hidrobiologica 3: 71–80.

Suárez-Morales, E., L. Segura & M. A. Fernández, 1986. Diversidad y abundancia del plancton en la Laguna de Catemaco, Ver., durante un ciclo anual. An. Inst. Cienc. del Mar y Limnol. UNAM. 13: 313–316.

Torres-Orozco B. R. E. & M. Estrada-Hernandez, 1997. Patrones de migración vertical en el plancton de un lago tropical. Hidrobiológica 7: 33–40.

Torres-Orozco B. R. E. & S. A. Zanatta, 1998. Species composition, abundance and distribution of zooplankton in a tropical eutrophic lake: Lake Catemaco, Mexico. Rev. Biol. Trop. 46: 285–296.

108

Torres-Orozco B. R. E., C. Jiménez-Sierra & A. Pérez-Rojas, 1996. Some limnological features of three lakes from Mexican neotropics. Hydrobiologia 341: 91–99.

Van de Velde, I., H. J. Dumont & P. Grootaert, 1978. Report on the collection of Cladocera from Mexico and Guatemala. Arch. Hydrobiol. 83: 391–404.

Zafar, A. R., 1986. Seasonality of phytoplankton in some South Indian lakes. Hydrobiologia 138: 177–187.

Hydrobiologia **467**: 109–116, 2002.
J. Alcocer & S.S.S. Sarma (eds), Advances in Mexican Limnology: Basic and Applied Aspects.
© 2002 *Kluwer Academic Publishers.*

Oligochaetes from six tropical crater lakes in Central Mexico: species composition, density and biomass

Laura Peralta[1], Elva Escobar[2], Javier Alcocer[1],* & Alfonso Lugo[1]

[1]*Limnology Laboratory, Environmental Conservation & Improvement Project, UIICSE, FES Iztacala, UNAM, Av. de los Barrios s/n, Los Reyes Iztacala, Tlalnepantla 54090, Estado de Mexico, Mexico Fax: +52-5277-1829; E-mail: jalcocer@servidor.unam.mx*
[2] *Unidad Académica Sistemas Oceanográficos y Costeros, Instituto de Ciencias del Mar y Limnología, UNAM, A. Postal 70-305, Mexico 04510, D.F., Mexico*
(*Author for correspondence)

Key words: Dero, high altitude, *Limnodrilus*, Nais, saline lakes, *Tubifex*

Abstract

The assemblage of littoral oligochaetes in six crater lakes in Central Mexico, was studied throughout a yearly cycle. To establish species composition, richness, density and biomass, 14 localities were sampled in the lakes. A total of eight species belonging to the families Naididae (five species), Tubificidae (two species), and Enchytraeidae (one species) were found. The dominant species, *Limnodrilus hoffmeisteri*, contributed with up to 99% in both abundance and biomass. Sediment organic matter is the most important environmental variable explaining the differences in density and biomass. Seasonal (dry and rainy seasons) changes were not significant for density and biomass. Higher density and lower biomass values characterized these lakes in contrast to other tropical and subtropical lakes worldwide. The small size of the dominant species *L. hoffmeisteri* was recorded in all lakes and explained the low biomass recorded in the area of study. The correlation between *L. hoffmeisteri* and four other species (*Dero (Dero) nivea, D. (D.) digitata, Nais variabilis* and *Tubifex tubifex*) was negative. The naidid species were positively correlated (>0.5) to each other.

Introduction

Little is known on the inland-water oligochaete fauna of Mexico, despite the efforts to understand the inland benthic invertebrate communities. The aquatic oligochaetes are known to be a rather 'difficult' group and are often referred to as 'worms' or 'unidentified Oligochaeta' in limnological studies. Published work on the oligochaete fauna of Mexico include Rybka (1898), Cernosvitov (1936), Cook (1974), and Harman & Loden (1978), and were carried out in the northern part of the country and along the Gulf of Mexico coast.

Few studies have provided accounts of the relevance of oligochaete species in lakes of the Mexican Plateau (Alcocer-Durand & Escobar-Briones, 1992; Alcocer et al., 1997, 1998). The number of aquatic oligochaete species reported for this area is small. This is more likely to be the result of the paucity of investigations than a scarcity of the fauna. To help

filling this gap, this study addresses the composition of the benthic oligochaetes in the littoral area of six crater-lakes of different salinity located in the southeastern portion of the Mexican Plateau. Data regarding the density and biomass were analyzed along with the environmental characteristics throughout a yearly cycle.

Area of study

The crater lakes Alchichica (ALC), Atexcac (ATE), Quechulac (QUE), La Preciosa (LAP), Aljojuca (ALJ) and Tecuitlapa (TEC) are located in the Cuenca Oriental (18° 56′ 51″–19° 43′ 25″ N and 97° 07′ 10″–98° 03′ 04″ W, at a mean altitude of 2312 m.a.s.l.) (Fig. 1). The climate is temperate dry with a dry summer and small temperature fluctuation in the northern Llanos de San Juan where ALC, ATE, QUE and LAP are

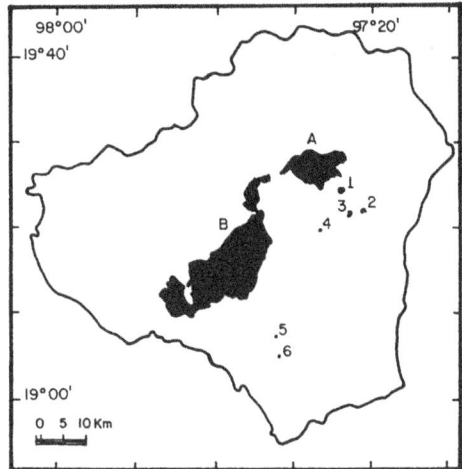

Figure 1. Geographic location of the crater lakes of Puebla, Mexico. (1 = Alchichica, 2 = Quechulac, 3 = La Preciosa, 4 = Atexcac, 5 = Tecuitlapa, 6 = Aljojuca). (Termini lakes: A = Tepeyahuako, B = Totolcingo).

getation. Submerged macrophytes *Ruppia maritima* L. are common in some sampling sites of ALC and *Potamogeton pectinatus* L. in ALJ, LAP and QUE. The emergent *Cyperus laevigatus* L. inhabits ALC, *Phragmites australis* (Cav.) Trin. ex Steud. and *Cyperus laevigatus* L. grows in ATE, *Typha domingensis* Pers. Emer in ALJ, *Elocharis montevidensis* Kunth and *Juncus andicola* Hook in TEC, *Juncus andicola* Hook in LAP, *Scirpus californicus* (Mey.) Steud in QUE and LAP (Ramírez-García & Vázquez-Gutiérrez, 1989). A benthic macrofauna rich in species inhabits the littoral area of these lakes (Alcocer, 1988; Alcocer et al., 1993), and the worms constitute up to 95% of it.

Materials and methods

After a careful examination of the shoreline of the six crater lakes, five (ALC1-ALC5), three (LAP1-LAP3, ALJ1-ALJ3), or one littoral sampling site (ATE1, QUE1, TEC1) per lake, were established according to lake size and heterogeneity. The sampling sites showed different habitat conditions (e.g., vegetation coverage and type, protection against wave action, and sediment texture, Table 1). A quarterly sampling program was carried out throughout a year (with a total of five sampling dates). At every station, triplicate bottom samples were taken with a 0.00225 m² in area, 0.15-m high Ekman grab. Bottom samples were sieved in the field and subsequently in the laboratory through a mesh size of 590 μm. The oligochaete fauna was sorted under a stereoscope, and identified using the identification keys published by Brinkhurst & Marchese (1991), Edmondson (1959), Hiltunen & Klem (1980) and Pennak (1978). Species identified were confirmed by taxonomic experts (R. O. Brinkhurst & M. Marchese). Organisms were dried at 60 °C to constant weight. Loss on ignition (biomass) after 1 h at 550 °C was used to determine the ash-free dry weight (AFDW).

A cluster analysis (agglomerate classification with complete linkage) was carried out based on Euclidean distances that measured the degree of affinity between the data sets. Data were first logarithmically transformed according to Kunz (1988) and Mirza & Gray (1981). The sampling sites were grouped into four clusters based on species composition, density and biomass (AFDW), using the statistics software *Statistica* (release 6.0, 1997). The four clusters were validated (Legendre & Legendre, 1998) against a Principal Component Analysis (PCA) considering the environ-

situated. The climate is subhumid temperate with summer rains in the southern portion of the Llanos de San Andrés, where ALJ and TEC are found. García (1988) recognizes two dominant seasons in the region of study: a cold dry season from November to April, and a warm rainy season from May to October.

The crater lakes have mean water temperatures >18 °C, an alkaline pH>8, and a high dissolved oxygen concentration (>72%) throughout the time of daylight (8:30–17:30 h), when sampling took place. ALC and ATE are saline (7.4 and 6 g l⁻¹, respectively), while QUE and ALJ (0.1 g l⁻¹) and LAP and TEC (1.0 g l⁻¹) are freshwater. The sediment, covered by a thin, superficial mud layer, is composed of sand and gravel, with low to medium concentrations of calcium carbonate (>0.1%), and medium to high concentrations of organic matter (>1.5%). Rooted submerged and emergent macrophytes as well as benthic algae (diatoms and filamentous chlorophytes and cyanobacteria) characterize the aquatic ve-

Table 1. Environmental variables of the littoral areas of **ALC**hicha, **ATE**xcac, **LA** Preciosa, **QUE**chulac, **ALJ**jojuca, y **TEC**uitlapa crater lakes. (Average±one standard deviation). (**DO**=dissolved oxygen, **TEMP**=temperature, **SAL**=salinity; **OM**=organic matter, **CO₃**=carbonates, **ST**=sediment texture, **EM**=emergent macrophytes, **SM**=submerged macrophytes, **BA**=benthic algae)

Variable	ALC	ATE	LAP	QUE	ALJ	TEC
pH	8.9–9.1	8.2–8.6	8.8–9.5	8.7–9.1	8.9–9.3	9.7–9.9
DO (mg l⁻¹)	6.5–12.3	5.4–8.4	5.5–12.8	4.1–7.7	3.7–9.1	6.5–12.1
DO (% Sat)	84–196	78–122	80–197	57–111	55–135	96–178
TEMP (°C)	18.3–24.9	18.7–21.7	17.4–24.5	15.5–19.5	19.4–23.4	26.3–21.9
SAL (g l⁻¹)	6.0–7.4	6.0	1.0	0.1	0.1	1.0
OM (%)	3–8	0–7.2	6–18	2–4	1–2	7–8
CO₃(%)	2–29	4–5	5–24	0.5–1	0.5–1	3–6
ST (ø)	0.2–2.3	0.9–1.1	0.1–2.6	0.2–0.3	0.2–0.3	2–2.2
EM (%)	0	50–60	0–50	80–100	0	0
SM (%)	25–100	70–75	0–75	0	20–25	90–100
BA (%)	0–25	20–25	25–75	10–25	50–60	0

mental variables listed in Table 1. Through PCA, we were able to recognize the main environmental variables responsible of the classification.

Box and whisker analysis was applied to test seasonal differences in density and biomass among the four biological clusters. A correlation analysis (Pearson coefficient, 95% confidence level) of the identified species was applied to reveal the relationships among the oligochaete taxa.

Results

Eight oligochaete species were recognized within the three families Naididae (*Dero (Aulophorus) furcatus* Müller, *Dero (Dero) digitata* Müller, *Dero (Dero) nivea* Aiyer, *Nais variabilis* Piguet and *Pristina aequiseta* Bourne); Tubificidae (*Limnodrilus hoffmeisteri* Claparède and *Tubifex tubifex* Müller), and Enchytraeidae (a single unidentified species).

According to abundance and frequency of occurrence in the lakes, *L. hoffmeisteri* and *D. (D.) nivea* were the dominant components (high abundance and frequency) of the oligochaete assemblage; whilst *T. tubifex* and *N. variabilis* were considered seasonal components (high abundance and low frequency) with abundance peaks in December and March. The other four species were classified as rare components due to low abundance and frequency throughout the sampling period.

The four clusters based on composition, density and biomass (linking distance=0.5) are shown in Figures 2a and b (Table 2). The cluster G1 (ALC4

Table 2. Mean annual density (ind.m⁻²) and biomass (mg AFDW.m⁻²) values of oligochaete clusters. (G1=ALC2-ALC5; G2=TEC1-QUE1; G3=LAP1-LAP2-LAP3-ATE1-ALC2-ALC3; G4=ALJ1-ALJ2-ALJ3-ALC1)

	Density (ind.m⁻²)		Biomass (mg AFDW.m⁻²)	
	Mean	s.d.	Mean	s.d.
G1	54 307	33 026	3157.9	1939.2
G2	8587	6653	538.9	459.9
G3	29 171	19 494	1679.8	1149.4
G4	19 295	13 067	1196.9	784.3

and ALC5) was characterized by the high density and biomass values (>45 000 ind.m⁻² and >2700 mg AFDW.m⁻²). The cluster G2 (TEC1 and QUE1) aggregated sampling sites with the lowest densities (7508 and 9594 ind.m⁻²) and lowest biomasses (498.3 and 579.1 mg AFDW.m⁻²). The remaining two clusters (G3 and G4) had intermediate density and biomass values. Densities of group G3 (LAP1, LAP2, LAP3, ATE1, ALC2 and ALC3) ranged between 25 532 and 63 111 ind.m⁻² and biomasses between 1571.4 and 1825.1 mg AFDW.m⁻². The densities in cluster G4 (ALJ1, ALJ2, ALJ3 and ALC1) fluctuated from 16 622 to 22 761 ind.m⁻², and biomasses from 1018.6 to 1329.1 mg AFDW.m⁻². Density and biomass values of oligochaetes (Fig. 3a and b) did not show significant seasonal differences ($p<0.05$). The PCA revealed organic matter in superficial sediments to be the most important environmental variable that structures the biological clusters (Fig. 2b).

Table 3. Seasonal (dry and rainy seasons) density (ind.m^{-2}) and biomass (mg AFDW.m^{-2}) values of the species identified from the crater lakes of Puebla, Mexico

	Density (ind.m^{-2})		Biomass (mg AFDW.m^{-2})	
	Dry	Rainy	Dry	Rainy
L. hoffmeisteri	28 305±26942	25 468±20818	1.66±1.55	1.51±1.23
T. tubifex	349±783	383±859	0.012±0.028	0.015±0.03
D. (D.) nivea	1015±1725	1023±1481	0.03±0.05	0.029±0.044
D. (D.) digitata	141±264	50±142	0.004±0.007	0.001±0.004
D. (A.) furcatus	0	14±47	0	0.0009±0.003
P. aequiseta	0	16±38	0	0.0002±0.0004
N. variabilis	114±386	82±281	0.002±0.006	0.001±0.004
Enchytraeidae	0	10±30	0	0.0001±0.0004

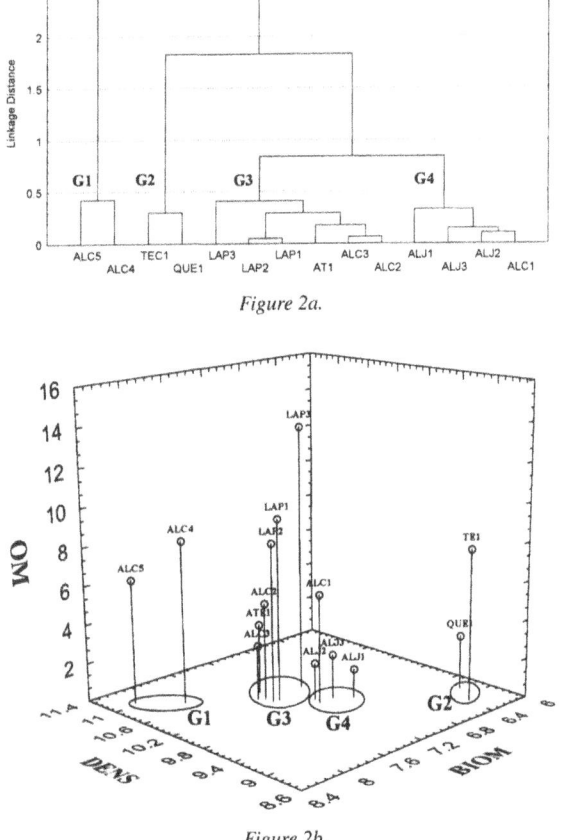

Figure 2a.

Figure 2b.

Figure 2. (a) Similarity dendrogram (Euclidean distances, complete linkage) of the 14 sampling stations. (b) Validation of the clusters through PCA analysis. (OM=sediment organic matter, DENS=density, BIOM=biomass).

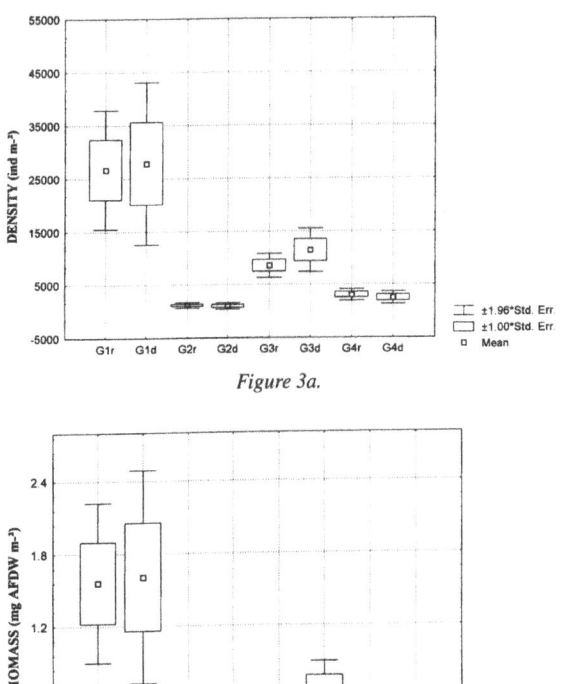

Figure 3a.

Figure 3b.

Figure 3. Box and whisker plots of (a) density (ind.m^{-2}) and (b) biomass (mg AFDW m^{-2}) seasonal variation of the biological clusters. G1–G4=groups identified by analyses in Figure 2. (r=rainy season, d=dry season).

Density and biomass values were mostly determined by the dominant species *L. hoffmeisteri* (Table 3). There was no statistically significant seasonal difference in density or biomass (*p*<0.05) in spite that *P. aequiseta*, *D. (A.) furcatus* and Enchytraeidae were only sampled during the rainy season.

The Pearson correlation analysis among oligochaete species showed a weak (<0.5) significant (*p*=0.05) negative correlation between the species *L. hoffmeisteri* and *D. (D.) nivea*, *D. (D.) digitata*, *N. variabilis* and *T. tubifex*, and a robust (>0.5) and significant (*p*=0.05) positive correlation between all naidid species.

Discussion

Numerous studies (Brinkhurst, 1967; Verdonschot et al., 1982; Verdonschot, 1984; Gluzman de Pascar, 1987; Coates & Stacey, 1994) have accounted for oligochaete species richness in lakes. Many of these studies attain from 10 up to 41 species at each lake (Hiltunen, 1967; Howmiller, 1974; Reynoldson, 1978; Särkkä & Aho, 1980; Timms, 1982; Lang, 1984; Slepukhina, 1984; Lafont, 1987; Särkkä, 1987; Lang & Reymond, 1992) which is notably higher than the total of eight species found in this study. Four out of the six crater lakes (ALC, LAP, TEC and ATE) were inhabited by only two species, while the lakes ALJ and QUE had seven and eight species, respectively. A low species richness was recorded similarly to other lakes (Bretschko, 1975; Osborne et al., 1976) in central Mexico. The low oligochaete diversity recorded in these lakes has been related to the preference of species to deeper habitats in contrast to littoral habitats (Lang, 1984; Martinez-Ansemil & Prat, 1984; Probst, 1987; Särkkä, 1987; Verdonschot 1984). Although small naidids could escape through a 590 μm mesh (Christer Erséus, pers. com.), our unpublished studies on the meiofauna of the same set of crater lakes showed low oligochaete density and diversity composed by the same species reported here.

Limnodrilus hoffmeisteri, the dominant species in the six crater lakes, showed a similar pattern in the six lakes studied by Osborne et al. (1976). Lang (1984) mentioned a relationship between oligochaete frequency and abundance, more frequent taxa reaching higher densities. The species *Tubifex tubifex* was a dominant component in Lake Vorderer Finstertaler (Bretschko, 1975), but a was less important in our lakes. Studies by Probs (1987) and Särkkä (1987) no-

ticed that while *L. hoffmeisteri* diminishes its density with depth, *T. tubifex* increases its density in deeper waters.

Oligochaete density values recorded in this study show higher values than those reported from other tropical and subtropical lakes of the world (Table 4). A consistent pattern seems to appear in which values tend to diminish in a latitudinal gradient. An exception has been recorded in the Toronto harbor in Lake Ontario, Canada, where densities noted for *T. tubifex*, *L. hoffmeisteri* and *Peloscolex multisetosus* (Brinkhurst & Cook, 1974) were twice ours.

The large amount of organic matter being dumped into Lake Ontario (McCall & Tevesz, 1982) promotes high tubificid densities. A similar pattern to the one we observed has been recorded elsewhere (Hiltunen, 1967; Johnson & Brinkhurst, 1971; Macioroswski et al., 1977; Lang, 1978). This pattern considers the occurrence of the three species *L. hoffmeisteri*, *T. tubifex* and *Peloscolex multisetosus*. The oligochaetes are able to increase in number with increasing organic matter, replacing other, less tolerant, benthic macroinvertebrates (Newrkla & Wijegoonawardana, 1987). However, few species are able to withstand the stress attributed to excessive organic input simplifying the community to few tolerant flourishing (Qi, 1987).

The biomass values in the area of study were consistent with those recorded in eutrophic lakes in Mexico (Alcocer, 1988) and Poland (Kasprzak, 1984). The biomass values in these lakes (Table 3) were lower than in other mesotrophic and eutrophic lakes (Šapkare & Točko, 1972; Brinkhurst & Cook, 1974; Okland, 1964 in Wetzel, 1975; Wetzel, 1975; Lafont, 1987). The biomass of benthic organisms increases with the trophic status based on a bottom up scheme, where the potential primary production will lead to high secondary production (Jonasson, 1964, 1965 in Jonasson, 1969; Wetzel, 1975; Barnes & Mann, 1980). This fact is explained by the size differences of the dominant species. Naidu & Naidu (1981) and Naidu et al. (1981) recorded that *L. hoffmeisteri* attains lengths of 25–28 mm and 0.6 mm in diameter; our specimens were only 10–19 mm long and 0.5–0.6 mm in diameter. The largest density and biomass values found in this study were probably related to the reproduction season and largest size of the dominant *L. hoffmeisteri*.

Some naidid species (e.g., *D. (A.) furcatus* and *P. aequiseta*) were absent for extended periods in the area of study. The littoral area of the crater lakes with abrupt changes in temperature and water level will have an effect in naidids shortening their sexual re-

Table 4. Oligochaete density and biomass values reported from lakes worldwide

Lake/country	Density (ind $^{-2}$)	Biomass (g DW m^{-2})	Trophic state	Reference
Tropical/Subtropical				
Banyoles, SPA	869	Not mentioned	Not mentioned	Rieradevall & Prat (1991)
				Rieradevall & Real (1994)
Spanish reservoirs	12 500	Not mentioned	Various	Prat (1976)
Dojran, YUG.	1795	Not mentioned	Eutrophic	Šapkarev (1975)
Ohrid, MAC.	1453	4.725	Not mentioned	Šapkarev & Točko (1972)
Lago Viejo, MEX.	10 229	1.09	Eutrophic	Alcocer (1988)
Lago Mayor, MEX.	2736	0.27	Eutrophic	Alcocer (1988)
Lago Menor, MEX.	385	0.02	Eutrophic	Alcocer (1988)
Alchichica, MEX.	38 247	0.77	Eutrophic	This study
La Preciosa, MEX.	26 975	0.59	Eutrophic	This study
Aljojuca, MEX.	18 747	0.15	Eutrophic	This study
Quechulac, MEX.	9583	0.06	Eutrophic	This study
Tecuitlapa, MEX.	7390	0.20	Eutrophic	This study
Atexcac, MEX.	34 240	0.55	Eutrophic	This study
Temperate				
Borrevan, NOR.	622	3.33	Eutrophic	Okland (1964) in Wetzel (1975)
Esrom, DEN.	1250	1.28	Eutrophic	Wetzel (1975)
Léman, SWI.	864	2.96	Meso-eutrophic	Lafont (1987)
Ontario, CAN.	81 000	45.50	Eutrophic	Brinkhurst & Cook (1974)
Zhechv, POL.	769	0.15	Eutrophic	Kasprzak (1984)

production period and inducing a rapid deposition of cocoons. Naidids are able to withstand long periods of environmental stress, leading to fewer individuals. In spite of a complete absence for extended periods, the naidid population recovers and regains dominance when more favorable environmental conditions are reached (Loden, 1981).

Organic matter content in sediment plays a major role in tubificid success and influences availability of bacteria, an important food source for oligochaetes, which in turn controls the relative abundance of oligochaetes. Resource partitioning by oligochaetes seems to take place through selecting specific bacteria (Brinkhurst, 1972 in Brinkhurst & Cook, 1974). The differences in salinity among lakes did not have an influence on the oligochaete species composition, density or biomass. The naidids are typical organisms of the littoral region characterized by abundant free-floating vegetation and drifted algae from blooms (Brinkhurst, 1967; Learner et al., 1978; Paoletti & Sambugar, 1984; Verdonschot, 1984; Särkkä, 1987). The abundant vegetation can explain the dominance of *D. (D.) nivea* right after *L. hoffmeisteri* in the crater

lakes of Puebla. Other naidid species like *D. (D.) digitata* and *P. aequiseta* seem to avoid organically enriched environments (Brinkhurst & Cook, 1974; Lafont, 1984), explaining their poor representation in these crater lakes.

In summary, the littoral oligochaete assemblages of the crater lakes studied show some of the lowest species richnesses in the world. Their density is, however, among the highest recorded worldwide. The reduced size of the dominant species (*L. hoffmeisteri*) explains the low biomass values. No significant seasonal differences were found. The organic matter content in sediments is the factor that determines density and biomass changes in the littoral area of these crater lakes.

Acknowledgements

The authors acknowledge the taxonomic support of Prof. R.O. Brinkhurst (Aquatic Resources Center) and Dr M. Marchese (INALI). CONACyT project 25430-T partially supported this study. M.R. Sánchez, L.A.

Oseguera and M.J. Montoya assisted in the field and laboratory.

References

Alcocer, J., 1988. Caracterización hidrobiológica de los lagos de Chapultepec, México. Master in Science Thesis, U.N.A.M., Instituto de Ciencias del Mar y Limnología. 86 pp.

Alcocer-Durand, J. & E. Escobar-Briones, 1992. The aquatic biota of the now extinct lacustrine complex of the Mexico basin. Freshwat. Forum 2: 171–183.

Alcocer, J., A. Lugo, S. Estrada, M. Ubeda & E. Escobar, 1993. La macrofauna bentónica de los axalapazcos mexicanos. Actas del VI Congreso Español de Limnología 33: 409–415.

Alcocer, J., A. Lugo, E. Escobar & M. Sánchez, 1997. The macrobenthic fauna of a former perennial and now episodically filled Mexican saline lake. Int. J. Salt Lake Res. 5: 261–274.

Alcocer, J., E. Escobar, A. Lugo & L. Peralta, 1998. Littoral benthos of the saline crater lakes or the basin of Oriental, Mexico. Int. J. Salt Lake Res. 7: 87–108.

Barnes, R. S. K. & K. H. Mann, 1980. Fundamentals of Aquatic Ecosystems. Blackwell, Oxford: 229 pp.

Bretschko, G., 1975. Annual benthic biomass distribution in a high-mountain lake (Vorderer Finstertaler See, Tyrol, Austria). Verh. int. Ver. Limnol. 19: 1279–1285.

Brinkhurst, R. O., 1967. The distribution of aquatic oligochaetes in Saginaw Bay, Lake Huron. Limnol. Oceanogr. 12: 137–143.

Brinkhurst, R. O. & D. G. Cook, 1974. Aquatic earthworms (Annelida: Oligochaeta). In Hard, C. & S. Fuller (eds), Pollution Ecology of Freshwater Invertebrates. Academic Press. New York: 389 pp.

Brinkhurst, R. O. & M. R. Marchese, 1991. Guía para la identificación de oligoquetos acuáticos continentales de Sud y Centroamérica, 2. Colección Climax No. 6, Argentina: 207 pp.

Cernosvitov, L., 1936. Resultats zoologiques de voyage de Mr Le Dr. J. Storkan au Mexico. Cekoslovenska Spolecnost Zoologicka Prague. Vestnik V. 3: 80–83.

Coates, K. A. & D. F. Stacey, 1994. Oligochaetes (Naididae, Tubificidae, Enchytraeidae and Alluroididae) of Guyana, Peru and Ecuador. Hydrobiologia 278: 79–84.

Cook, D. G., 1974. The systematics and distribution of marine Tubificidae (Annelida: Oligochaeta) in the Bahia de San Quintin, Baja California, Mexico, with description of five new species. Bull. South. Calif. Acad. Sci. 73: 126–140.

Edmonson, W. T., 1959. Fresh-water Biology, Wiley, New York: 1248 pp.

García, E., 1988. Modificaciones al sistema de clasificación climática de Köppen. E. García. Mexico. 217 pp.

Gluzman de Pascar, C., 1987. Aquatic oligochaeta in some tributaries of Rio de la Plata, Buenos Aires, Argentina. Hydrobiologia 144: 125–130.

Harman, W. J. & M. S. Loden, 1978. A re-evaluation of the Opistocystidae (Oligochaeta) with descriptions of two new species. Proc. biol. Soc. Wash. 91: 453–462.

Hiltunen, J. K. & D. J. Klemm, 1980. A guide to the Naididae (Annelida: Clitellata: Oligochaeta) of North America. U.S. Environmental Protection Agency. EPA/670/4-73-001. Cincinnati: 41 pp.

Hiltunen, J. K., 1967. Some oligochaetes from Lake Michigan. Trans. am. Microsc. Soc. 86: 433–454.

Howmiller, R. P., 1974. Some Naididae and Tubificidae from Central America. Hydrobiologia 44: 1–12.

Johnson, M. G. & R. O. Brinkhurst, 1971. Associations and species diversity in benthic macroinvertebrates of Bay of Quinte and Lake Ontario. J. Fish Res. Bd Can. 28: 1683–1697.

Kasprzak, K., 1984. The oligochaetes (Annelida, Oligochaeta) in a lake and a canal in the agricultural landscape of Poland. Hydrobiologia 115: 171–174.

Kunz, I., 1988. El uso de la estadística para la construcción de clasificaciones y regionalizaciones. Serie Varia 1(11): 34 pp.

Jonasson, P. M., 1969. Bottom fauna and eutrophic. In National Academy of Sciences. Eutrophication: causes, consequences, correctives. N. A. S., Washington: 274–305.

Lafont, M., 1984. Oligochaete communities as biological descriptors of pollution in the fine sediments of rivers. Hydrobiologia 115: 127–129.

Lafont, M., 1987. Production of Tubificidae in the littoral zone of lake Léman near Thonon-les-Bains: A methodological approach. Hydrobiologia 155: 179–187.

Lang, C., 1978. Factorial correspondence analysis of oligochaeta communities according to eutrophication level. Hydrobiologia 57: 241–247.

Lang, C., 1984. Eutrophication of Lakes Léman and Neuchâtel (Switzerland) indicated by oligochaete communities. Hydrobiologia 115: 131–138.

Lang, C. & O. Reymond, 1992. Reversal of eutrophication in Lake Geneva: evidence from the oligochaete communities. Freshwat. Biol. 28: 145–148.

Learner, M. A., G. Lochhead & B. D. Hughes, 1978. A review of the biology of British Naididae (Oligochaeta) with emphasis on the lotic environment. Freshwat. Biol. 8: 357–375.

Legendre, P. & L. Legendre, 1998. Numerical Ecology. Developments in Environmental Modelling 20. Elsevier, Amsterdam: 853 pp.

Loden, M. S., 1981. Reproductive ecology of Naididae (Oligochaeta). Hydrobiologia 83: 115–123.

Maciorowski, A. F., E. F. Benfield & A. C. Hendricks, 1977. Species composition, distribution and abundance of oligochaetes in the Kanawha River, West Virginia. Hydrobiologia 1: 81–91.

Martinez-Ansemil, E. & N. Prat, 1984. Oligochaeta from profundal zones of Spanish reservoirs. Hydrobiologia 115: 223–230.

McCall, P. L. & M. J. Tevesz (eds), 1982. Animal–Sediment Relation. The Biogenic Alteration of Sediments. Plenum Press, New York: 336 pp.

Milbrink, G., 1978. Indicator communities of oligochaetes in Scandinavian lakes. Verh. int. Ver. Limnol. 20: 2406–2411.

Mirza. F. B. & J. S. Gray, 1981. The fauna of benthic sediments from the organically enriched Oslofjord, Norway. J. exp. mar. Biol. Ecol. 54: 181–207.

Naidu, K. V., K. Kalpana & K. S. Kumar, 1981. Aquatic oligochaeta from among the roots of *Eichhornia crassipes* Solms. Hydrobiologia 76: 103–112.

Naidu, K. V. & K. A. Naidu, 1981. Some aquatic oligochaetes of the Nilgiris, South India. Hydrobiologia 76: 113–118.

Newrkla, P. & N. Wijegoonawardana, 1987. Vertical distribution and abundance of benthic invertebrates in profundal sediments of Mondsee, with special reference to oligochaetes. Hydrobiologia 155: 227–234.

Osborne, J. A., M. P. Wanielista & A. Yousef, 1976. Benthic fauna species diversity in six central Florida lakes in summer. Hydrobiologia 48: 125–129.

Paoletti, A. & B. Sambugar, 1984. Oligochaeta of the middle Po River (Italy): principal component analysis of the benthic data. Hydrobiologia 115: 145–152.

Pennak, R. W., 1978. Freshwater Invertebrates of the United States. Wiley. New York: 803 pp.

116

Prat, N., 1976. Bentos. In Margalef, R., D. Planas, J. Armengol, A. Vidal, N. Prat, A. Guiset, J. Toja & M. Estrada (eds), Limnologia de los Embalses Españoles. Dir. Gral. Obras Hidraúl., Ministerio de Obras Públicas, Madrid, 123: 260–289.

Probst, L., 1987. Sublittoral and profundal Oligochaeta fauna of the Lake Constance (Bodensee–Obersee). Hydrobiologia 155: 277–282.

Qi, S., 1987. Some ecological aspects of aquatic oligochaetes in the lower Pearl River (People's Republic of China). Hydrobiologia 155: 199–208.

Ramírez-García, P. & F. Vázquez-Gutiérrez, 1989. Contribución al estudio limnobotánico de la zona litoral de seis lagos cráter del estado de Puebla. An. Inst. Cienc. del Mar y Limnol. Univ. Nal. Autón. México. 16: 1–16.

Reynoldson, T. B., 1978. Observation on the typology of some Alberta lakes with special reference to their oligochaete faunae. Verh. int. Ver. Limnol. 20: 190–191.

Rieradevall, M. & M. Real, 1994. On the distribution patterns and pollution dynamics of sublittoral and profundal oligochaeta fauna from Lake Banyoles (Catalonia, NE Spain). Hydrobiologia 278: 139–149.

Rieradevall, M. & N. Prat, 1991. Benthic fauna of Banyoles Lake (NE Spain). Verh. int. Ver. Limnol. 24: 1020–1023.

Rybka, J., 1898. Contribution a la morphologie et la classification du genre Limnodrilus Claparede. Mem. Soc. Zool. France 11: 376–392.

Šapkarev, J., 1975. Seasonal and annual variation of the population density and biomass of the bottom-fauna in the deepest waters of lake Doiran, Macedonia. In Salanki, J. & J. E. Ponyi (eds), Limnology of Shallow Waters. Akademiai Kiado, Budapest: 247–254.

Šapkarev, J. & M. Toþko, 1972. Dynamics of the biomass of bottom fauna from Ohrid Lake, Macedonia. Verh. int. Ver. Limnol. 14: 494–504.

Särkkä, J., 1987. The occurrence of oligochaetes in lake chains receiving pulp mill waste and their relation to eutrophication on the trophic scale. Hydrobiologia 155: 259–266.

Särkkä, J. & J. Aho, 1980. Distribution of aquatic Oligochaeta in the Finnish Lake District. Freshwat. Biol. 10: 197–206.

Slepukhina, T. D., 1984. Comparison of different methods of water quality evaluation by means of oligochaetes. Hydrobiologia 115: 183–186.

Timms, B. V., 1982. A study of the benthic communities of twenty lakes in the South Island, New Zealand. Freshwat. Biol. 12: 123–138.

Verdonshot, P. F. M., 1984. The distribution of aquatic oligochaetes in the fenland area of N.W. Overijssel (The Netherlands). Hydrobiologia 115: 215–222.

Verdonshot, P. F. M., M. Smies & A. B. J. Sepers, 1982. The distribution of aquatic oligochaetes in brackish inland waters in the SW Netherlands. Hydrobiologia 89: 29–38.

Wetzel, R. G., 1975. Limnology. Saunders. Philadelphia: 743 pp.

Hydrobiologia **467**: 117–122, 2002.
J. Alcocer & S.S.S. Sarma (eds), Advances in Mexican Limnology: Basic and Applied Aspects.
© 2002 *Kluwer Academic Publishers.*

Fish abundance and trophic structure from the commercial catch in Lake Patzcuaro, Mexico

César A. Berlanga-Robles[1], Juan Madrid-Vera[2] & Arturo Ruiz-Luna[1]

[1]*CIAD-Unidad Mazatlan, P.O. Box 711, Mazatlan, Sin. 82010, Mexico*
Tel: +69-880157; Fax: +69-880159; E-mail: cesar@victoria.ciad.mx
[2]*CRIP-Mazatlan, Av. Sabalo Cerritos s/n, Mazatlan, Sin. 82010, Mexico*

Key words: endemism, fish assemblages, trophics, exotic species, tropical lakes, fish production, Lake Patzcuaro, Mexico

Abstract

The composition, abundance and trophic structure of the fish community in Lake Patzcuaro is analyzed, using data from eight samples taken from the commercial fishery acquired between August 1990 and April 1991. A total of 65 767 individuals, caught with gillnets selective to all fish and size, were analyzed. Eleven species (8 native and 3 exotic) were recorded during the study. Zooplanktivorous were the most representative species, amounting to 91% of the total number. Also, they were the most characteristic species (36.4%) when feeding habits were considered, followed by carnivore and omnivore (27.3% each), and herbivore (9.1%). Three species groups and three time-associated groups were detected by means of classification methods. Lake Patzcuaro's ichthyofauna is to be considered as a transitional community that has not reached equilibrium, particularly after the introduction of exotic species and local extinction of native species. However, Lake Patzcuaro has the greatest fishery production of all tropical lakes in the Central Plain of Mexico and it is mainly sustained by native species. The differential use of feeding resources by the fish species and the diversity in their habits must be taken into account to design strategies to maintain the fish structure (avoiding new introduction of exotic species) and the fish production in Lake Patzcuaro.

Introduction

The diversity of any community responds to historic, regional and unique events of speciation and dispersion, as well as contemporary and continuous events of predation, competition, adaptation and environmental local events. They contribute to increase (by speciation) or to reduce (by extinction), the species diversity (Sale, 1980; Ricklefs, 1987).

In epicontinental aquatic communities, it could be expected that species richness is modulated by contemporary rates of colonization and extinction, as proposed by the McArthur & Wilson theory, but when the means of species dispersion is exclusively aquatic, species richness responds much better to short periods of colonization and further interactions among the co-existing species (Barbour & Brown, 1974). Regional and historic processes play an extra role in limiting species abundance in lake fish communities. Most of the tropical lakes outside Africa, display low fishing yields (Melack, 1976). This low production appears as a result of bottlenecks in the trophic chain, more than species interactions, local environmental fluctuations or even by fish overexploitation.

Some species from lotic origins are not adapted completely to lentic conditions and the whole potential of the trophic chain is underutilized. When the geological history of the lake is associated with regional processes, related with its dimensions, the lakes in the vicinity and stability in environmental conditions over a long period of time, endemic fauna, completely adapted to lentic conditions and occupying most of the positions in the trophic scale, could increase. This is true for African lakes, which show the highest natural fishing yields for tropical lakes in the world (Fernando & Holcik, 1982).

Lake Patzcuaro's ichthyofauna, with a high endemic component, strongly related with tectonic and vulcanic events that happened in the Mesa Central (Central Plateau) of Mexico since the later Tertiary (de Buen, 1947; Alvarez, 1972; Barbour, 1973a), has been modified by local, contemporary events. The introduction of exotic species (centrarchids cyprinids, and cichlids) since 1933, along with overexploitation of fishing resources and strong environmental perturbations (Flores et al., 1992; Berlanga, 1993; Chacón, 1993; Orbe & Acevedo, 1998), have had similar effects to regional and historic processes. This situation has created new conditions in the structure and dynamics of the fish associations in Lake Patzcuaro, that possibly reflect similar responses of fish species to the perturbations (Berlanga et al., 1997).

This work describes the fish composition in Lake Patzcuaro and analyses their abundance and trophic structure, discussing the role played by different natural and antropogenic processes to define the actual fish structure of Lake Patzcuaro.

Materials and methods

Data were gathered from eight samples from the commercial fishery in Lake Patzcuaro, taken between August 1990 and April 1991. Lake Patzcuaro is a tropical, polymictic and eutrophic lake, located within an endorheic basin, in northern-central part of Michoacan, Mexico, between 19° 32′–19° 42′ N and 101° 32′–101° 43′ W, at 2035 meters above sea level (Anonymous, 1990).

The fishing devices used are gillnets constructed with different mesh sizes (1.2–11.6 cm), that are jointly selective for all fish species and commercial sizes. Fish from the total catch were separated and the total mass by species was recorded, excepting the 'charales' group, which included three species of the genus *Chirostoma*, unable to be identified at species level in the fishing grounds. The proportion of the three species was later estimated, with two 1-kg samples taken in September and November 1990. The *Chirostoma* species were identified using the keys proposed by Barbour (1973b). All the recorded fish species were arranged in a taxonomic order following the proposal of Nelson (1994) for teleostean fish.

The relative abundance (number and mass) by species was estimated after standardizing the fishing effort, dividing the catch by the number of gillnets employed. The number for charales was estimated after their proportion in the samples was evaluated and extrapolated to the total catch. The standardized number was used for further analysis.

To describe variations of temporal patterns in the fish community and search for fish assemblages, a correlation analysis with the standardized fish weight was made using Spearman's correlation coefficient. Also temperature data, from Anonymous (1990) and Chacón & Muzquiz (1991), were used. After that, a classification method was applied, using Euclidean distance as measure of similitude and the unweighted paired group average (UPGMA) as the clustering algorithm (Pielou, 1984). The data matrix with 8 months (columns) and 11 species (rows) was used in R-mode and Q-mode to define the species and samples (months) relationships. Two dendrograms were outputed, and a nodal analysis was done with both results.

Elton's pyramid, that relates number of individuals to their size and trophic level, was the model used to describe the trophic structure of the fish community (Colinvaux, 1980). Fish size was recorded individually, and histograms were constructed per species with amplitude intervals of 40 mm, calculating the relative frequency by length interval. Frequencies were added by length interval to construct the pyramid with the distribution of organisms by size. A similar method was used with number of individuals by trophic level, based on information of feeding habits provided by Rosas (1976), Nepita (1993) and Orbe & Acevedo

Table 1. Checklist of Lake Patzcuaro's ichthyofauna

Order Cypriniformes
Family Cyprinidae
1. *Algansea lacustris* Steindachner, 1985	Acúmara
2. *Cyprinus carpio*[b] (Linnaeus, 1765)	Carpa común

Order Cyprinodontiformes
Family Goodeidae
3. *Alloophorus robustus* (Bean, 1902)	Chehua
4. *Goodea atripinis luitpoldi* (Jordan, 1879)	Tiro
5. *Neophorus diazi* Meek, 1902	Choromu

Order Atheriniformes
Family Atherinidae
6. *Chirostoma attenuatum* (Meek, 1902)	Charal prieto
7. *Chirostoma estor* Jordan, 1879	Pescado blanco
8. *Chirostoma grandocule* (Steindachner, 1894)	Charal g"uero
9. *Chirostoma patzcuaro* (Meek, 1902)	Charal pinto

Order Perciformes
Family Centrarchidae
10. *Micropterus salmoides*[b] (Lacépède, 1822)	Lobina negra o trucha

Family Cichlidae
11. *Oreochromis niloticus*[b] (Linnaeus, 1766)	Mojarra

[a] According to the taxonomic arrangement proposed by Nelson (1994).
[b] Indicates exotic species.

Table 2. Standardized data of mean abundance (kg) of fish from Lake Patzcuaro, Mexico from 8 monthly samples (August 1990–April 1991). Data are arranged for total abundance and for three seasonal groups obtained by cluster analysis[a]

	Total		Group 1		Group 2		Group 3	
	Mean	S.D.	Mean	S.D.	Mean	S.D.	Mean	S.D.
A. lacustris	28.2	30.4	2.7	0.71	24.0	–	63.5	7.3
Chirostoma spp.	42.5	23.2	22.9	4.33	50.0	–	66.1	15.0
C. estor	3.9	1.2	4.3	0.14	2.5	–	4.0	1.7
G. a. luitpoldi	4.3	4.5	3.7	1.24	1.8	–	6.0	7.7
A. robustus	2.3	1.5	2.9	1.37	0.7	–	2.8	1.9
N. diazi	0.2	0.2	0.2	0.25	0.1	–	0.3	0.2
C. carpio	6.8	3.5	4.9	3.30	5.2	–	9.8	2.2
M. salmoides	2.0	2.0	2.9	2.47	1.3	–	1.0	1.2
O. niloticus	20.4	12.7	28.5	12.89	8.1	–	13.6	6.6

[a] S.D. Standard deviation. Group 1, August to November 1990. Group 2, January 91. Group 3, February to April 1991.

(1998) for the fish species. The omnivorous group were not considered regarding its uncertain position in the trophic chain (Colinvaux, 1980).

Results

From the commercial catch, 1763 kg of fish were analyzed identifying eleven species, belonging to eight genera and five families. A total of 65 771 fish specimens were estimated after the standardization. The *Chirostoma* species were the most representative of the fish fauna in Lake Patzcuaro and, in addition to *Chirostoma estor*, at least three species of the genus were identified from the two 1-kg samples obtained during September and November 1990. The analysis of the proportion of the *Chirostoma* species found in both samples did not show statistical differences between samples (Wilcoxon; $n=5$, $t=7.0$, $Z=0.3$, $P=0.89$).

Most of the fish species (8) are native to Lake Patzcuaro; the three others were introduced during this century. The systematics of the species caught is displayed in Table 1, indicating their common name and origin (local or introduced).

Fish abundances, seen as standardized weight (mass/net number), indicate that the 'charal group' are the most important species, followed by *Algansea lacustris* and *Oreochromis niloticus* in this order. Most of the species reach captures above 1.0 kg/net, excepting *Neophorus diazi*, which was the least abundant species (Table 2).

Associations between species, evaluated throughout the correlation analysis ($P<0.5$) were significant only for the charales group and *A. lacustris* ($r=0.95$) and also for *O. niloticus* with *Cyprinus carpio* ($r=-0.77$) and with *Micropterus salmoides* ($r=0.74$).

Temperature obtained significant values of the correlation coefficient with the species *O. niloticus* (0.8) and with the charal group (−0.75).

Once the classification analysis was performed, three clusters were detected, at 50% of the maximun euclidean distances, for species and dates respectively (Fig. 1). The first species group includes *A. lacustris* and the charales group. The second group gather most of the species of Lake Patzcuaro's icthyofauna, with the choromu *N. diazi* and *M. salmoides* as the nearest species. *C. estor* and *Alloophorus robustus* were almost at the same level of distance as the former group. This group encompasses the species *Goodea atripinis luitpoldi* and *C. carpio*. Finally, *O. niloticus* was discriminated from the other two groups.

From a temporal viewpoint, there is a group with four months (August–November, 1990), a three month group (February–April, 1991) and again, an separated component (January, 1991). The first group represents the rainy season, mainly occurring during August and September, with lake surface temperatures around 21 °C. The second group includes the winter months, with exception of January which appears as an isolated component. From February to April the lake surface temperature is maintaned at 17 °C, but January is the coldest period, with an average of 15 °C at the surface of the lake.

The nodal analysis relates the abundance of each group with the temporal groups, emphasizing the relationship with temperature, specially those of *O. niloticus* (warm months) and the charales group (coldest months).

Regarding fish composition as a function of their feeding habits, from the 65 771 individuals, 2% were carnivorous, mainly feeding on fish, represented by the species *A. robustus*, with 646 individuals, *C.*

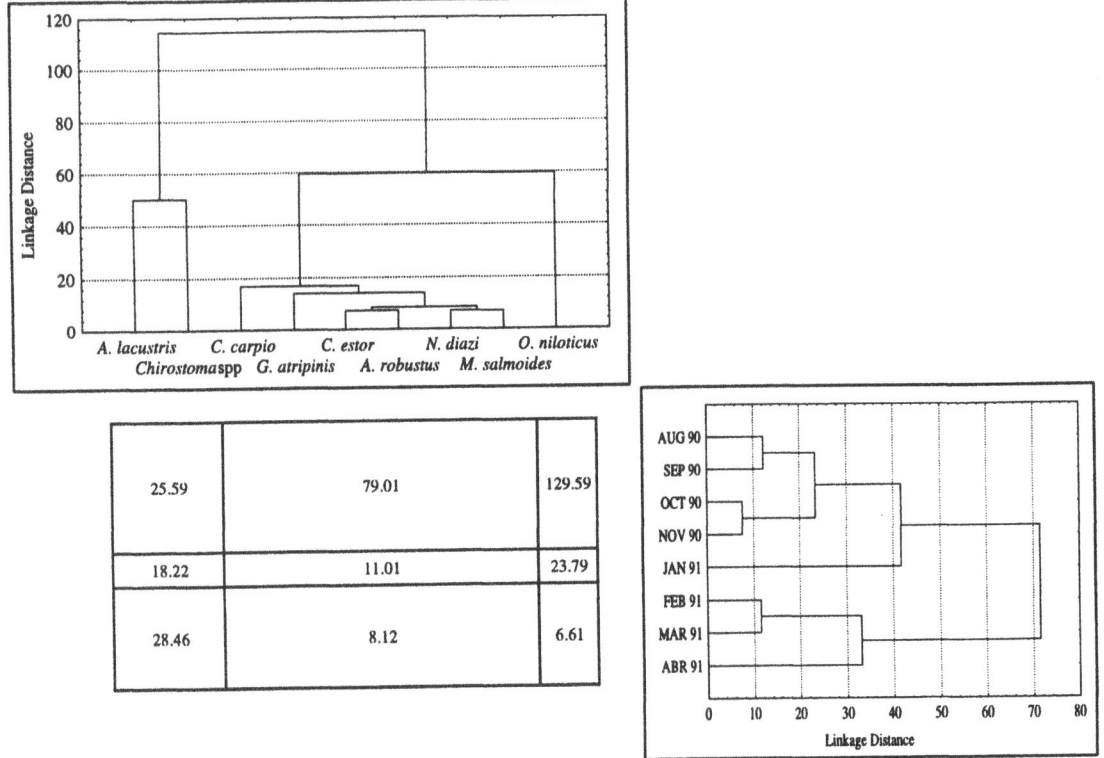

25.59	79.01	129.59
18.22	11.01	23.79
28.46	8.12	6.61

Figure 1. Species and samples (time) classification by means of Cluster analysis using Euclidean distance as measure of similitude and Unweighted Paired Group Average (UPGMA) as the clustering algorithm. The species groups (1–3) and sample groups (a–c) were defined with 50% of Euclidean distance. The nodal analysis (inner table) describes the relation between mean abundance of species groups with the temporal groups.

estor (597 ind.), *M. salmoides* (90 ind.). Most of the catch (91%) was represented by four zooplanktivorous: *Chirostoma grandocule, C. attenuatum, C. patzcuaro,* that jointly amounted 59 831 individuals, and *N. diazi* with 125 individuals. From the total number, only 2% belongs to one herbivorous species (*G. atripinis luitpoldi*) with 1299 individuals, and 5% is for omnivorous fish, represented by three species *A. lacustris* with 1602 individuals, *O. niloticus* (1510) and *C. carpio* (72).

Because most of the whole catch is represented by zooplanktivore, Elton's pyramid with the trophic levels is not included, and was only constructed using length intervals (40 mm) against the fish number. The small species, basically the charales group, again represent the larger proportion of the fish, *C. grandocule* being the most abundant species comprising up to 56% of the group. Considering that, data were log-transformed previous to Elton's pyramid drawing (Fig. 2). After 120 mm-length, the proportion among length groups decreased monotonically.

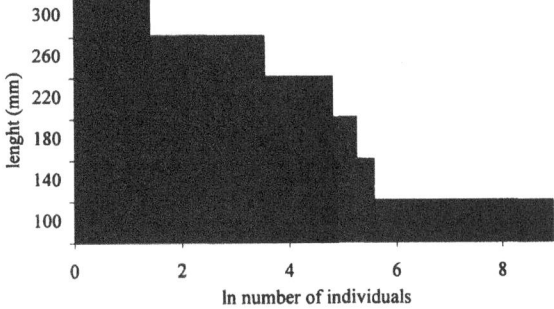

Figure 2. Elton's pyramid of the fish community in Lake Patzcuaro relates the number of individuals (ln) to their size.

Discussion

Lake Patzcuaro's species richness is defined as an assemblage of eight native and three introduced species, where the native *C. grandocule* is the most abundant. *Skifia lermae* and *Allophorus dugesi*, other native species of Goodeidae that had been previously reported as part of Patzcuaro's ichthyofauna (Martin del Campo, 1940; de Buen, 1944; Osorio-Tafall, 1944;

Rosas, 1976), were not recorded during this study, neither were *Ctenopharingodon idellus* nor *Tilapia rendalli*, which were unsuccessfully introduced during the 1970s.

This assemblage is characterized by a high degree of endemism because *A. lacustris, N. diazi, C. grandocule* and *C. patzcuaro*, 50% of the native species and more than 36% of the fish assemblage, are endemic to Patzcuaro (Alvarez, 1970; Barbour, 1973b; Barbour & Miller, 1978). In general terms, *Algansea, Chirostoma* and Goodeidae are endemic to Central Mexico.

This endemism has been explained as a consequence of the geological history of the Central Plateau of Mexico and also by the interactions among the fish species. During the Tertiary and the early Pleistocene, there was high energy tectonic and vulcanic activity that modified natural drainage, opening new routes for species dispersion and closing some others (de Buen, 1941; Barbour, 1973a). Also, the differences in size of the fish species allowed a differential selection for food, allowing coexistence among them.

Thus, the charales (*Chirostoma* spp.) occupy different areas in the neritic and limnetic zones and predominantly they inhabit different areas of the lake (Berlanga, 1993). The presence of *C. estor* was minimum or null when charales were caught, despite the fact that it feeds on charales and other small fish. This is due to the fact that recruitment of *C. estor* occurs when other *Chirostoma* are in an adult stage and their populations are not mixed. After that the size of *C. estor* is enough to avoid the gillnets used in the charales fishing. These natural characteristics must be considered to plan an adaptive fishery management (closed areas and seasons, fishing devices design), regarding the diversity in size and habits of these species.

The introduction of exotic species modified the species richness of Lake Patzcuaro, and changed the trophic structure. Before the inclusion of exotic species, 20% of the fish composition was represented by carnivore, 50% zooplantofage, 20% omnivore and 10% herbivore. After that, and considering the local extinction of *Skifia lermae* and *Allotoca dugesi*, the balance changed, increasing the carnivorous and omnivorous species proportion up to 27%. The zooplantofage amounted to 37% and the herbivore to 9%.

Differential use for feeding resources is common to fish assemblages (Ross, 1986). In Lake Patzcuaro, before the introduction of *M. salmoides* in 1993, the carnivorous species did not show feeding interferences because the diet of *C. estor* is based on charales and *A. robustus* feeds on goodeids. Once *M. salmoides* was introduced, the differential use of feeding resources changed, because *M. salmoides* also feeds fundamentally on charales (Lizarraga & Tamayo, 1989). Under this scenario, competition for food between both species has been suggested, only taken into account diet overlapping, but not food abundance (Contreras & Escalante, 1984). On the other hand, de Buen (1941) proposed coexistence between both species for a scenario where there is not significant insufficiency of feed in the lake.

The output Elton's pyramid, shows that abundance of small fish, seen as prey, was greater that predator's abundance, condition that could allow niche overlapping, but not exclusion or displacement of the interacting species (Ross, 1986).

Where the ecosystem's conditions limit the number of species that can be carried, it is not possible to introduce new species unless compensatory processes (e.g. local extinction) occur. This is not true for communities in transition, where the species number has not reached the equilibrium or for those that always are far from it. In both cases, the communities are not in saturation and it could be possible to introduce new species, with no exclusion of any other by competition for food (Ricklefs, 1987; Ault & Johnson, 1998).

Zooplanktivorous seem to be the most successfull species in the lake, linking the primary production with the upper trophic levels, allowing a sustainable fishery production that averages around 120 kg·ha^{-1} year^{-1} (Lizarraga & Tamayo, 1989; Orbe & Acevedo, 1998), greater than that of some Asian tropical lakes sustained by native species, but lesser than some others that include exotic species from African origin (Berlanga, 1993). This is not the case for *O. niloticus* in Lake Patzcuaro, were natural conditions have not been propitious for the population development of this African species.

The evidence provided by species clustering and the structure of Elton's pyramid for the fish in Lake Patzcuaro, lead to the conclusion that there are no bottlenecks in its trophic dynamics, resulting in high fishery production levels. The use of classification techniques revealed a main assembly of six species, native and introduced, that are mainly related by their catch stability, with little variation throughout the year. Only the charales *Chirostoma* spp. and *A. lacustris* displayed a seasonal pattern, negatively related to the surface temperature of the lake. Also the catch of *O.*

niloticus seems to be influenced by temperature, but in this case the relation is direct.

Conclusions

Lake Patzcuaro's ichthyofauna could be seen as an unsaturated or transitional community that has not reached its equilibrium in species number, largely modified by the introduction of exotic species and local extinction of native species. Fishery production in Lake Patzcuaro is the highest for tropical lakes in the Central Plateau of Mexico and it is mainly sustained by native species, but after the successful introduction of exotic species, the lake's fishery production has increased. Even that, the increase in catch is not large enough to approve the introduction of other exotic species, particularly when has been demostrated that *O. niloticus*, *C. idellus* or *T. rendalli* were inadequately selected.

The addition of new species has modified the structure of fish composition, but is not satisfactory evidence to demonstrate that this is the cause of native species extinction, that could be happen as a consequence of overfishing, environment degradation or by the sinergic combination of several factors.

The differential use of feeding resources by the fish species and the diversity in their habits must be taken into account when designing fishery strategies to maintain or increase fishery production in Lake Patzcuaro.

References

Anonymous, 1990. Rendimiento pesquero potencial de grandes embalses. BIOTEC, Siglo 21. México. 190 pp.

Alvarez, del V. J., 1970. Peces mexicanos. SECOFI, Inst. Nac. Inv. Biológico-Pesqueras, México: 166 pp.

Alvarez, del V. J., 1972. Ictiología michoacana V. Origen y distribución de la ictiofauna de Michoacán. An. Esc. Nac. Cienc. Biol. Mex. 19: 155–161.

Ault, R. T. & C. R. Johnson, 1998. Spatially variations in fish species richness on coral reefs: habitat fragmentation and stochastic structuring processes. Oikos 82: 354–364.

Barbour, C. D., 1973a. A biogeographical history of *Chirostoma* (Pisces: Atherinidae): a species flock from the Mexican plateau. Copeia 3: 533–556.

Barbour, C. D., 1973b. The systematics and evolution of genus *Chirostoma* Swainson (Pisces Atherinidae). Tulane Stud. Zool. Bot. 18: 97–141.

Barbour, C. D. & J. H. Brown, 1974. Fish species diversity in lakes. Am. Nat. 108: 473–489.

Barbour, C. D. & R. R. Miller, 1978. A revision of the Mexican cyprinid fish genus *Algansea*. Miscellaneous Publications Museum of Zoology, University of Michigan, Ohio. 155: 1–72.

Berlanga, R. C. A., 1993. Contribución al conocimiento de las comunidades de peces del Lago de Pátzcuaro, Michoacán. Professional thesis, Facultad de Ciencias, UNAM, México: 91 pp.

Berlanga, R. C. A., A. Ruiz, M. R. Nepita & J. Madrid, 1997. Estabilidad y diversidad de la composición de peces del lago de Pátzcuaro, Michoacán, México. Rev. Biol. Trop. 45: 1553–1558.

Colinvaux, P., 1980. Introducción a la ecología. LIMUSA, México: 679 pp.

Contreras, B. S. & M. A. Escalante, 1984. Distribution and known impacts of exotic fishes in Mexico. In Stauffer, J. R. (ed.), Distribution, Biology and Management of Exotic Fishes. John Hopkins Univ. Press, London: 102–130.

Chacón, T. A., 1993. Lake Patzcuaro, Mexico: Watershed and water quality deterioration in a tropical high-altitude Latin American lake. Lake and Reserv. Manage. 8: 37–47.

Chacón, T. A. & I. E. Muzquiz, 1991. Síntesis limnológica del Lago de Pátzcuaro, Michoacán, México. Univ. Mich. San Nicolás de Hidalgo, México: 48 pp.

de Buen, F., 1941. El *Micropterus (Huro) salmoides* y los resultados de su aclimatación en Pátzcuaro. Rev. Soc. Mex. Hist. Nat. 2: 69–78.

de Buen, F., 1944. Limnobiología del lago de Pátzcuaro. An. Inst. Biol. Mex. 15: 261–312.

de Buen, F., 1947. Investigaciones sobre ictiología mexicana I. Catálogo de los peces de la Región Neártica en suelo mexicano. An. Inst. Biol. Mex. 18: 257–348.

Fernando, C. H. & J. Holcik, 1982. The nature of fish communities: a factor influencing the fishery potential and yields of tropical lakes and reservoirs. Hydrobiologia 97: 127–140.

Flores, R., V. Magallanes & E. Mestre, 1992. Evaluación de las técnicas para el control de la erosión. Comisión Nacional del Agua, SARH, México. 127 pp.

Lizarraga, Y. & P. Tamayo, 1989. Análisis de la producción pesquera del Lago de Pátzcuaro, período 1980–1987. Informe de Labores 1986–1988, CRIP-Pátzcuaro, Inst. Nal. Pesca, Mexico: 49–70.

Martin del Campo, R., 1940. Los vertebrados de Pátzcuaro. An. Inst. Biol. Mex. 11: 481–492.

Melack, J. M., 1976. Primary productivity and fish yields in tropical lakes. Trans. am. Fish. Soc. 105: 575–580.

Nelson, J. S., 1994. Fishes of the World. John Wiley & Sons, New York: 523 pp.

Nepita, V. M. R., 1993. Hábitos alimenticios de tres especies de godeidos del Lago de Pátzcuaro, Mich., Mex. Professional thesis, Escuela de Biología, UMSNH, Mexico: 32 pp.

Orbe, M. A. & J. Acevedo G., 1998. El Lago de Pátzcuaro. In de la Lanza & J. L. García-Calderón (eds), Lagos y Presas de México. Centro de Ecología y Desarrollo A. C., México: 89–108.

Osorio-Tafall, B. F., 1944. Biodinámica del Lago de Pátzcuaro. I Ensayo de interpretación de sus relaciones tróficas. Rev. Soc. Mex. Hist. Nat. 5: 197–227.

Pielou, E. C., 1984. The Interpretation of Ecological Data. John Wiley & Sons, New York: 263 pp.

Ricklefs, R. E., 1987. Community diversity: relative roles of local and regional processes. Science 235: 167–171.

Rosas, M. M., 1976. Peces dulce-acuicolas que se explotan en México y datos sobre su cultivo. CEESTEM No. 2, México: 135 pp.

Ross, S. T., 1986. Resource partitionning in fish assemblages: a review of field studies. Copeia 1986: 352–388.

Sale, P. F., 1980. The ecology of fishes in coral reef-fish communities. Oceanogr. Mar. Biol. Ann. Rev. 18: 367–421.

Hydrobiologia **467**: 123–131, 2002.
J. Alcocer & S.S.S. Sarma (eds), Advances in Mexican Limnology: Basic and Applied Aspects.
© 2002 *Kluwer Academic Publishers.*

Spatial and temporal variation patterns of a waterfowl community in a reservoir system of the Central Plateau, Mexico

Julieta Barragán Severo[1], Eugenia López-López[1] & Kathleen Ann Babb Stanley[2]

[1]*Laboratorio de Ictiología y Limnología, Departamento de Zoología, Escuela Nacional de Ciencias Biológicas, Instituto Politécnico Nacional, Prol. de Carpio y Plan de Ayala, Col. Sto. Tomás, México, D.F. 11340, México*
[2]*Laboratorio de Vertebrados Terrestres, Departamento de Biología, Facultad de Ciencias, UNAM. México, D.F. 04510, México*

Key words: species richness, habitat use, Mexican Central Plateau, migratory birds, waterfowl, wetlands

Abstract

The reservoirs studied are located on a very important migratory flyway in the Mexican Central Plateau. The survey included the waterfowl, macrophytes and physico-chemical parameters of three localities visited on ten occasions during a 1-year period. A total of 23 waterfowl species were observed: 6 transients, 10 winter visitants, and 7 residents. A period of increased richness was detected in winter, when migratory species arrive at the reservoirs. A canonical correspondence analysis showed segregation patterns of the species for the use of the space. Migratory ducks occurred in Umécuaro, where there is a larger variety of macrophytes and eutrophication occurs. Loma Caliente was the area with less eutrophication and was not preferred by ducks. *Gallinago gallinago* and *Ardea herodias* were the species with the highest frequency of occurrence, while the Effluent was the preferred site of resident species. Although the reservoirs studied are small, they are suitable for a significant number of species to be considered as a waterfowl protection and conservation zone.

Introduction

A large number of waterfowl are migrants and are adapted to wetlands, ecosystems that are highly dynamics, with seasonally available resources. These migratory movements, whether nonstop or with intermediate stopping points, take place between the breeding and wintering zones, and depend on habitat conditions. It has been estimated a total of 61 million individuals of aquatic birds for North America, of them 10% winter in Mexico. These birds arrive in Mexican territory by three flyways (Pacific, Central, and Gulf). The estimated number of individuals for each flyway is 2 100 000; 1 900 000; and 2 100 000, respectively (SEMARNAP, 1995).

Those arriving by the central flyway cross the Mexican Central Plateau, and confront a semiarid region with few natural lakes. In order to satisfy the human demand for water in this area, it has been necessary to build artificial reservoirs, ponds, and livestock watering holes, which have become an alternative source of refuge for waterfowl. These bodies of water, as well as the natural ones, have been affected to different degrees by the various activities of people. Those located in lower areas and close to human settlements have been more extremely affected, since they receive urban and industrial wastewaters, which directly or indirectly affect the aquatic fauna.

The role of artificial reservoirs as alternative habitats for migratory and resident birds can be examined through a systematic survey of bird arrivals, limnological conditions, and distribution patterns of species in space and time. In North America certain chemical, physical, morphometrical and biological lake factors have shown direct relationships to bird abundance and diversity (Hoyer & Canfield, 1994). In connected ponds in Alaska, Murphy et al. (1984) found that the anatids prefer bodies of water with high levels of eutrophication.

There are few studies of bird populations in artificial reservoirs in Mexico (Carrera & Canales, 1985; Cisneros, 1985; Babb, 1988), and even rarer there are few studies of seasonal variations in the physical and chemical conditions of bodies of water, and waterfowl response to them (Huerta et al., 1985). This study analyzes the spatial and temporal variations of waterfowl in two reservoirs on the Mexican Central Plateau, and examines the relationship between limnological factors and the composition of bird populations.

Study area

The Río Grande de Morelia basin is located in the southern part of the Mexican Central Plateau, in the state of Michoacan, in an area highly disturbed by human settlements and industrial developments, which have brought a decrease in the natural habitat of the aquatic flora and fauna. The basin is fed by the Río Tirio (see Fig. 1); in its upper portion there are the Loma Caliente and Umécuaro reservoirs, at 2175 and 2170 m above sea level, respectively. They were built in serie to control floods, and their water is used for irrigation and to supply adjacent towns. The Loma Caliente effluent is the tributary of Umecuaro; this effluent forms a flood plain that is covered by aquatic vegetation and grasses.

The reservoirs have a surface area of 7 km^2, and receive a mean of 1000–1200 mm annual precipitation, with 16–18 °C mean annual temperature. The climate is temperate subhumid with summer rains, beginning in June, and reaching their climax in August, September and October. The dry season last from November to May.

Methods

Monthly observations, each one of three days of duration, were made on 10 occasions, beginning in August, 1992 and ending in May, 1993. Fixed-radius observation points and transects, in accordance with Hutto et al. (1986) and Hoyer & Canfield (1990), respectively, were chosen at three sites (see Fig. 1): E-1 Loma Caliente, E-2 Effluent, and E-3 Umécuaro. Observations were made twice a day: 6:00 h, and 15:00–18:00 h. Taxonomic determinations were established using Peterson & Chalif's (1989) and National Geographic Society (1987) field guides. Pettingill's (1985) classification was used to place birds in the different migratory or resident categories. The nomenclature and systematic ordering of the species is that of the American Ornithologist Union (AOU, 1993).

Traditional methods were used to record temperature, pH, dissolved oxygen, conductivity, and Secchi depth at each study site (Wetzel & Likens, 1979). Water samples were taken in order to quantify total dissolved solids, nitrates and phosphates using a Hach DREL/2000 spectrophotometer. Aquatic macrophytes were collected at each observation site. The trophic status of each locality was calculated with Carlson's (1977) trophic state index (TSI).

A matrix of species by site was produced from the avian fauna composition results, and a second matrix, of factors by site, from the environmental factors results. Both matrices were log-transformed (log $10(x + 1)$), and a canonical correspondence analysis (CCA) was performed using the ANACOM statistical program (de la Cruz, 1994).

Results

Avian fauna composition

A total of 23 aquatic bird species were recorded, of which 30.4% are resident and 69.6% migratory. Species occurrence varied by month and by site (see Fig. 2), with the Effluent being the site with the lowest species richness; it also lacked transients species. In Umécuaro and Loma Caliente, species richness is highest during the winter, with the addition of migratory species 57.9% and 61.53% of the total species, respectively. Breeding activity was observed in *Podilymbus podiceps* (Linnaeus), *Gallinula chloropus* (Linnaeus), *Fulica americana* Gmelin, and *Jacana spinosa* (Linnaeus) (see Table 1).

In Loma Caliente 13 species were identified, belonging to eight families (see Table 1). The family best represented was the Ardeidae, with four species, one of which, *Ixobrichus exilis* Gmelin, was exclusive to this reservoir. *P. podiceps* and *F. americana* were observed on the entire lake surface. *G. chloropus, J. spinosa* and *I. exilis* were sighted only in the central part of the reservoir, where there is a zone of emergent and floating aquatic vegetation. The rest of the species were found on both the central vegetation and the reservoir shores.

A total of 21 species, belonging to 10 families, were identified at Umécuaro (see Table 1). The most common family was the Anatidae, with six species, one of which is resident. Five of the seven species exclusive to this reservoir are Anatidae. Most of the species were mainly observed towards to the tail end of the reservoir. *Ardea alba* (Linnaeus) uses the dry trees in the middle part of the lake only as a roosting area.

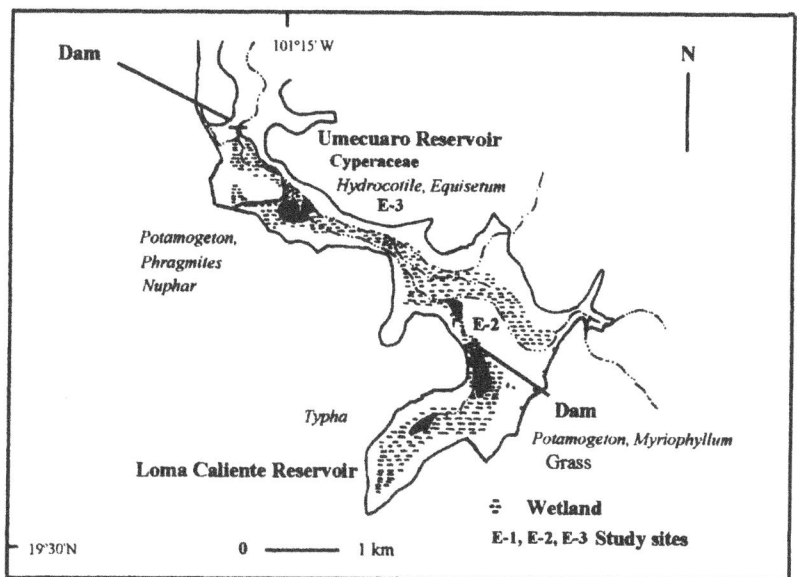

Figure 1. Study area location. E-1 Loma Caliente, E-2 Effluent, E-3 Umecuaro.

The Effluent sheltered five species. *Rallus limicola* Vieillot was exclusive to this site; this bird is evasive and though it was heard throughout the study period, it was only sighted once. Species like *Anas platyrhynchos diazi* Ridgway, and limicolous species, such as *Porzana carolina* (Linnaeus), *G. chloropus* and *J. spinosa*, were seen only from March to May.

It is worth noting that four of the species sighted (17.4% of the total) fall into one or another of the protected categories, according to Mexican Official Regulation (NOM, 1994). *Anas discors* Linnaeus and *Aythya affinis* (Eyton) are species of special interest, *Ardea herodias* Linnaeus is considered rare, and *I. exilis* has an endangered status.

Limnological conditions in the reservoirs

The bodies of water studied are shallow environments whose maximum depth ranges from 3 to 5 m in the dry and wet seasons, respectively. The Effluent's zone has an abundant aquatic vegetation with species of *Typha*, *Sagittaria*, *Phragmites*, and *Potamogeton*. A wide marshy area develops at this site during the low-water period in the dry season. In Loma Caliente, the shoreline is fringed with species of *Potamogeton*, *Myriophyllum*, and several Gramineae, while in the central area there is a zone with *Typha* sp., and *Nuphar* sp. In Umécuaro, in addition to *Potamogeton* and *Nuphar*, there are several species of *Equisetum*, *Phragmites*, *Hydrocotyle*, and Cyperaceae (see Fig. 1).

Table 1. Avian fauna composition of the sites studied E-1 Loma Caliente, E-2 Effluent, E-3 Umécuaro (W – Winter visitor T – Transient; R – Resident)

Especies	(E-1)	(E-2)	(E-3)	W	T	R
Podilymbus podiceps	x		x			x
Ixobrichus exils	x				x	
Ardea herodias	x		x	x		
Ardea alba	x		x	x		
Egretta thula	x		x			x
Egretta caerulea			x		x	
Anas plathyrhynchos diazi		x	x			x
Anas discors			x	x		
Anas cyanoptera			x		x	
Anas clypeata			x	x		
Anas crecca			x		x	
Ahythya affinis			x		x	
Rallus limicola		x				x
Porzana carolina		x	x	x		
Gallinula chloropus	x	x	x		x	
Fulica americana	x		x			x
Charadrius vociferus	x		x	x		
Jacana spinosa	x	x	x			x
Actitis macularia	x		x	x		
Gallinago gallinago	x		x	x		
Larus delawarensis			x		x	
Pandion haliaetus	x		x	x		
Ceryle alcyon	x		x	x		

Due to their height above sea level, the study sites exhibit a marked seasonal variation in temperature. Minimum water temperatures ranged from 8.5 to 10 °C in January and February, and the maximum was 23 °C in August. No thermal stratification was detected. Dissolved oxygen ranged from 6.3 to 12.4 mg/l in September and March, respectively; Loma Caliente exhibited the widest fluctuations (see Table 2). In both reservoirs, a drop in the concentration of dissolved oxygen towards the bottom was noticed during the summer months, a condition that is typical of eutrophic bodies of water. Water pH ranged from 5.9 to 7.8 in September and December, respectively; Umécuaro had the widest variations with a marked tendency towards acidification at the bottom.

Water transparency registered the lowest values, 0.65 m, during spring and the highest, 2.2 m, at the end of the wet season in October. Conductivity reached its lowest values, 15 μS/cm, at the end of winter and the highest, 53 μS/cm, during the wet season. Nitrates and phosphates were lowest at the end of winter, with values non detectable for both nutrients, and highest during the rainy season, with values of 2.6 and 4.8 mg/l, respectively.

As indicated by the trophic state index (TSI), the values displayed by the reservoirs correspond to a mesoeutrophic condition (see Table 2); nevertheless, it is possible to note that Umécuaro is slightly more eutrophic. With respect to seasonal variation, an increase in trophic status was found towards the end of the dry season and beginning of the wet period.

Canonical correspondence analysis

In the CCA diagram, spatial and seasonal variations can be seen, which are related to physical and chemical factors (see Fig. 3). In the upper quadrants appear the study sites in the spring and summer, correlated with higher temperatures and increased rainfall. The upper left quadrant corresponds to Loma Caliente, which is correlated with high oxygen and nitrogen values. In the upper right quadrant are the Effluent and Umécuaro, which are correlated with greater conductivity (see Fig. 3a and Table 3).

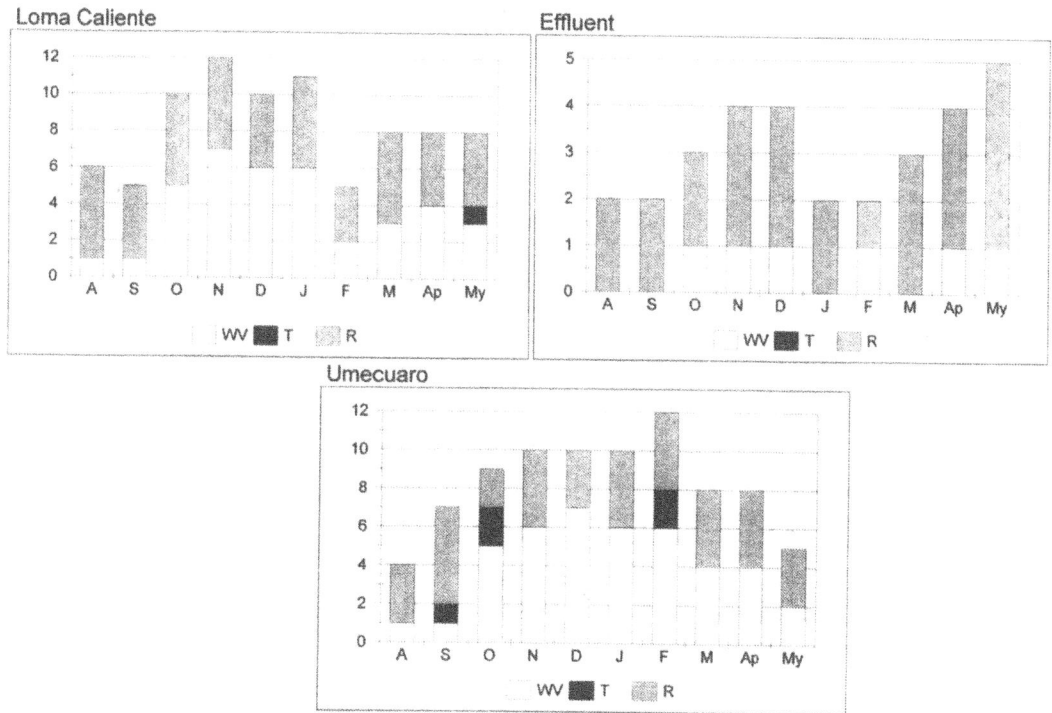

Figure 2. Monthly fluctuation of specific richness of the waterfowl in the sites studied. WV – Winter Visitors; T – Transient species; R – Resident species.

In the lower left quadrant, all three study sites are found during the fall, correlated with low temperatures and high pH values. In the lower right quadrant, all three study sites appear clustered in winter, with the lowest temperatures and high conductivity, phosphorus and Secchi depth values.

From the avian fauna composition, we were also able to detect patterns of species association and use of space and time (see Fig. 3b). The presence of *J. spinosa* and *G. chloropus*, both residents, was characteristic in Loma Caliente during the hot period. *I. exilis*, a migratory species, was sighted only once in the month of March; it was possibly a transient individual on its way back north. During the same hot period in Umécuaro, there was a single sighting of *Egretta caerulea* (Linnaeus), a migratory species that is probably one of the first to fly south. In the Effluent, *R. limicola* was observed, and at both sites, *Anas platyrhynchos diazi*.

During the fall, species such as *Gallinago gallinago* (Linnaeus) and *Charadrius vociferus* Linnaeus, which were common to Umécuaro and the Effluent, are associated. *A. herodias* and *Pandion haliaetus* (Linnaeus) were present in Loma Caliente and Umécuaro.

In the wintertime, the presence of different species of winter migratory Anatidae was common: *Anas crecca* Linnaeus, *A. discors*, *Anas clypeata* Linnaeus, *Anas cyanoptera* Vieillot, *Aythya affinis*, *Actitis macularia* (Linnaeus), *Larus delawarensis* Ord, and *Ceryle alcyon* (Linnaeus). Species like *F. americana* and *Egretta thula* (Molina) are residents that share the study sites, and appear near the origin in the diagram (see Fig. 3b).

The CCA allowed us to determine in a clear manner a marked spatial segregation of waterfowl, as well as a seasonal arrival in the reservoirs. We were able to establish the periods when winter migratory species and those that are returning to their place of origin occur (October through March), as well as the April through October period, when resident species predominate.

Discussion

Since winter migratory and resident species, some of them included in the NOM (059-Ecol-1994), predominate in the aquatic bird population of the reservoir system, the latter may be considered an important arrival and refuge area for avian fauna in spite of its

Table 2. Physical and chemical characteristics of the sites studied. TSI – Trophic State Index, '–' data no recorded

Parameters	Loma Caliente				Effluent				Umecuaro			
	Min	Max.	Mean	σ	Min.	Max.	Mean	σ	Min.	Max.	Mean	σ
Temperature (°C)	8.5	22	16.3	4.6	10	23	16.3	3.9	9	23	17.5	4.2
Oxygen (mg l^{-1})	6.3	12.4	8.3	2.0	6.3	12	8.0	1.6	6.4	11.9	8.0	1.4
pH	6.5	7.4	6.8	0.3	6.5	7.1	6.6	0.3	5.9	7.8	6.8	0.5
Secchi depth (cm)	62	220	113	42	–	–	–	–	37	150	100	30
Conductivity (μS cm^{-1})	15	52	24.7	9.7	18	52	35	11.28	30	53	42.4	9.54
Nitrate (mg l^{-1})	0	2.2	0.98	0.68	0	2.64	0.66	0.7	0	1.98	0.7	0.58
Orthophosphates (mg l^{-1})	0.01	2.2	1.11	0.81	0.4	4.8	1.6	1.47	0	4.0	1.35	1.9
TSI	54.5	61.9	59.9	1.1	–	–	–	–	58	64	60.35	0.96

small size. This assertion becomes evident when the species richness of the locality is compared with that of larger bodies of water in the same biogeographical region, such as the Laguna de Chapala and the Yuriria reservoir, which have lake surface areas of 8860 km^2 (Limón & Lind, 1990), and 12 km^2, respectively. A total of 45 species have been recorded in these two lake areas (Babb, verbal communication, 1988), while at our locality, 23 species were sighted in an area of only 7 km^2. This circumstance may be due to the inaccessibility of the study places, since part of the area is fenced in for use as pasture and agricultural fields, and it is also far from large urban centers, so that it has been affected and disturbed to a fairly low degree. Then again, it could also be due to the fact that in spite of the degree of eutrophication, the study sites are not markedly contaminated because they do not receive any wastewater discharge.

During the study period, seasonal and spatial patterns were detected which reflect resource partition. With regard to seasonal variation, an increase in species richness during the winter was particularly evident at all three sites. This phenomenon is related to the geographic location of the sites, which are on an important migratory flyway in Mexican territory. Once the migratory species leave, the remaining resident species also display patterns of spatial resource use.

As far as spatial patterns are concerned, a marked segregation was observed. The Umécuaro reservoir evidently had the highest species richness, and it was also the preferred arrival site of various Anatidae species. This body of water was characterized by a larger variety of aquatic macrophytes, which provide greater habitat heterogeneity for the avian fauna. In accordance with Hoyer & Canfield (1990), in a study on the influence of limnological factors on species richness

Table 3. Multiple correlation coefficients of environmental factors and coordinates of the sites and species studied

Axis I	Axis II	
0.828	0.704	Multiple correlation coefficients Factors
0.253	0	Temperature
0.073	0.111	Oxygen
−0.189	0.077	pH
−0.194	−0.1	Transparency
−0.029	−0.107	Conductivity
0.013	−0.103	Nitrate
−0.006	−0.196	Orthophosphate
0.181	−0.166	Precipitation

and abundance of birds in 33 Florida lakes, they pointed out that although these birds can be shown to be influenced by many factors, the more relevant ones are the trophic status and the aquatic macrophytes, since they are correlated with an increase in the number of species. In similar way, Hoyer & Canfield (1994), adding 13 more lakes to their previous data, examine trophic status, lake morphology, and macrophytes, and find a close correlation between greater trophic status and increase in species richness and abundance of birds. In this same study, they emphasize bird-macrophyte relations at the species level, such as that of *Aythya collaris* (Donovan), which is only found in lakes where the macrophyte *Hydrila verticillata* occurs.

Another important difference in the environmental characteristics of Umécuaro is the higher degree of eutrophication, which is manifested in the CCA by a higher degree of mineralization. Moreover, according to the study by López & Dávalos-Lind (1998) on algal growth potential, Umécuaro was identified as having

a)

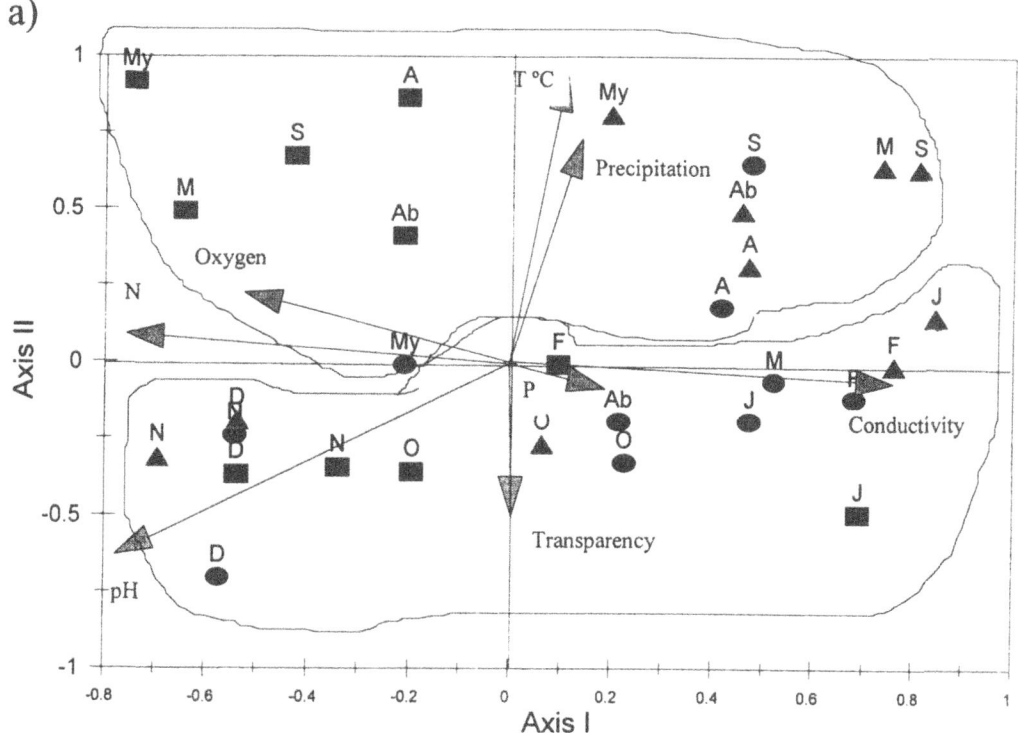

b)

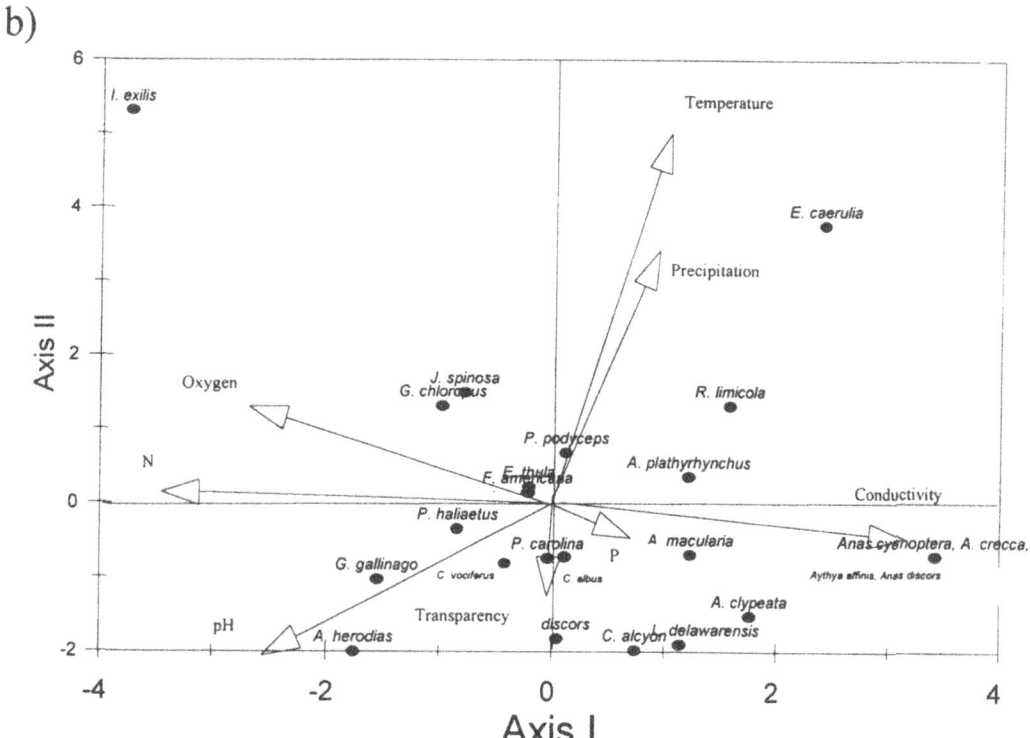

Figure 3. Diagram of the canonical correspondence analyses. (a) Physical and chemical factors and sites studied. + Loma Caliente, ▲ Effluent, ● Umecuaro, A – August; S – September; N – November; D – December; E – January; F – February; M – March; Ab – April; My – May. (b) Waterfowl species and physical and chemical factors.

130

the greatest nutrient enrichment in the reservoirs system. With regard to the avian fauna composition, this study showed that Anatidae preferred the site with more eutrophication and a larger variety of macrophytes. In this respect, there seems to be a high correlation between ponds that are hydrologically connected and have high nutrient (nitrogen and phosphorus) levels, and the greater use of these sites by ducks, as has been suggested by Murphy et al. (1984) for connected ponds in Alaska. In reference to this, Blanco et al. (1996) found a relation between Anatidae requirements and river productivity in the Middle Tajo Valley. In Loma Caliente, the species that showed higher correlations with respect to habitat were *J. spinosa, I. exilis* and *G. chloropus*. This body of water had a lower trophic status than Umécuaro, and according to the CCA, a lower degree of mineralization and higher levels of oxygenation. The habitat requirements that have been recorded for *I. exilis* are environments with abundant emergent vegetation (Stotz et al., 1996), a condition that is common in Loma Caliente. *J. spinosa* and *G. chloropus* are typically found in environments with floating and emergent vegetation, and therefore in waters with a certain degree of eutrophication.

As far as the Effluent was concerned, it sheltered resident species, *R. limicola* being exclusive to it. The latter has been reported from wetlands invaded by tule (Stotz et al., 1996), a typical condition at this site particularly during the low-water period in the dry season. Only one winter migratory species visited this site.

Conclusions

We were able to detect the existence of various patterns of spatial and temporal segregation among the waterfowl and the reservoirs studied, which reflect that the different requirements of the birds are met by these limnologically variable sites. Anatidae showed a preference for the Umécuaro reservoir, the site with the most mineralization and the largest variety of macrophytes. We were also able to assess the importance of these small bodies of water in the arrival of migratory birds, which require adequate conditions for rest, refuge, and/or feeding to foster a good breeding season. In this sense, we can point out that bodies of water located far from large cities and less, such as Loma Caliente and Umécuaro reservoirs, are excellent areas for waterfowl rest and arrival. This, together with the fact that they shelter species which have been placed in one or another of the protected categories, makes

them a particularly important area to protect in order to preserve the habitat of avian fauna.

Acknowledgements

The authors wish to express their gratitude to the following colleagues: María Eugenia López Islas, Dora Luisa Apáez, Fernando Urbina, and Edmundo Díaz Pardo, for their collaboration during field work; Eduardo Soto, Joel Paulo, and José Angel Serna, who took water samples and collected macrophytes in the reservoirs studied, and to Oscar Polaco for their comments to the manuscript.

References

American Ornithologist Union (AOU), 1983. Check-list of American Birds, 6th edn. American Ornithologists Union, Washington DC, U.S.A. 877 pp.

Blanco, G. A., A. Acha, J. Cuevas, P. Ruiz & T. Velasco, 1996. Fenología de la reproducción y productividad de anatidas en ríos del Valle Medio del Tajo. Ardeola. 43: 31–39.

Carlson, R., 1977. A trophic state index for lakes. Limnol. Oceanogr. 22: 361–369.

Carrera, J. & G. E. Canales, 1985. Aves acuáticas migratorias (Anatidae) de Coahuila. Memorias del I Simposio Internacional de Fauna Silvestre. México: 873–883.

Cisneros, T., 1985. Mini-hábitat, estrategia para la conservación del pato mexicano. Memorias del I Simposio Internacional de Fauna Silvestre. México: 957–965.

De La Cruz, G., 1994. ANACOM. Sistema para el Análisis de comunidades. Versión 3.0. CICIMAR. Instituto Politécnico Nacional, México. 99 pp.

Hoyer, M. V. & D. E. Canfield, Jr., 1990. Limnological Factor Influencing bird Abundance and Species Richness on Florida Lakes. Lake and Reservoir Management 6: 133–144.

Hoyer M. V. & D. E. Canfield, Jr., 1994. Bird abundance and species richness on Florida lakes: influence of trophic status, lake morphology, and aquatic macrophytes. Hydrobiologia 297/280: 107–119.

Huerta, L.A., C. T. Chávez & C. J. M. Chávez, 1985. Plan de Manejo y Desarrollo para la conservación y uso público de las comunidades de aves acuáticas del Ex-Lago de Texcoco. SEDUE, The Wild Soc. Méx. Memorias del Primer Simposio Internacional de Fauna Silvestre, México, D. F.: 678–708.

Hutto, R. L. S., M. Pletschet & P. Hendricks, 1986. A fixed-radius point method for nonbreeding and breeding season use. The Auk. 103: 593–602.

López, L. E. & L. Dávalos-Lind, 1998. Algal growth potential and nutrient limitation in a tropical river–reservoir system of the Central Plateau México. Aquatic Ecosystem Health and Management 1: 345–351.

Limón, M. G. J. & O. T. Lind, 1990. The management of Lake Chapala (México): Considerations after significant changes in the water regime. Lake and Reservoir Management 6: 61–70.

Murphy, S. M., B. Kessel & L. J. Vining, 1984. Waterfowl Populations and Limnologic Characteristics Of Taiga Ponds. J. Wild. Mgmt. 48: 1156–1163.

National Geographic Society, 1987. Field Guide To The Birds Of North America. National Geographic Society, 2nd edn., Washington. 463 pp.

Norma Oficial Mexicana. NOM-059-Ecol-1994. Que determina las especies y subespecies de flora y fauna silvestres terrestres y acuáticas en peligro de extinción, amenazadas, raras y las sujetas a protección especial, y que establece especificaciones para su protección. Diario Oficial del lunes 16 de mayo. Tomo CDLXXXVIII No. 10:1–60 México, D. F.

Peterson, R. T. & E. L. Chalif, 1989. Aves De México. Guías De Campo. Diana, México: 473 pp.

Pettingill, O. S. Jr., 1985. Ornithology In Laboratory And Field. Academic Press, Orlando: 403 pp.

SEMARNAP, 1995. Mortandad de Aves Acuáticas En La Presa De Silva, Guanajuato. Informe Técnico. Procuraduría Federal De Protección Al Ambiente. Subprocuraduría De Recursos Naturales. México: 167 pp.

Stotz, D. F., J. Fitzpatrick, T. Parker & D. Moskovits, 1996. Neotropical Birds, Ecology And Conservation. Univ of Chicago Press, Chicago: 481 pp.

Wetzel, R. & G. Likens, 1979. Limnological Analyses. Saunders Co. Boston, U.S.A.: 357 pp.

Hydrobiologia **467**: 133–139, 2002.
J. Alcocer & S.S.S. Sarma (eds), Advances in Mexican Limnology: Basic and Applied Aspects.
© 2002 *Kluwer Academic Publishers.*

Structure of a pond community in Central Mexico

E. Escobar-Briones[1], A.M. Cortez-Aguilar[2], M. García-Ramos[3], L.M. Mejía-Ortíz[4]
& A.Y. Simms-Del Castillo[5]
[1]*Unidad Académica Sistemas Oceanográficos y Costeros, ICML-UNAM, 04510 D.F., Mexico*
E-mail: escobri@mar.icmyl.unam.mx
[2]*Dolphin Discovery, 77500 Cancún, Q. Roo, Mexico*
[3]*Lab. de Producción Acuícola, FMVZ, UNAM, 04510 D.F., Mexico*
[4]*Lab. PHI: UAM-Xochimilco, Fisiología y Comportamiento Animal*
[5]*Pasgrado en Sistemas y Recursos Acuáticos, Facultad de Ciencias, UNAM*

Key words: benthos, model, phytoplankton, zooplankton

Abstract

The community structure of a tropical pond located in Mexico City was described. Primary producers were diverse and abundant and macrophyte components provide a broad habitat diversity. Major biotic components in the water column were algae of the genera *Microcystis*, *Scenedesmus* and *Chlamydomonas*. Mats of *Cyclotella* were located in deeper, shaded areas associated with the bottom. Three species of emergent macrophytes characterize the large primary producers. The zooplankton was dominated by protozoa, which were more abundant than rotifers and Crustacea. Dominant invertebrate species of the pond bottom were insect larvae, gastropod mollusks, oligochaete annelids and nematodes. Vertebrates (fish, amphibians, reptiles) feed on this variety of biotic components. A flow diagram depicted the major biological interactions and suggested a complex trophic structure.

Introduction

Ponds have been considered ornamental components in public parks and gardens, yet are of major interest to limnologists. These small bodies of water allow studies with a holistic view of the structure and function (Williams et al., 1997). The general classification system describes 'ponds' as water bodies of anthropogenic origin, with shallow depth, and commonly without stratification (Rigler & Peters, 1995). Due to their shallowness, and relatively high surface to volume ratio, ponds often have water temperatures similar to the local atmosphere (Lampert & Sommer, 1997). Light, turbidity, and mixing of the water column control the trophic status of ponds. The import of nutrients into the pond, originated from runoff in the catchment area, is the principal source of energy that, combined with light, enhances high primary production. Ponds show abundant vascular plant growth, which varies in mass according to depth and form of the basin (Williams et al., 1997).

The information available on pond assemblages in the tropics is limited to few studies (e.g., Alcocer et al., 1988; Alcocer & Lugo, 1995), not directed towards the understanding of their structure and the processes that occur in them. In contrast, studies focused on ponds in temperate latitudes have a longer tradition (e.g., Collinson et al., 1995; Everard et al., 1999).

The objective of this study is to describe the structure of a high altitude pond located at tropical latitude. A conceptual model based on the composition and potential interactions is herein suggested to help future management studies.

Study area

The current study took place in a pond, built in 1975, located in the Botanical Garden of the Universidad Nacional Autónoma de México (UNAM) campus in Mexico City. It has an area of 42.3 m^2, a volume of 16.9 m^3 and is located at 19° 19′ 4″ N, 99° 11′

134

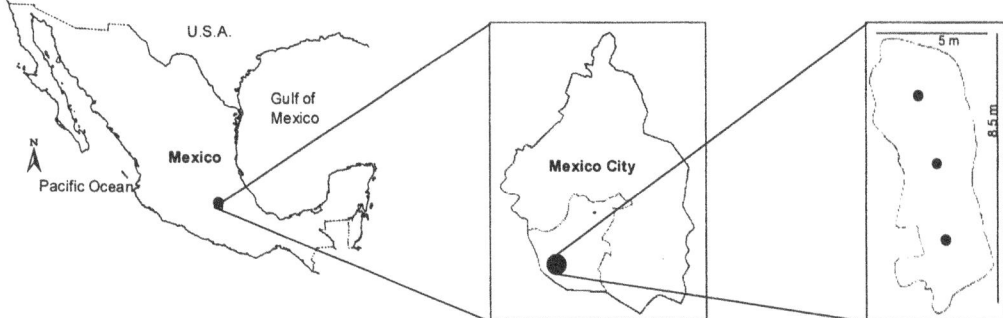

Figure 1. Location of the area of study.

99″ W and 2300 masl (Fig. 1). Its shape is irregular, depth varies from 0.35 to 0.49 m, and it is fed by an intermittent inflow in the northern section. The pond is surrounded by deciduous vegetation; marginal vegetation is absent. The surface is partially covered by free-floating and rooted aquatic vascular plants that are limited to the central portion of the pond. Light incidence on the surface varies throughout the day, generating differential distribution of algae, promoting a varied number of microhabitats.

Materials and methods

Selected physical and chemical variables were analyzed; subsurface water temperature, and dissolved oxygen concentration were measured with a YSI oxygen meter model 5100, pH was determined with an YSI 63 pH/temperature meter. Total suspended solids (TSS) were estimated by filtering five 10 ml samples of subsurface water through 0.7 μm GF/F pre-combusted filters (dia.=4.7 cm). The concentration values of the TSS were determined by the difference in weight of the filter before and after the filtration.

A set of three quadrants (area per quadrant=0.01 m^2) was used to randomly sample the macrophytes. The taxonomic composition of the plant material was identified using Riemer (1984). Aerial coverage of macrophytes (in units of m^2) was determined by means of photographs and grids of the pond surface. Phytoplankton samples were obtained from 0.5 l surface water filtered through a 20 μm mesh. The collected phytoplankton material was fixed in 4% lugol-acetate solution and identified using González (1988) and Ortega et al. (1994). The phytoplankton biomass was estimated based on the chlorophyll concentration following Vollenweider (1974).

Microphytobenthic organisms were sampled in triplicate with hand cores (Area=7×10^{-4} m^2); taxa were identified with the keys cited above.

Zooplankton samples were obtained by filtration of 5 l of three sub-superficial water samples with a net with mesh size opening of 0.5 mm, and fixed with a 10% solution of buffered formaldehyde. Taxonomic groups were identified using Edmonson (1959), Pennak (1978), Streble & Krauter (1987), Thorp & Covich (1991), and were quantified with a stereoscopic microscope.

A core (dia.=5 cm) of superficial sediment was collected to determine the organic matter content in the top 2 cm. The sediment was combusted at 450 °C for an hour in a furnace and the organic matter determined by the difference in weight of the sediment before and after combustion.

Benthic macrofauna samples were collected in triplicate with a hand corer (area=12.5×10^{-2} m^2) that was randomly launched in order to obtain 20 replicates from the sediments. The fauna, retained after sieving through 1 mm mesh, was quantified and classified. The benthic components were identified to genus levels to homogenize the taxonomic list of all sampled groups. The hand netting of aquatic macroinvertebrates omitted sampling at sites within the vegetated bottom. The density of frogs and turtles was evaluated by *in situ* direct counting. Hand nets were used to collect fish; the catch specimens were identified following Alvarez (1970) and Capula (1990). The gut contents of fish were analyzed following the gravimetric method (Hyslop, 1980).

A flow diagram was conceptualized to explain the pond trophic interactions by using Stella II modeling language (from High Performance Systems Inc.). Three subsystems (water column, benthos, and detritus) were suggested in this flow diagram based on the different processes that characterize each subsystem. The trophic compartments were boxes that depict subsystems. Each subsystem was represented by different taxa. Each trophic compartment was interdependent from the other compartments within each subsystem and among all three subsystems. All three subsystems were shown here in a cross-linked fashion. The flow was pulsed by the processes recognized in the three subsystems of the pond being these photosynthesis, predation and detritivory. This conceptual diagram suggested a hypothesized model of how the pond functions and that should be tested in future experimental and field studies.

Results

Environmental factors

Subsurface water temperature varied from 13.1 ± 0.3 °C to 18.8 ± 0.6 °C, the minimum values were found at the pond's bottom. Dissolved oxygen concentration was 6.79 ± 2.14 mg O_2 l^{-1} throughout the water column. The TSS values were 0.23 ± 0.13 mg ml^{-1} dry weight. The average pH was 6.9 ± 0.6. The organic matter content in sediment ranged from 3.3 to 5.8% in the pond peripheral zone and attained values of 6.3–8.7% at the center of the pond at the base of the macrophytes.

Phytoplankton

Fourteen phytoplankton taxa were recognized (Table 1), of which *Microcystis* was the most common followed by *Scenedesmus* which contributed to >50% of the algal density. The chlorophyll concentration of the water column varied spatially from 0.32 to 2.13 mg Chl m^{-3}. The largest values occurred in shaded areas of the pond. Chlorophyll *a* had the lowest concentration (0.04–0.08 mg Chl_a m^{-3}), followed by chlorophyll *b* (0.20–0.42 mg Chl_b m^{-3}) and chlorophyll *c* (0.08–1.63 mg Chl_c m^{-3}).

Aquatic macrophytes

Three forms of emergent vegetation were present in the pond, *Nymphaea* had the largest biomass (7.20 g wet weight m^{-2}), with a coverage of 6.55 m^2; this species dominated the central part of the pond with 84% of relative dominance. *Ceratophylum* and *Naja* were moderately represented in zones of lower light

Table 1. Average density values of each taxonomic group identified in the pond

Taxonomic group	Density
Phytoplankton	cells.ml^{-1}
Microcystis sp.	822
Scenedesmus sp.	680
Chlamydomona sp.	390
Navicula sp.	100
Nitzchia sp.	50
Pandorina sp.	47
Synedra sp.	40
Gomphonema sp.1	33
Gomphonema sp.2	27
Pediastrum sp.	23
Fragilaria sp.	20
Tetrastrum sp.	17
Scenedesmus sp.2	17
Mastigasphaera sp.	7
Zooplankton	**Individuals.ml^{-1}**
Protozoans	
Volvox sp.	227
Gymnodinium sp.	333
Chilidonella sp.	53
Vorticella sp.	33
Rotifers	
Philodina sp.	79
Embata sp.	40
Trichocerca sp.	67
Brachionus sp.	60
Cladocerans	
Daphnia sp.	24
Copepods	3
Vertebrates	**Individuals.m^{-2}**
Poeciliopsis sp.	1.8
Rana sp.	0.05
Trachemys sp.	0.02
Benthos	
Macrophytes	**Individuals.m^{-2}**
Nymphaea sp.	7
Ceratophylum sp.	5
Naja sp.	21
Microphytobenthos	**Individuals.m^{-2}**
Cyclotella sp.	11353
Navicula sp.	1297
Anabaena sp.	757
Synedra sp.	450
Anacystis sp.	217
Benthic fauna	**Individuals.m^{-2}**
Physella sp.	260
Limnodrillus sp.	136
Zigoptera sp.	294
Pseudosuccinea sp.	119
Tipulidae	366
Nematoda	112

(0.49 g wet weight m^{-2} and 0.6 g wet weight m^{-2}, respectively, Table 1).

Microphytobenthos

The microphytobenthos was composed by five genera of algae. *Cyclotella* were the most abundant, while *Anacystis* was the least abundant (Table 1). The benthic cyanobacteria *Anabaena* dominated the central portion of the pond.

Zooplankton

The taxonomic composition of the zooplankton included four genera of Protozoa, four of Rotiferan, one of each Cladocera, and Copepoda (Table 1). The crustaceans had the lowest density (27 ind ml^{-1}) in the pond in contrast to other faunal components, and had the lowest species richness (2 taxa). The Protozoa showed the highest density (646 ind ml^{-1}) followed by the Rotifera (246 ind ml $^{-1}$).

Macrobenthos

The fauna consisted of insect larvae of the family *Tipulidae* and *Zigoptera*, snails of the genera *Physella* and *Pseudosuccinea*, oligochaetes and nematodes. The most abundant were the insect larvae (660 ind m^{-2}) followed by the snails (379 ind m^{-2}) with less abundance (Table 1).

Vertebrates

Fish of the genus *Poeciliopsis* were dominant, additional to frogs of the genus *Rana* and turtles of the genus *Trachemys*. The diet of *Poeciliopsis* was comprised of benthic invertebrates (65%), zooplankton (20%) and algae identified from the water column (15%). The frogs occurred only in the pond area covered by *Nymphaea*.

The flow diagram

The water column, the detritus and the benthos were considered as the major subsystems in the pond. The phytoplankton, the zooplankton and the vertebrates were the three trophic groups that were linked and characterized the water column. Macrophytes and organic matter content in sediment define the detritus box that was linked with the water column through the export of biogenic carbon. The patches of microphytobenthos and the benthic fauna were the benthic components that comprise two trophic levels. The benthos box was linked to the detritus box through detritivory and was linked to the water column through predation by fish on the benthic invertebrates (Fig. 2).

Discussion

The phytoplankton biomass in the tropics is an important component that pulses the interactions in the water column. The uneven distribution of submerged vegetation and the variability of the solar radiation are the main factors affecting the heterogeneity of habitats (Scheffer, 1990) in both the horizontal and vertical gradients of the pond. A strong spatial variation along the horizontal axis has been recorded in tropical ponds, resulting in a large variety of microhabitats (Margalef, 1983). The phytoplankton biomass in tropical water bodies covers a wide range. As an example, shallow soda lakes attain chlorophyll *a* concentrations of 500 mg m^{-3} (Melack, 1981) to 20 mg m^{-3} (Payne, 1986). Many shallow water bodies sometime support standing crops of phytoplankton comparable to the highest values known for terrestrial communities and hypertrophic aquatic environments (Lampert & Sommer, 1997).

The dissolved oxygen concentration in the pond is homogeneous characteristic of shallow environments; the small variations are related to the consumption of the heterogeneously distributed diverse organisms and the mineralization of detritus (Lampert & Sommer, 1997). As in other shallow water bodies, the absence of anoxic conditions are attributed to mixing and photosynthesis providing the benthic dwellers with oxygen (Payne, 1986). Few areas, those with poor water exchange, remain hypoxic at night. An example is the bottom with large densities of macrophytes. Locations close to the outfall in diverse aquatic systems have shown oxygen gradients that can be easily disrupted by mineralization throughout the day and in distance from the outfall (Harper, 1992). The small variation in pH observed can be attributed to patches of Cyanobacteria and macrophytes on the pond bottom. Irradiance promotes the primary production in ponds that constitutes the base of the trophic chain (Maitland, 1990). The largest input of imported organic matter to the aquatic system arrives from the deciduous vegetation surrounding the pond. The emerged and submerged vegetation provide the immediate largest amount of biogenic carbon generating detritus within the pond. They may also accomplish the key func-

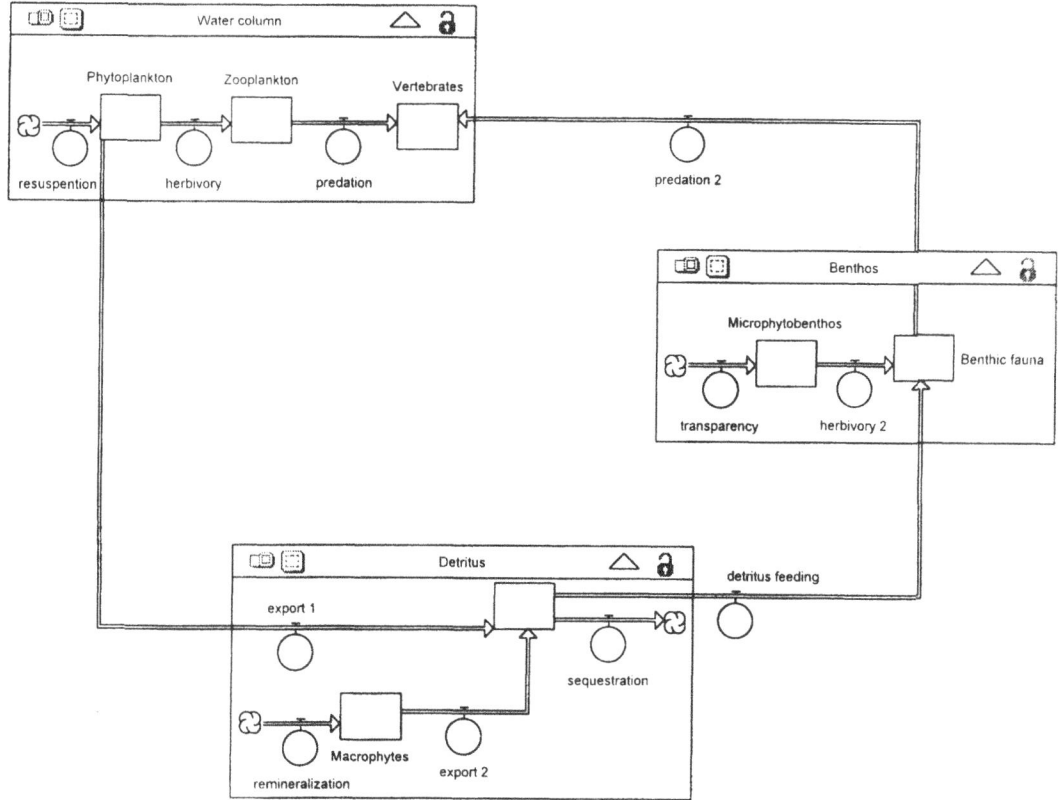

Figure 2. Conceptual diagram of the biological trophic subsystems and compartments in the pond.

tion of stabilizing fresh water ecosystems (Scheffer, 1990), and possibly provide refuge for diverse fauna. The size and depth of the pond seem to be relevant variables to be considered in the pond productivity (Russell-Hunter, 1970).

As with phytoplankton, the zooplankton assemblage from tropical shallow bodies of water indicates a similar pattern in the variety of species. Some open water tropical lakes host in up to 89 rotifer species, while Cladocera may vary from 49 to 65 species. There are in contrast only two or three species of cyclopoid copepods per lake in the tropics. Usually ponds tend to have the highest number of species on a more local scale (Collinson et al., 1995). The pond zooplankton is a mixture of littoral and limnetic species. The low number of zooplankton species encountered in this pond can be attributed to the fluctuating regime of the filling and the small size of the pond as well as to the fish predation pressure. Mean annual standing crops are highly variable 44 mg dw m^{-3} (Payne, 1986) to 559 mg dw m^{-3} (Burgis, 1974). The number of spe-

cies of crustacean zooplankton is not high in contrast to other shallow sites at different latitude and altitude.

Tight coupling between the water column components and the benthic components in the pond community is suggested in the transfer of biogenic carbon within the shallow aquatic environment. Deposition of biogenic carbon into the detritus pool promotes the linkage between water column and benthos. The organic material in the sediment of shallow bodies of water has proven to be low, between 3.7 and 16.9% dry weight, that is significantly lower when compared with temperate ponds (Estèves, 1983). This large variability has been recorded in another urban lake in Mexico by Alcocer et al. (1988) and has suggested a rapid remineralization of organic material both in the water column and the sediment.

Communities from tropical marine pond sediments do not appear to be particularly diverse (Lampert & Sommer, 1997). This has been attributed to the physical and chemical instability of the environment. The bottom fauna over most of the pond bed is poor in species, largely consisting of midge larvae, which are

138

temporary residents of the pond bed. Several studies show however that amongst marginal vegetation the diversity increases dramatically as a wide variety of insects appear (Green et al., 1976; Payne, 1986). Our study was limited to sediment cores and the diversity of benthic fauna was notably reduced. Similar low taxonomic composition has been reported in urban lakes in Mexico City (Alcocer & Lugo, 1995). The vegetation, which provides shelter for these marginal communities, is rarely itself diverse and tends to be made of a few species.

The Stella diagram shows that the pond is pulsed by environmental factors such as turbulence and mixing that determine the origin and nature of the biogenic carbon. Each subsystem is short with a maximum of three trophic levels each, however the entire system is linked, enhancing the length of the food web in the pond. As in other tropical shallow water bodies two complementary pathways for energy utilization within the pond were identified. Refractory organic matter, whose residence time is dependent on the cleaning schedule of the pond, may define the sequestration of biogenic carbon for less extended periods than deeper lakes. A comparison of the taxonomic composition, the abundance, the pH and the depth with other ponds (Rast et al., 1989; Harper, 1992) places the site of study within a meso-eutrophic state.

Conclusions

Due to their size, ponds provide excellent sites for research and formal studies with a holistic view of the structure and function. The relevance of these small bodies of water relies in their value for wildlife. Limited information exists on pond structure and the processes that define the interactions among trophic components in tropical ponds. A large abundance and richness of primary producers characterized the pond of the Botanical Garden. Dominant components of the pond community include planktonic protozoa followed by rotifers and benthic insect larvae. All of these groups support trophically the fish and other vertebrates that inhabit the pond. The major pathways of the biogenic carbon flow are the primary producers and detritus. Wind mixing and predation are respectively the bottom-up and the top-down controls of the pond community structure. The pond is suggested a complex system according to the conceptual flow diagram.

Acknowledgements

The Ecology course of the School of Sciences, UNAM graduate program granted support for field and laboratory work. Dr R. Bye, Institute of Biology, UNAM, allowed the access to the pond in the Botanical Garden.

References

Alcocer, J. & A. Lugo, 1995. The urban lakes of Mexico City (Lago Viejo de Chapultepec). Lake Line 15: 14–15, 31.

Alcocer, J., E. Kato, E. Robles & G. Vilaclara, 1988. Estudio preliminar del efecto del dragado sobre el estado trófico del Lago Viejo de Chapultepec. Rev. Int. Cont. Amb. 4: 43–56.

Alvarez, J., 1970. Peces mexicanos. Claves para la determinación de especies en los peces de las aguas continentales mexicanas. Instituto Nacional de Investigaciones Biológico Pesqueras, México, D.F.: 166 pp.

Burgis, M. J., 1974. Revised estimates for the biomass and production of zooplankton in Lake George, Uganda. Freshwat. Biol. 4: 535–541.

Capula, M., 1990. Amphibians and Reptiles. MacDonald Orbis, London: 2567 pp.

Collinson, N. H., J. Biggs, A. Corfield, M. J. Hodson, D. Walker, M. Whitfield & P. J. Williams, 1995. Temporary and permanent ponds: an assessment of the effects of drying out on the conservation value of aquatic macroinvertebrate communities. Biol. Conserv. 74: 125–133.

Edmonson, W. T. 1959. Freshwater Biology. Wiley & Sons, New York: 1248 pp.

Estèves, A. F., 1983. Levels of phosphate calcium, magnesium and organic matter in the sediments of some Brazilian reservoirs and implications for the metabolism of ecosystems. Arch. Hydrobiol. 96: 129–138.

Everard, M., B. Blackham, K. Rouen, W. Watson, A. Angell & A. Hull, 1999. How do we raise the profile of ponds? Freshwat. Forum. 12: 32–43.

González, Y. A., 1988. El plancton de las aguas continentales. OEA. Monografía 33: 85–109.

Green, J., S. A. Corbet, E. Watts & O. B. Lan, 1976. Ecological studies on Indonesian lakes. Overturn and restratification of Ranu Lamongan. J. Zool. Lond. 180: 315–354.

Harper, D., 1992. Eutrophication of Freshwater. Chapman and Hall, New York: 327 pp.

Hyslop, E. J., 1980. Stomach contents analysis – A review of methods and their applications. J. Fish Biol. 17: 411–429.

Lampert, W. & U. Sommer, 1997. Limnoecology: The Ecology of lakes and streams. Oxford University Press, New York: 382 pp.

Maitland, P. S., 1990. Biology of freshwater. Blackie, New York. 432 pp. Margalef, R., 1983. Limnología. Omega, Barcelona: 1010 pp.

Melack, J., 1981. Photosynthetic activity of phytoplankton in tropical African soda lakes. Hydrobiologia 81: 71–85.

Nogrady, T., 1983. Succesion of planktonic rotifer populations in some lakes of the East African Rift Valley, Kenya. Hydrobiologia 98: 45–54.

Ortega, M. M., J. L. Godinez & S. G. Ollivia, 1994. Ficología de México: Algas continentales. AGT Editor, S.A., México: 215 pp.

Payne, A. I., 1986. The Ecology of Tropical Lakes and Rivers. John Wiley & Sons. 301 pp.

Pennak, R. W., 1978. Fresh-water Invertebrates of the United States. Wiley Interscience, New York: 803 pp.

Rast, W., V. H. Smith & J. A. Thornton, 1989. Characteristics of eutrophication. In Ryding, S. & W. Rast (eds), The Control of Eutrophication of Lakes and Resorvoirs. UNESCO and The Panthenon Publishing Group, Paris: 37–63.

Riemer, D. N., 1984. Introduction to freshwater vegetation. Avi. Publi. U.S.A.: 356 pp.

Rigler, F. H. & R. H. Peters, 1995. Science and Limnology. In Kinne, O. (ed.), Excellence in Ecology. Ecology Institute, Oldendorf: 239 pp.

Russell-Hunter, W. D., 1970. Productividad acuática: introducción o aspectos básicos de la oceanografía biológica y de la limnología. Acribia, España: 273 pp.

Scheffer, M., 1990. Multiplicity of stable state in freshwater systems. Hydrobiologia 61: 475–486.

Thorp, J. E. & A. P. Covich, 1991. Ecology and classification of northamerican freshwater invertebrates. Academic Press, San Diego: 911 pp.

Vollenweider, R. A., 1974. A Manual of Methods for Measuring Primary Production in Aquatic Environments. Blackwell, Oxford: 225 pp.

Williams, P. J., J. Biggs, A. Corfield, G. Fox, D. Walker & M. Whifield, 1997. Designing new ponds for wildlife. British Wildlife 8: 137–150.

Hydrobiologia **467**: 141–157, 2002.
J. Alcocer & S.S.S. Sarma (eds), Advances in Mexican Limnology: Basic and Applied Aspects.
© 2002 *Kluwer Academic Publishers.*

On the dynamical response of Lake Chapala, Mexico to lake breeze forcing

A.E. Filonov

Department of Physics, University of Guadalajara, Apdo. Postal 4-040, Guadalajara, 44421, Jal., Mexico
Tel./Fax: 52-3-619-82-92. E-mail: afilonov@ccip.udg.mx

Key words: Lake Chapala, breezes, temperature and current measurements, internal waves

Abstract

Fluctuations in the atmospheric characteristics, as well as variations in the water level of Lake Chapala are discussed. Field measurements of the atmospheric characteristics and lake level during December 1996 through January 1997 are described; using spectrum analysis of synchronous time series. The findings suggest that the variability is due to the diurnal cycle of atmospheric elements. Lake breeze circulation plays an important role in the area of Lake Chapala; since it was registered in 83% of the data. Periodic fluctuations in atmospheric pressure and wind generate significant seiche amplitudes in the lake, with the periods of about 6 h. With the help of a simple model, the seiche parameters are estimated. The amplitude of one-nodal seiches on one of the edges of the lake; is on average equal to 18 mm. This wave should generate currents of approximately 0.012 m s^{-1} at the lake's centre in the area of the nodal line. The experimental results on the thermal regime and circulation of Lake Chapala are discusssed as well. Surface temperature variations were registered at the eastern part of the lake. In all cross-sections, typical spatial variations of 3 °C were registered, over a distance of 100–300 m. A bouy station registered movements of an internal thermal front in the body of the water. The leading edge of the front was accompanied by intense internal waves, in the form of internal KdeV solitones. The front near the buoy station was produced by the movement of a warm body of water travelling from the shallow eastern part of the lake and trigered by morning breeze.

Introduction

Lake Chapala is the largest lake in Mexico and the third largest in Latin America. It is located at approximately 20°N, 103°W, 1500 m above sea level. It measures 75 × 25 km, and has an average depth of 6 m and a maximum depth close to 11 m (see Fig. 1). The lake's tributary and effluent systems are the Lerma and Santiago rivers. There is a mountain chain along its southern and northern shores. The Lerma–Chapala watershed system has a surface area of approximately 47 000 km^2 (Sandoval, 1994). The average precipitation in this area is about 750 mm a^{-1}, which can decrease to 300 mm a^{-1} in drought years and increase to 1200 mm a^{-1} in wet years. There is 1400–1600 mm of surface evaporation per year, which exceeds by a factor of 2 the average precipitation (Jaurégui, 1995; Filonov et al., 1998). The precipitation has short-term fluctuations (from 2 to 7

years) due to El-Niño and changes of solar activity with periods of 11, 22, 45 years or greater. Inter-annual variability of precipitation causes significant temporal fluctuations in the lake's level, area and volume (Filonov & Tereshchenko, 1997).

Local winds and breeze play a crucial role for the lake's dynamics. They develop daily, as a result of the large difference in temperature between the lake's surface and the surrounding land of scattered vegetation cover. Thus the land is quickly heated and cooled during day and night. Day time breeze is more intense than night time breeze and its direction is towards land. The speed of the lake breeze wind in the day time reach 8–10 m s^{-1} and it strongly increases evaporation from the lake surface (Filonov, 1998).

The lake plays an essential role in the economy of Mexico. About 14 million people live in the area of its watershed (Sandoval, 1994). During the past decades, the lake has become polluted with anthro-

Table 1. Daily fluctuation parameters of the meteorological characteristics and anomalies of lake level for Lake Chapala

	Mean value	Average maximal values for the month	Average minimal values for the month	Absolute maximum	Absolute minimum
Air temperature (°C)	18.2	21.7	13.4	25.2	10.5
Humidity, (%)	68.7	82.3	51.5	100	41.7
Atmospheric pressure, bm	848.3	852.1	846.8	854.5	843.7
Wind speed (m s^{-1})	1.3	3.6	0	16.3	0
Wind direction (grad)	165–345	–	–	–	–
Lake level anomalies (mm)	±6.3	+9.1	−9.7	+37	−36

Table 2. Values of the square function of the coherence, of the phase differences and the mean square amplitudes of the daily (a) and semidaily (b) harmonic of the hydrometeorological characteristics of the region of Lake Chapala

	a							b						
	Air temperature	Humidity	Atmospheric pressure	Wind speed	Wind direction	Lake level anomalies	Mean square amplitude	Air temperature	Humidity	Atmospheric pressure	Wind speed	Wind direction	Lake level anomalies	Mean square amplitude
Air temperature		48	142	47	350	210	5.0		182	124	90	74	—	1.6
Humidity	0.987		12	106	343	—	16.1	0.891		65	105	347	—	5.1
Atmospheric pressure	0.983	0.959	0.965	91	152	71	3.1	0.923			33	43	65	3.1
Wind speed	0.739	0.667	0.654		308	150	1.3	0.902	0.847	0.820		345	188	1.0
Wind direction	0.968	0.980	0.884	0.805		218	11.3	0.633	0.844	0.738	0.753		181	3.4
Lake level anomalies	0.781	0.743	0.812	0.773	0.718		1.9	0.486	0.384	0.772	0.658	0.683		—

pogenic agents, which enter through discharges from the Lerma River and neighboring areas. This pollution causes an intense growth in water lilies and tule. Furthermore, the level of the lake has decreased due to water usage for agriculture and domestic needs of the cities (Filonov et al., 1998). All this reduces its attractiveness as a recreation and tourist area, therefore, the Mexican government has taken measures to protect the dam and the lake from the negative consequences of anthropogenic actions. Among these measures is the development of research projects on the thermal-hydrodynamics of the lake, in order to study the distribution of the pollutants and thus select the necessary measures to decrease the evaporation of the lake surface. In the this paper, the experimental results of atmospheric parameters, level and thermodynamic regime of the Lake are discussed.

Character of the lake breeze circulation

Time series and their processing

The time series of the atmospheric characteristics were used for a more detailed analysis of the breeze circulation and its connection with the variations in lake level, which were measured with the help of an automatic meteorological station Weather-Monitor II, located at the climatic station at Chapala. The sampling interval

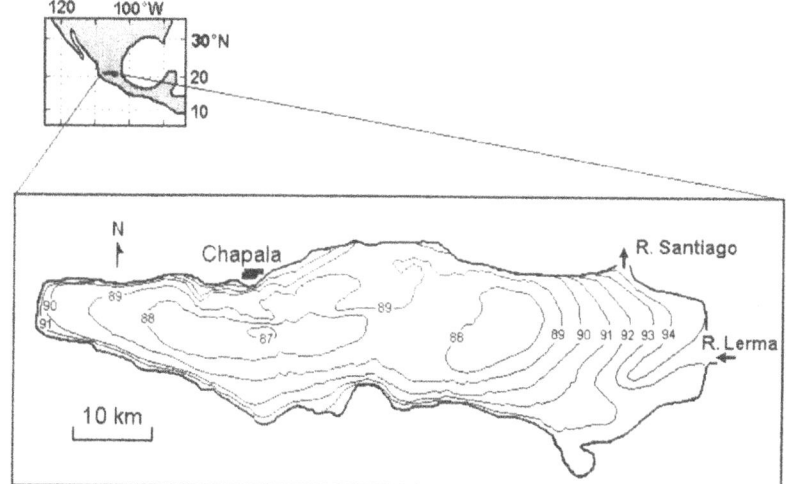

Figure 1. Batymetrical map of Lake Chapala. Depth is given in meters in relation to the 87 m isobath.

of measurement was 1 h. Synchronous with the atmospheric measurements, the time series of the lake level fluctuations with same sampling interval were received by the CNA (Comisión Nacional del Agua). The spectra of the amplitudes was estimated from the time series (Jenkins & Watts, 1969; Konyaev, 1990):

$$C_x(\omega) = \int_0^T x(t) \cdot \exp(-i2\pi\omega t)dt, \qquad (1)$$

where $x(t)$ is the time series; T is the total length of the series; and ω is the frequency. The autoperiodogram $S_{xx}(\omega)$ and cross-periodogram $S_{xy}(\omega)$ are defined as

$$S_{xx}(\omega) = \frac{1}{T}C_x(\omega) \cdot C_x^*(\omega), \qquad \text{and} \qquad (2)$$

$$S_{xy}(\omega) = \frac{1}{T}C_x(\omega) \cdot C_y^*(\omega) = P_{xy}(\omega) - iQ_{xy}(\omega). \qquad (3)$$

Here $P_{xy}(\omega)$, $Q_{xy}(\omega)$ are the real and imaginary parts of the crossed periodogram; and (∗) marks is the complex conjugate of the spectra. The spectral estimates were obtained by smoothing the frequencies of the periodograms:

$$\hat{S}_{xx}(\omega) = \int_{-\infty}^{\infty} S_{xx}(\omega')Z(\omega - \omega')d\omega', \qquad (4)$$

where $Z(\omega)$ is a smoothing function. The estimates of the coherence function were also calculated for pairs

of time series:

$$C_{0_{xy}}^2(\omega) = \frac{|\hat{S}_{xy}(\omega)|^2}{\hat{S}_{xx}(\omega) \cdot \hat{S}_{yy}(\omega)} \qquad (5)$$

and the phase difference

$$\Delta\varphi_{xy}(\omega) = \arctan(\hat{Q}_{xy}(\omega)/\hat{P}_{xy}(\omega)). \qquad (6)$$

For the structure of the time series of the wind velocity and direction of groups I and II, the method of rotary components (Gonella, 1972; Mooers, 1973) was used. To this end we find the estimates of the 'clockwise' spectra,

$$S_-(\omega) = \frac{1}{8}[S_{uu}(\omega) + S_{vv}(\omega) - 2Q_{uv}(\omega)] \qquad (7)$$

and 'anticlockwise' spectra

$$S_+(\omega) = \frac{1}{8}[S_{uu}(\omega) + S_{vv}(\omega) + 2Q_{uv}(\omega)] \qquad (8)$$

where $S_{uu}(\omega)$ and $S_{vv}(\omega)$ are the autospectra, and $Q_{uv}(\omega)$ is the quadrature spectrum of the zonal and meridional components ($u(t)$ and $v(t)$ respectively) of the wind velocity vector. The orientation of the major axes of the elliptic hodographs

$$\Theta(\omega) = \arctan\{2P_{uv}(\omega)/(S_{uu}(\omega) - S_{vv}(\omega))\} \qquad (9)$$

and their stability

$$E^2(\omega) = \frac{(S_{uu}(\omega) + S_{vv}(\omega))^2}{(S_{uu}(\omega) + S_{vv}(\omega))^2 - 4Q_{uv}^2(\omega)} - \\ - \frac{4(S_{uu}(\omega) \cdot S_{vv}(\omega) - P_{uv}^2(\omega))}{(S_{uu}(\omega) + S_{vv}(\omega))^2 - 4Q_{uv}^2(\omega)}, \qquad (10)$$

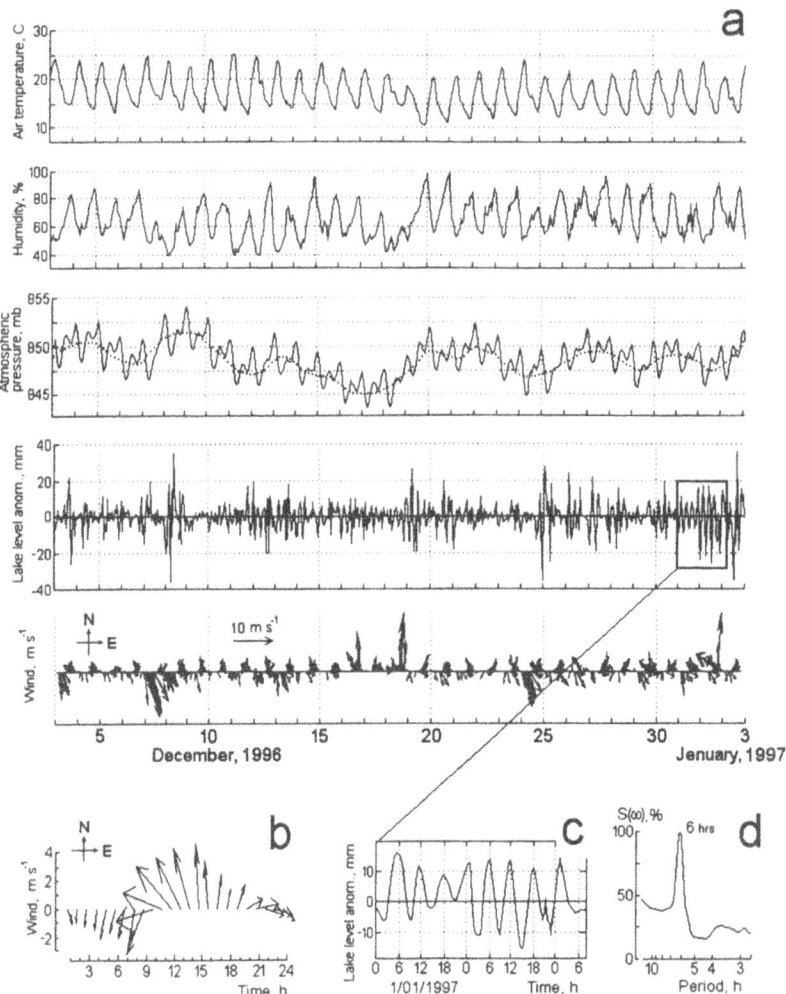

Figure 2. (**a**) Time series of fluctuations of the meteorological characteristics (measured by automatic meteorological station Weather-Monitor II, located at Chapala's climatic station) and fluctuations in the anomalies of lake level. The lower panels show the daily course of average wind for 1 month (**b**), almost pure sine fluctuations in lake level are described in (**c**) as well as their spectrum are given (**d**).

were also calculated, where $P_{uv}(\omega)$ is the co-spectrum of $u(t)$ and $v(t)$.

The confidence intervals were calculated at each level for all spectral estimates; from standard algorithms described in the literature (Jenkins & Watts, 1969; Konyaev, 1990). The number of degrees of freedom v, was found as $v = 2\alpha(2F + 1)$ where α is the number of independent segments of realization in which the spectral estimates were averaged, and F is the half-width of the filter that is used to average the periodograms.

The spectral characteristics of the daily weather fluctuations

Previous studies have shown; that the main processes contributing to climatic conditions and water circulation of Lake Chapala are the local winds and lake breeze (Filonov et al., 1998). The basic power source of the lake breeze circulation is the daily course of the temperature caused by the change in incoming heat to the terrestrial surface, and to the atmosphere during day time. Interactions between the land, water and atmospheric system are a very complex system to study. The district around Lake Chapala is mountainous, with valleys of various spatial orientations. Acting as pulses, with daily periodicity, the thermal energy does not remain at a fixed frequency. It is redistrubuted in a complex way as fluctuations at different frequencies (Scorer, 1978).

Figure 2 shows the fluctuation in the time series of air temperature, relative humidity, atmospheric pressure and wind. Such series were measured by the automatic meteorological station during December 1996 through January 1997. The descriptive statistical analysis is presented in Table 1 where daily periodicities are presented. The curves have no correct 'sine wave' form, though their significant modulation is visible at the expense of fluctuations of lower frequencies. The spread of daily fluctuations of air temperature near the lake reaches, on the average, 8.3 °C. For some days, for example, 11 December 1996, it was 12 °C. Daily fluctuations in relative humidity do not have such a regular character as those in temperature. Variation in humidity might be explained by two phenomena. First, is the amount of clouds above the lake which causes a significant difference in humidity between the land and the lake. Second, is the lake breeze direction and speed. Nevertheless, its daily course is well expressed: in the afternoon; with an increase of temperature; the air humidity is reduced to 40–50%, and at night it increases up to 80–100%. In the atmospheric pressure fluctuations, even the synoptic fluctuations of the 5–7-day period (see dashed line in a Fig. 2) are visually well allocated, as well as lake breeze fluctuation of daily and semidaily periods, with approximately equal amplitudes (see spectrum 3, in Fig. 3).

Lake breeze circulation above Lake Chapala develops from the background daily course of meteorological characteristics. These are always present, even when the lake breeze is absent. Therefore, to separate the lake breeze component from the usual daily course of fluctuations in air temperature, humidity and atmospheric pressure is not an easy task, nor is its statistical analysis. Speed and direction of wind are the most suitable variables when studying breeze presence, because lake breeze circulation develops on background variations in the field of pressure, thus having small horizontal gradients and is not a baric wind. They are not 'contaminated' with daily course and low-frequency fluctuations (Burman et al., 1983). Let's consider that there is a lake breeze circulation, when the following conditions are realized. There should not be more than 30% cloud cover, wind speed at height of the weathercock is no more than 3 m s^{-1}, during the day the wind is driven to the land and during the night to the lake (Burman, 1969).

In Figure 2b, the average value for December 1996 is given as the daily course of wind velocity

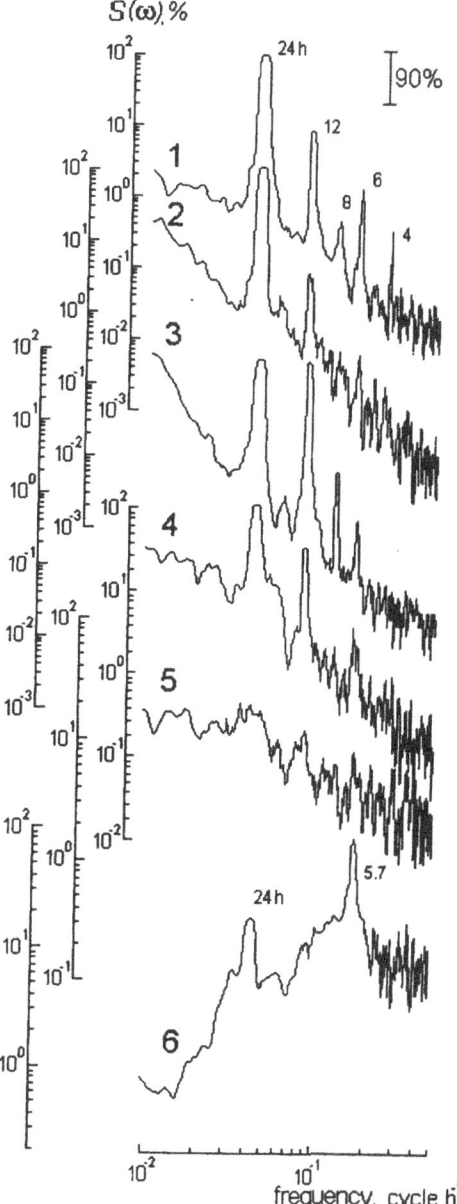

Figure 3. The normalized power spectra of fluctuations in the meteorological characteristics: air temperatures (1), relative humidity (2), atmospheric pressure (3), 'clock-wise' (4) and 'anticlock-wise' (5) wind vector components, and anomalies in lake level (6). Vertical line in the spectrum indicates the 90% confidence limit.

(for each hour of the day at Chapala's climatic station. Clearly, the wind had a lake breeze character. At night time it was weak and had a direction toward the lake. The maximum value of 2 m s^{-1} of night breeze was reached at 04:00 h, and then it decreased. After 08:00 h, the wind was amplified and turned clockwise.

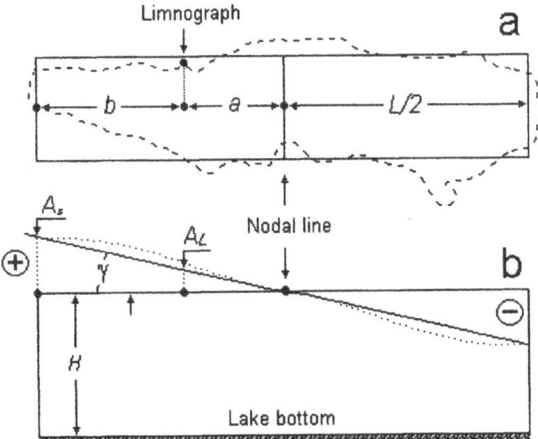

Figure 4. Model of the form of Lake Chapala, used for calculationg some of the parameters for the one-nodal seiches. A plan view (a) and side view (b).

Between 09:00 h and 10:00 h, it reached a speed of 3 m s⁻¹ and was directed parallel to the coast of the lake, to the west. From 14:00 to 15:00 h the wind reached a maximum of 4 m s⁻¹ and was landward directly to the north. From 17:00 to 24:00 h, the wind was eastward and had a speed of just 1 m s⁻¹.

A wind vector diagram (Fig. 2a) shows that the lake breeze circulation on Lake Chapala did not develop every day. For some days under the influence of the synoptic scale features, the wind is sharply amplified consequently affecting lake breeze development. For such days (7, 17–19 and 24–26 December), the atmospheric pressure was increased, and the wind reached 10–15 m s⁻¹ and had an almost constant direction. In December 1996, 83% of cases were days with lake breeze.

To find out specific information about the variability in the analyzed meteorological characteristics, the auto and cross-spectra were calculated for all time series. The normalized (in the percentage of the maximum value) spectra of analyzed time series are given in Figure 3. A daily peak prevails in the spectra of temperature and relative humidity. In the spectra there are also statistically valid peaks with periods of 12, 8, 6 and 4 h. They demostrate 'the cascade' mechanis of transfer of energy acting with daily periodicity from the Sun, to fluctuation with smaller amplitudes and periods (and apparently with the smaller linear sizes). For the atmospheric pressure spectra, the spectral density peaks have identical amplitudes for the daily and semi-daily periods. It is caused by the daily course of atmospheric pressure, which can be seen as

two peaks in Figure 12(3). In time these peaks correspond to the moments of the maximum temporal gradients of temperature, that is the increase or decrease of temperature. On the contrary, at the moments when temperature of the air reaches maximum or minimum values, the atmospheric pressure above the lake reaches minimum values. In the rotary spectra of wind fluctuations the daily peak is well expressed only in the 'clock-wise' spectrum. It confirms the assumption, that in the area of Lake Chapala, daily fluctuations of wind are basically caused by the day time breeze, at which the turn of the wind velocity is clockwise (Gonella, 1972).

The analysis of coherence for pairs data sets has shown that the confidence interval was greater than 95% for daily and semi-daily periods. The matrices of square coherence, the difference of phases and sizes of mean square amplitudes for harmonics at these periods are given in Table 2. As can be seen, daily fluctuations of the meteorological characteristics are related to each other almost linearly, but with different phase shifts, whose physical sense is easily explained. For example, as a rule, a change in wind direction causes immediate change in the relative humidity ($\Delta\varphi \approx 0$ grad.). On the contrary, fluctuations of temperature, humidity as well as wind speed and direction always occurred at an opposite phase ($\Delta\varphi \approx 180$ grad.).

Results of the analysis have shown that thermodynamic processes in the area of Lake Chapala with periods from 1 h to 1 d are caused by processes of local circulation (lake breeze and valley winds), whose primary source is the daily course of temperature. Mean square amplitudes for daily harmonic have larger amplitudes than semi-daily fluctuations for all analyzed meteorological characteristics. There is aclose connection between rates of change of temperature and the intensity of local circulation. At night, wind speed decreases to nearly zero due to the thermal stability, and vertical movement decreases. Consequently, this results in weak night time evaporation and moisture collects at the surface of the lake and land. In the morning the stable layer of the atmosphere near the ground collapses under the influence of insulation, and wind speed quickly amplifies (Riehl, 1979). The classical mechanism of daily circulation in the atmosphere; above the lake should be strongly broken as the lake breeze and valley winds develop. The mountains surrounding of the lake make the breeze flow along its axial line. Various circulating processes of smaller intensity created by valley winds should be imposed on.

The stability factor **E** and orientation Θ, of the major axes of elliptic orbits calculated for daily and semi-daily harmonics are nearly equal to $E_{24}=0.994$, $\Theta_{24}=173-353°$ for daily harmonics; and $E_{12}=0.773$, $\Theta_{12}=82-262$ degrees for semi-daily harmonics. The result means, that on average, the elliptic hodographs of the daily and semi-daily vector speed fluctuations of wind have high stability. Concerning the daily hodographs, the major axis is oriented north–south, while for the semi-daily it is oriented east–west. The statistical reliability of shorter fluctuations of the meteorological characteristics available in the spectra in Figure 3 (periods 8, 6 and 4 h) have low coherence, which can attest to the presence of nonlinear relations between these fluctuations in the atmosphere above the lake.

Lake level fluctuation

Level fluctuations are always present in natural waters (also in the seas) and have been well described (for example, Bergamasco & Gacic, 1996; Parsmar & Stigebrandt, 1997). The synchronous fluctuation of wind and lake level anomalies, which were measured in Cecember 1996 at the Chapala climatic station, indicates the continual occurence of well-expressed seiches (see Fig. 2). Fluctuations in the lake level are irregular in time. On average, they do not exceed ±10 mm. However, for separate days they reached ±35 mm. Sometimes, over short intervals of time, they looked like free harmonic fluctuations with a period of approximatly 6 h, as shown in Fig. 2c,d.

Cross analysis with wind and atmospheric pressure fluctuations has shown that sharply amplified seiches always after pressure increases which are caused by synoptic processes. A pulse of pressure acting on Lake obviously causes an inclination of its level on one of its parts (because of the large linear size of the lake). And then this pulse generate long free gravitational waves causing horizontal currents and level fluctuations (Filatov, 1983). Undoubtedly, the lake breeze also causes fluctuations of lake level with daily periodicity, causing a 'wind tide' on the coast in the afternoon and at its centre in the night. The frequency spectra (Fig. 3) shows that the energy transmitted to the water mass by atmospheric pressure (and wind), acts on it, basically, with synoptic periodicity and also with periodicities of 24 and 12 h. The Lake acquires this energy which eventually results in free lake level fluctuations, with periods of 5.7 h as well as forced fluctuations ('wind tide') with periods of 24 h. The mean square amplitudes of free fluctuations was estimated as 3.5 mm and the amplitude of the forced fluctuation was estimated as 1.3 mm. Fluctuations in lake level occur with one or several nodal lines, which are determined by the lake's bottom topography. The fluctuations gradually decay under the action of the frictional forces. The basic equations describing seiches follow (LeBlond & Mysak, 1978):

$$\frac{\partial u}{\partial t} - fv = -c^2\frac{\partial \eta}{\partial x} + \tau_x, \tag{11}$$

$$\frac{\partial v}{\partial t} - fu = +c^2\frac{\partial \eta}{\partial y} + \tau_y, \tag{12}$$

$$\frac{\partial u}{\partial x} + \frac{\partial v}{dy} + \frac{\partial \eta}{\partial t} = 0, \tag{13}$$

where x and y are the horizontal coordinates, u and v are the corresponding components of the vertically-integrated current, t is a time, $\eta(x,y)$ is the free surface perturbation, f is the Coriolis parameter, $c = \sqrt{gH}$ is the speed of propagation of long gravity waves, g is the gravitation constant, H is the average depth of water, τ_x, τ_y are the horizontal components of the force which causes seiches.

Solving Equations (11)–(13) can be accomplished only by numerical methods because their solution depends on the lake form. Let's make some estimates on the seiches without fully solving the given system of the equations. In long, narrow and shallow lakes ($H/L \ll 1$, L is the length of the lake), such Lake Chapala, it is possible to neglect geostrophic effects and vertical acceleration. Then the solution for the current speed u (v, the cross component, is very small), the free surface perturbation, and period of the seiches can be approximated under the following formulas (Graf & Mortimer, 1978; Filatov, 1983):

$$u = u_{max} \sin(n\pi x/L) \sin(2\pi t/\tau_s), \tag{14}$$

$$\eta = A_s \cos(n\pi x/L) \cos(2\pi t/\tau_s), \tag{15}$$

$$\tau_s = 2L/(n\sqrt{gH}). \tag{16}$$

u_{max} is the maximal current through a nodal line; A_s is the seiche amplitude at the edges of the lake; n is the number of seiche nodes. Equation (16) is known in the literature by the name of 'Merian formula' (LeBlond & Mysak, 1978).

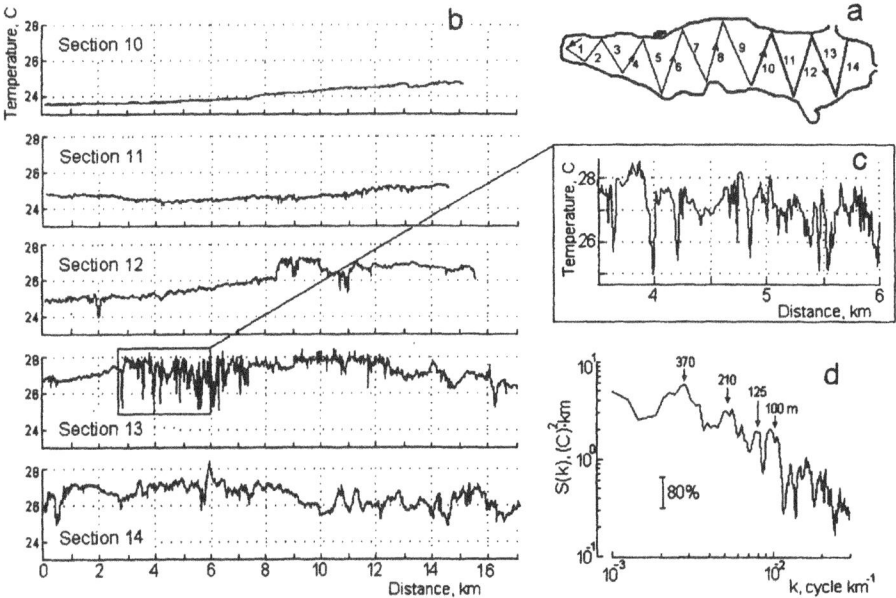

Figure 5. (a) Number and distribution of transects. Towing made from the 18–19 of September 1996. (b) Surface temperature at transects 10–14. (c) The section that especially recorded intense fluctuations in temperature and their spatial spectrum (d). The vertical line in the spectrum indicates the 80% confidence limit.

Let's estimate some parameters of the seiches in Lake Chapala, using the almost pure harmonic fluctuations (Fig. 2c) having an average heigh (the double amplitude) of 15 mm and a period of 6 h. These fluctuations should undoubtedly cause one-nodal seiches. The estimation of the period of such a seiche using Formula (16) gives a value of 5.4 h. Though this value is slightly less than the one described above by spectrum analysis. However, it can be used to calculate the data below.

Let's consider that the lake has a rectangular form, in length 75 km and a constant depth H=6 m. The first nodal line for such geometry should occur at the centre of the lake. Assuming a lake level inclination caused by seiches, a straight line is used as shown in Figure 4. To estimate the seiche amplitude at the western edge of the lake. First the angle of inclination of the free surface is: $\tan \gamma = (A_L/a)$. Here, $a = 15.5$ km and refers to the distance from the centre of lake to the point of arrangement of the limnograph (look at Fig. 4); $A_L = 7.5$ mm; is the seiche amplitude at the point of measurement. A_S represents the seiches amplitude at the western edge of the lake. From here we shall find, that $A_S = \tan \gamma \cdot L/2 = \tan \gamma \cdot (a+b) \simeq 18.3$ mm, where $f = L/2 - a = 22$ km is the linear distances, as shown in Figure 4a.

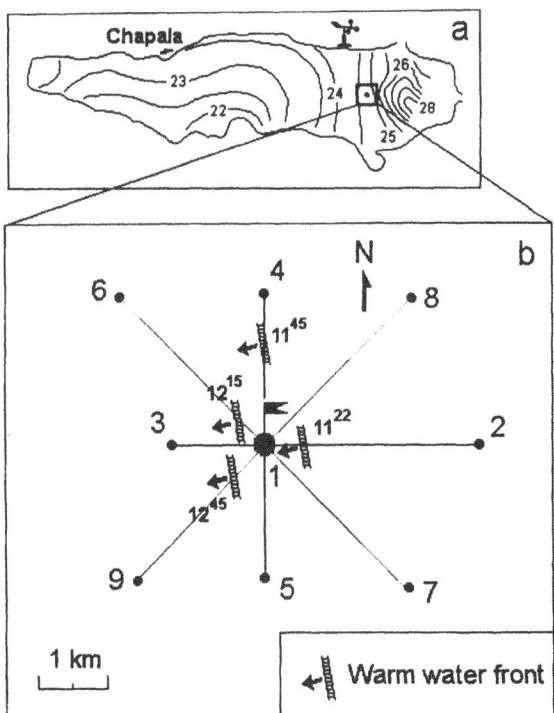

Figure 6. (a) Distribution of Lake Chapala's surface temperature. It represents the results on 18 and 19 of September 1996. With the help of a symbol the location of meteorological station is shown. (b) The measurement scheme of surface temperature in the micropolygon on 26 September, 1996.

The inclination level caused by seiches moves every 5.4 h a volume of water equal to $v = (L/2 \cdot A_S)/2 \approx 341$ m^3, for example in the west part of the lake. The displacement of such volume corresponds to one meter of the lakes width. Hence, the speed of current from the surface to the bottom in the area of a nodal line can be calculated as $u = v/(\tau_S/4)/H \simeq 0.012$ m s^{-1}. This value is comparable with the value of the drift current in Lake Chapala (Simons, 1984; Filonov & Tereshchenko, 1999b). On this basis the first mode seiches should cause significant fluctuations in level in the western and the eastern parts of the lake, and intensive horizontal currents in its central part. Here we use the very simplified model of the form of Lake Chapala. Although simplistic, the model is of great value for researchers wanting to study currents and pollution dynamics in the lake.

Thermal structure and lake circulation

Experimental details

In September 1996, researchers from the University of Guadalajara carried out an experiment at Lake Chapala. The aim of the experiment was to measure the processes of vertical and horizontal mixing, as well as the processes of horizontal circulation (Filonov & Tereshchenko, 1997). During the experiment, vertical temperature profiles were measured, using a CTD SBE-19. Measurements of the currents from the surface down to the bottom were also taken at 16 points uniformly distributed. Surface temperature measurements with an accuracy of 0.001 °C were also made using a towed CTD (Filonov et al., 1996). The survey consisted of 14 cuts perpendicular to the coast, that uniformly covered the lake by zigzagging from west to east (see Fig. 5a).

To investigate temporaral temperature variability in the easternn part of the lake, measurements were taken at 5.8 m using autonomous SBE-16 instruments (the accuracy of the measurement is similar to the SBE-19). The instruments were placed at 1.3 and 5.1 m below the water surface. The data were recorded for 8 h with a discretization of 15 sec. At the same time, temperature profiles were made near the water bouy from the boat every 5 min with a resolution of 0.05 m. In addition, horizontal currents were measured every meter from the surface to the bottom. To study surface temperature fluctuations in more detail, a fast survey was conducted close to the water bouy on 26 September, 1996 with the help of a SBE-19 (Fig. 6). The survey consisted of two crosswise sections forming an star and the angles between rays is equal to 45° extending from the centre (see Fig. 6).

Basic features of thermal structure and Lake circulation

Some results concerning the thermodynamic regime of Lake Chapala have been described by several authors (Filonov et al., 1998; Filonov & Treshchenko, 1999a, 1999b). Such results can be briefly summarized as follows. Measurements have shown that Lake Chapala (in spite of its shallowness) has a quite complex thermodynamic regime and a significant vertical stratification. The temperature difference between the surface and the bottom at the centre of the lake was 0.5–1 °C, while at the eastern part it was 2–3 °C. The warmest water was located in the eastern part of the lake, where the shallow waters (0.5–2 m) occur. There turbid waters from the Lerma River are quickly heated. Thus given calm periods, forming a sharp thermocline forms with gradients from 1–2 °C m^{-1}. Waters from the central and southern parts of the lake are 1–2 °C cooler than waters from the northern shore (see Fig. 6a). At the central, western and southern parts of the lake, the vertical temperature gradients are rather weak, reaching just 0.05–0.2 °C m^{-1}. However, in the eastern shallow waters, temperature gradients in the top surface layer (1–2 m thick) can reach unusually high magnitudes, up to 3 °C m^{-1}. This high stratification eventually interferes with both gas and momentum exchange between surface and bottom waters, thus resulting in water stagnation.

Measurements carried out in the early morning hours, when winds were absent, showed the presence of cyclonic circulation with speeds ranging from 0.02–0.03 m s^{-1} at the surface and 0.01–0.02 m s^{-1} in the deeper layers. The lake's mean circulation with weak eastward wind has enhanced the westward currents,and stronger eastward compensation flows in the central part of the lake (Simons, 1984; Filonov & Tereshchenko, 1999b). Together they form a cyclonic circulation in the north part of the lake and anticyclonic circulation in the southern part. It also leads to the formation of a current that moves warm surface water from the eastern region along the northern shore to its western part, and causes downwelling at the northern boundary of the lake and upwelling at the southern boundary where the cool water centre was observed.

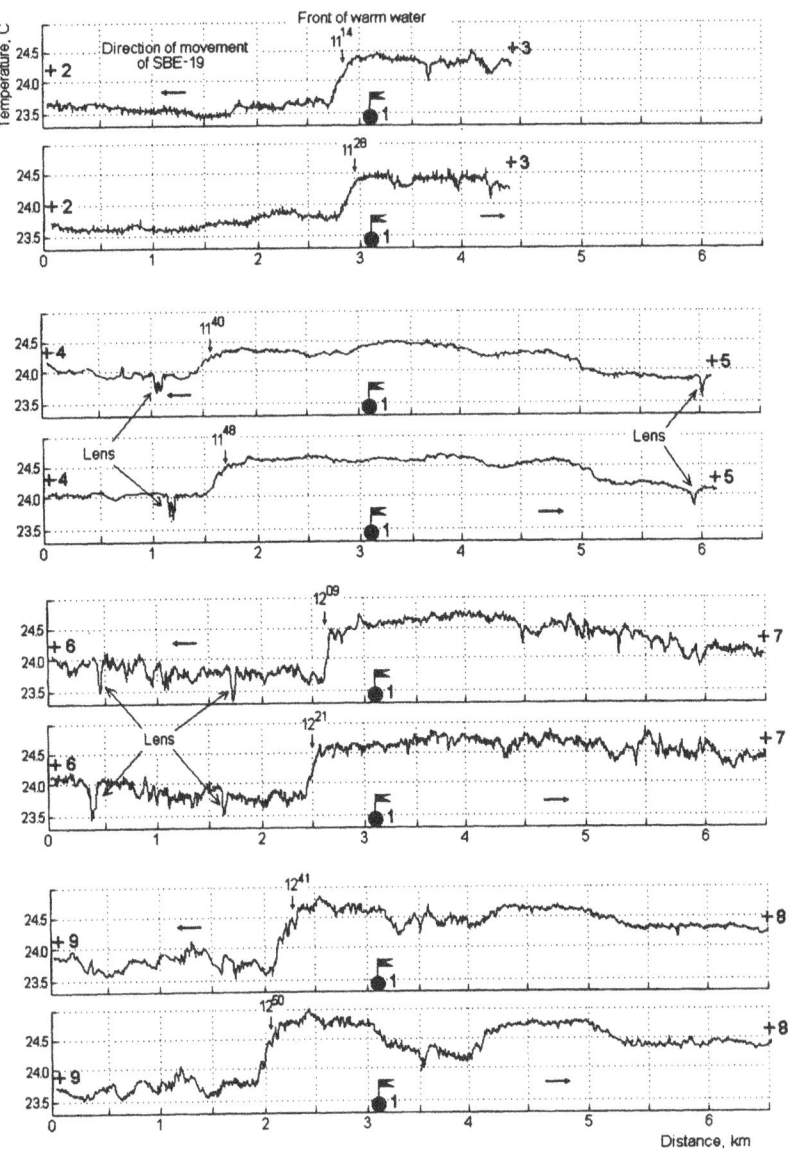

Figure 7. Spatial fluctuations of surface temperature in the micropolygon on 26 September, 1996. Bold points with a tag show water buoy location, relative to the sections of the temperature field. The cross symbol represents both starting and ending points of the appropriate sections of surface temperature. The horizontal arrows show towage direction. The vertical arrows show the location of the warm front on various sites in the micropolygon; the time of the front crossing is given beside it.

Speeds of depth-integrated currents change from 0 to 0.01 m s^{-1}, depending on the region of the lake. Results from instrumental measurements turned out to be almost equal to results of model calculations. Differences in speed values occur due to the fact that, generally, the speed of the currents in the subsurface layer are always greater than the vertically integrated speed. Close to noon, there is usually a land breeze circulation over the lake. During these periods, the current speed of the thin subsurface water layer influenced by the wind, was significant. Its magnitude, based on the transport speed of water lilies with a submerged root systems of 0.4–0.5 m, was estimated to be 0.2–0.3 m s^{-1}.

Towed CTD measurements showed a peculiarity in the distribution of surface temperature that had not been observed before. In the eastern part of the lake, at transects 12, 13 and 14, spatial variation of

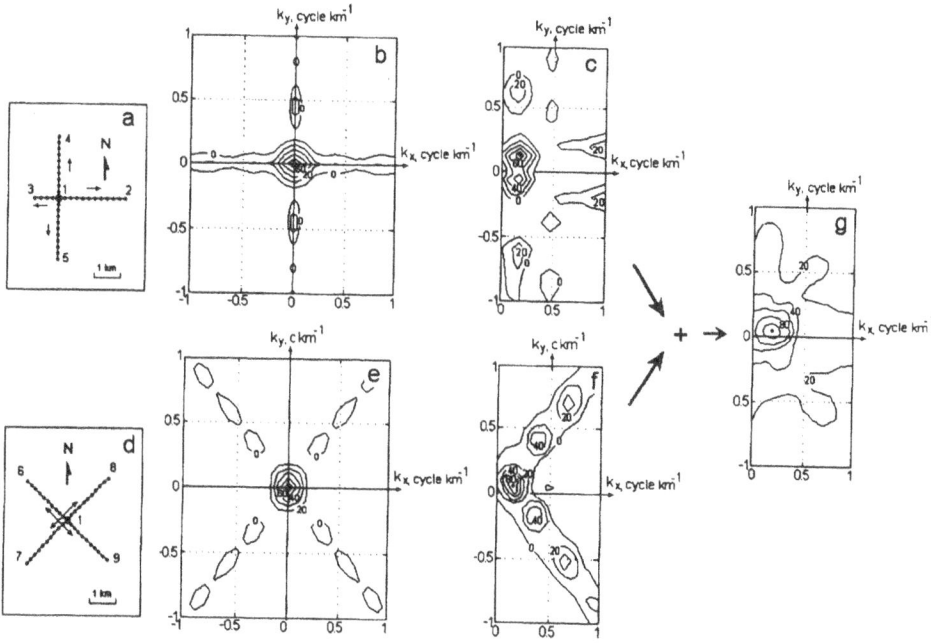

Figure 8. Sectional forms used in the spectra calculation (a,d) and spectral windows (b,e). A two-dimensional normalized spectrum from surface temperature fluctuations, particularly from the micropolygon's data of the two crosswise section pairs (c,f) and their average spectrum (g). The relief in the spectral density is shown as a percentage of the maximum value.

1–2 °C were recorded. The largest temperature variations reached 3 °C in the form of cool water patches (Fig. 5) only 50–100 m wide. The most intense group of patches was recorded at transect 13, which is shown in the spatial spectrum (Fig. 5d) indicating that the temperature patches of the group had a spatial scale of 100–400 m. The physical nature of these spatial variations of the surface temperature will be analysed below.

Temperature fluctuations in the eastern part of the lake

Spatial fluctuations

The micropolygon over which the measurements of the detailed structure of spatial fluctuations of surface temperature were made was located in the eastern part of the lake, with an average depth of 6 m (Fig. 6a). The micropolygon is represented by lines of approximately 3 km length. At its centre the water bouy was established, from which two crosswise sections (in opposing directions) with the towed CTD were made (see Fig. 6b). During these measurements, the following weather conditions prevailed: clear sky, westward wind up to 7–9 m s^{-1} and waves of approximately 0.3–0.4 m height. The measurements on cuts totaling 45 km length were made over 1 h 52 min (beginning at 11:13 h, ending at 13:05 h), with an average towing speed for the temperature sensor of 24 km h^{-1} (6.7 m s^{-1}). The boat position was recorded by means of a GPS.

Based on results from the micropolygon (Fig. 7) the surface temperature field has an inhomogenious spatial distribution. On all sections, the temperature front divides two water masses distinguished by temperature differences of up to 1 °C and an interface width of 200–300 m. Curved lenses of cool water have widths ranging from 50 to 100 m and horizontal temperature gradients of up to 0.01 °C m^{-1}. The thermal front; found in the micropolygon; will be named the 'warm water front'. This front was crossed by the towed sensor on each section (Fig. 6b), this allowed estimation of its horizontal speed and examination of its evolution. The speed of the warm front at the different sites of the micropolygon changed from 0.12 m s^{-1} (sections 2–3), to 0.18 m s^{-1} (sections 6–7). On average, its speed was 0.15 m s^{-1}. All sections focused along the direction of the warm front (sections 2–3, 6–7 and 8–9) recorded significant spatial temperature fluctuations with lengths from a hundred meters up to several kilometers, as well as fluctuations as short as tens meters. There are lenses of cool water with

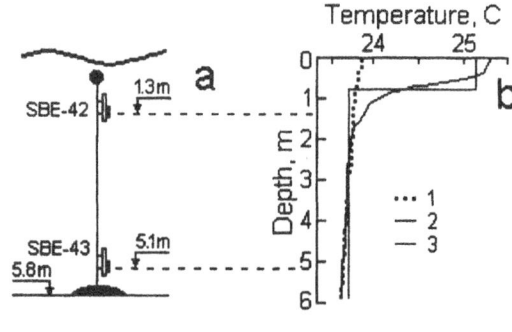

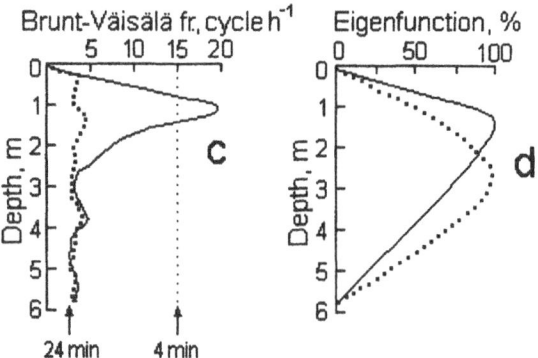

Figure 9. (a) The scheme of the mooring; (b) vertical structures of temperature; (c) Brunt–Väisälä frequency and (d) engenfunction for linear internal waves of the first mode at the period of 4 min. 1 – Curve obtained on vertical sounding at 8 h; curve 2 at 13 h of local time on 26 September, 1996 and curve 3, two-layer fluid approximation.

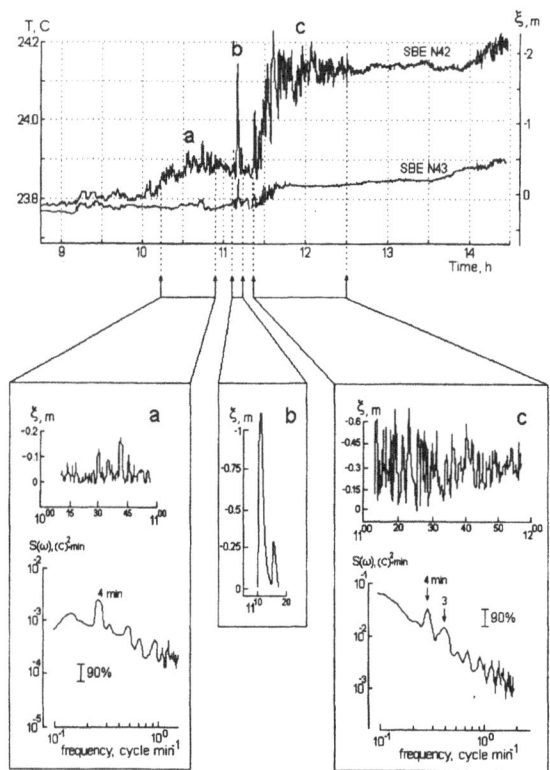

Figure 10. Temporal fluctuations of temperature at the mooring. Letters mark groups of internal waves. At the bottom part of the figure, these groups of waves are shown in expanded scale, together with the frequency spectra.

width up to 50 m and temperatures less than 0.5 °C in comparison with the backgound. On sections parallel to the front, the pattern of spatial fluctuations in temperature differs strongly. The curves look smooth and have no short components. This results suggests that all temperature fluctuations were due to a wind having westward direction. The similar of structure suggests wind generation at the initial stage of its development (LeBlond & Mysak, 1978).

The two pairs of the cross-shaped spatial sections made in the micropolygon during a short interval of time, enable estimation of the two-dimensional spectrum of the superficial temperature field. The following estimator (Konyaev, 1990) was used:

$$S(k_x, k_y) = \text{Re}\left[\overline{C(k_x) \cdot C(-k_y)}\right], \qquad (17)$$

where the bar at the top denotes ensemble averaging. Re denotes the real part of the complex expression in the square brackets. Expressions:

$$C(k_x) = \int_L \xi(x, 0) \cdot \exp(-i2\pi k_x x)dx \qquad (18)$$

and

$$C(-k_y) = \int_L \xi(0, y) \cdot \exp(-i2\pi k_y y)dy \qquad (19)$$

are the amplitude spectra of the sections of the field $\xi(x,0)$ and $\xi(0,y)$; L is the length of the spatial section; x, y are the space coordinates; k_x, k_y are the wave number components or spatial frequency (cycle km^{-1}). Several pairs of crosswise sections permit finding the wave length and its two possible directions (because of the central symmetry of a spectrum).

To calculate the two-dimensional spectrum, spatial sections of the temperature field were smothed by a bandpass filter. Fluctuations with lengths less than 200 m and more than 6 km were removed from the analysis. Then sections of the temperature field were transformed into discrete spatial values by cubic spline imterpolation (see Fig. 8a,d). After that, two estimations of the two-dimensional spectrum (see Fig. 8c,f) and their average (see Fig. 8g) were calculated. For the analysis of the acheived results from each pair of sections, their spectral windows (i.e., the response of the computational algorithm to a spatial harmonic of infinite length and unit amplitude) were also calculated (see Fig. 8b,c). It can be recognized that the two-dimensional spectrum has only one peak whose width exceeds the width of the main peak for both spectral windows. That variability in the temperature field was caused by several fluctuations of similar length and direction. This group of fluctuations was focused in the wind direction, i.e., westward. Their average length was about 4.7 km (0.213 cycle km^{-1}).

Temporal fluctuations

Analysis of the temporal temperature fluctuation shows significant variability in the lake's top layer (see Figs 9 and 10). As can be seen, it was possible to record the transport of the layer's internal thermal bore with instruments, which generates intense internal waves. At the beginning of the observation period (06:35 h, September 26, 1996), the area of the water bouy was weakly stratified and had a drop in temperature of only 0.3 °C between the surface and the bottom. Starting at 10:00 h, the instruments recorded a warm water lens no thicker than 1.5 m. There was a group **a** in Figure 10 of internal waves at the front part of the bore, which is the group of precursory waves that generally precede the principal group of internal waves which in turn are generated by the internal bore that moves in shallow water (Thorpe, 1971; Turner, 1973; Ostrovsky & Stepanyants, 1989; Thorpe et al., 1995; Thorpe, 1998). On average, these waves had periods of approximately 4 minutes (see Fig. 10a). Their amplitudes did not exceed 0.5 m as estimated with the formula:

$$\Delta \xi_z(t) = \xi_z(t_1) - \xi_z(t_0) = [T(t_1) - T(t_0)]/(\overline{dT/dz}),$$
$$(20)$$

where $\xi_z(t_0)$, $\xi(t_1)$ are the vertical displacements of the water layers at level z at times t_0, t_1, which cause the fluctuations in the temperature $T(t_0)$, $T(t_1)$; $\overline{dT/dz}$ is the average vertical temperature gradient during the time period t_1–t_0.

At 11:10 h, a solitary nonlinear internal wave (**b** in Fig. 10) as in a Korteveg–de Vries (KdeV) soliton with an amplitude of 1.5 m passed through the water bouy location. It lasted only 4 min. Approximately 10 min

later, an abrupt forward front of the warm water bore began to develop, this was accompanied by long series of internal waves (group **c** in Fig. 10) with amplitudes of 0.5–1 m, and which lasted almost 1 h. This group of internal waves consists of periodic fluctuations from 3–4 min. As a result of the onset of the warm bore, the water temperature at 1.3 m rose by 0.3 °C and nearly 2 °C at the surface.

To estimate the variability in the vertical stratification close to the water bouy, the Brunt–Väisälä frequency was calculated:

$$N^2(z) = \left(\frac{g}{\rho}\right) \cdot \left(\frac{\partial \rho}{\partial z}\right),$$

z is depth, $\rho = \rho(z)$ density of lake water, which was calculated under the formula:

$$\rho = \rho_o(1 - 1.96 \cdot 10^{-6}(T - 289)(T - 4^2)/ \quad (21)$$
$$(T + 68.1) \cdot 10^3),$$

which is usually used for shallow lakes (Filatov, 1983). The vertical temperature structure and the Brunt–Väisälä frequency for soundings made before and after the passage of the internal bore are given in Figure 9b,c. It can be seen that in the beginning of the experiment the lake water mass was poorly stratified meaning that the Brunt–Väisälä frequency did not exceed 2–3 cycles h^{-1}. After the passage of the internal bore in the water bouy, a sharp thermocline was formed. The maximal values of the Brunt–Väisälä frequency were near 20 cycle h^{-1} at 1.5 m depth.

Linear internal wave parameters in group **C** were calculated by the numerical solution of the wave equation (LeBlond & Mysak, 1978, Konyaev & Sabinin, 1992):

$$\frac{d^2W}{dz^2} + k^2\frac{N^2 - \omega^2}{\omega^2 - f^2}W = 0, \quad (22)$$

where W is vertical velocity and $W=0$ at $z=0$ and $z=H$; ω is the internal wave frequency; f is the Coriolis parameter; k is the wave number. The internal waves calculation were made for periods of 24 min (2.5 cycles h^{-1}). The vertical structure data recorded before and after the passage of the thermal bore, were used for the calculation. Appropriate eigenfunctions are given in Figure 9d. The figure shows that the eigenfunction maximum was displaced from 3 to 1.5 m, that is, on the depth of the maximum Brunt–Väisälä frequency. The wave number and phase speed for the first mode were found just for periods of 4 min. They

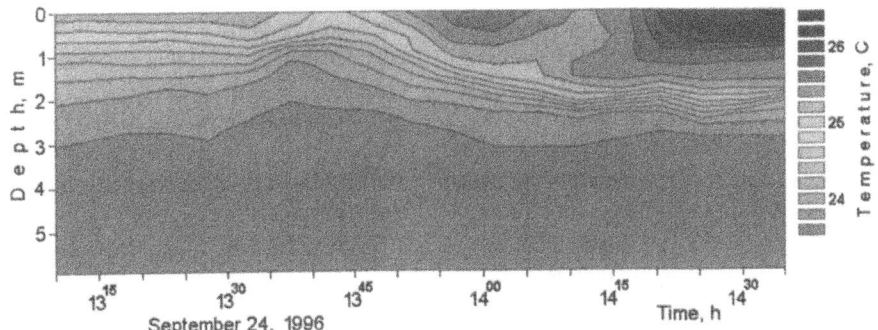

Figure 11. Temperature fluctuations at mooring, measured every 5 min using a SBE-19 Profiler.

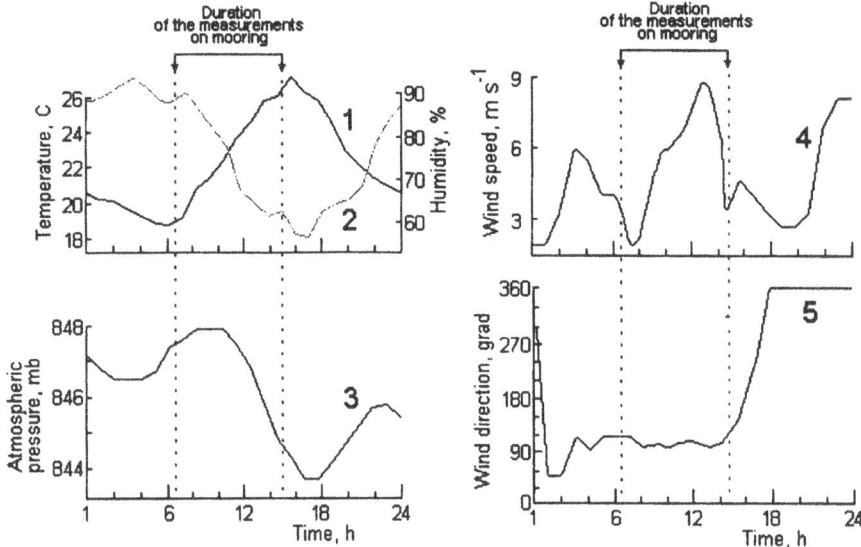

Figure 12. Fluctuations in the meteorological characteristics on 26 September, 1996. (1) air temperature, (2) relative humidity, (3) atmospheric pressure, (4) wind speed and (5) wind direction.

had a length of 14 m, and a phase speed of approximately 0.058 m s⁻¹, which was four times slower than the speed of the thermal bore, by which the measured group of waves **C** was determined. In fact, this group of waves could be distributed in the narrow waveguide within a thickness of approximately one meter, which are essentially formed by high values of the Brunt–Väisälä frequency at the lake's top layer after the intrusion of the thermal bore.

Since our measurements were made in shallow water, the non linear KdeV model was used to estimate the phase speed of a solitary internal wave (see fluctuation b in Fig. 10). The KdeV equation is (Hunkins & Fiegel, 1973; Ostrovsky & Stepanyants,

1989; Pelinovsky, 1996; Grimshaw, 1998):

$$\frac{\partial \xi}{\partial t} + C_0 \frac{\partial \xi}{\partial x} + \alpha \xi \frac{\partial \xi}{\partial x} + \beta \frac{\partial^3 \xi}{\partial x^3} = 0, \qquad (23)$$

where ξ is vertical displacement, C_0 is the linear phase speed, t is time and x the horizontal coordinate. For the model we used a two-layer approximation of the vertical stratification from the micropolygon. At the moment of intrusion of the thermal bore, the thickness of the upper and lower layers are h_1 and h_2 and densities are ρ_1 and ρ_2. For this case the expressions for phase speed follow from the equation (23) (Kakutani & Yamasaki, 1978; Volzinger et al., 1989):

$$C_0 = \left(g \frac{(\rho_2 - \rho_1)}{\rho_2} \frac{h_1 h_2}{h_1 + h_2} \right)^{1/2}, \qquad (24)$$

Table 3. Variables for a two-layer fluid (on a micropolygon)

Number of layer	Thickness of layer (m)	Temperature, (°C)	Density of layer (kg m^{-3})
1	0.8	25.2	997.02
2	5.2	23.3	997.49

$$C = C_0 \left\{ 1 + \frac{a}{2} \left(\frac{h_1 - h_2}{h_1 h_2} \right) \right\}, \qquad (25)$$

where C is the phase speed of the internal soliton, and a is the amplitude. It also follows, that $a>0$ for $h_1 > h_2$ and $a<0$ for $h_1 < h_2$ (Kakutani & Yamasaki, 1978; Holloway, 1987). The parameters of a two-layer fluid for the micropolygon are given in Table 3. Analysis of such data yields estimates of phase speeds $C_0 = 0.061$ m/s, and $C = 0.12$ m/s. (Note: this value of phase speed is close to that obtained by the above numerical estimate ($C_0 = 0.058$ m s^{-1}). As expected, the phase speed of the internal soliton is higher than that of a linear wave. Therefore, it should 'escape' faster from the vertical wall of the thermal bore, its generation area.

Repeated vertical soundings of the temperature (every 5 min), carried out during 90 min at the end of the experiment, documented the intrusion of another thermal bore, with warmer water than in the first bore. Its intrusion is appreciable in the last part of the temperature series recorded by the SBE-42 (see Figs 10 and 11). In total, for 1 h (from 13:30 h on September 24, 1996) the vertical gradient of temperature at 1.75 m rose from 0.4 to 2.8 C m^{-1}, and the surface temperature from 25.2 to 26.3 °C.

The internal bore was caused by the onset of the water lenses (patches) close to the water buoy site, moving westward, with speeds of 0.1–0.20 m s^{-1} from a region with significantly shallower depths, where the early morning lake breeze enhanced internal wave generation. This usually occurred from 09:00 h. until 10:00 h. At this time the wind close to the water bouy reached a speed generation of 5–8 m s^{-1} in a steady westward direction (see Fig. 12). At the beginning of the experiment, the current at the water bouy was close to zero. However the current quickly developed as the lake breeze started to build up. By 11:00 h the westward current reached the lake's surface with a speed of 0.1–0.2 m s^{-1}. Moreover, the current speed decreased exponentially with depth. At 4 m the speed did not exceed 0.01–0.02 m s^{-1}. By 15:00 h, a floating

aquatic lily island (*Eichhornia ctassipes*) and bulrush (*Typha latifolia*), with dimensions of approximately 50 × 50 m approched the instrumentation, thus threatening the equipment and the experiment had to be stopped.

The structure of surface temperature patches was evidently caused by the following mechanisms. The predominantly westward early morning lake breeze that blows from the shore and, in this part of the lake, causes the displacement of the thin (only 1–1.5 m) warmer surface water layer from the wide shallow region of the lake bordering the mouth of the Lerma River towards the cooler central part. A horizontal and vertical Langmuir-type circulation system of converging and diverging flows is generated in the upper layer; they initially disrup the warm water layer and then continues as lenses (patches) separated and extended by the wind, which due to the instability of their displacement, generate in their periphery groups of intense internal waves that transmit heat and kinetic energy from the wind towards the interior of the lake (Thorpe, 1992). Since the lake breeze circulation occurs over the entire lake throughout the year, it can be conjectured that it plays an important role in the mechanisms of vertical and horizontal mixing of the water mass.

Conclusions

A great deal of results presented here are unique for Lake Chapala. Altough this research was carried out for a short period of time, on everage it reflects the characteristics of dynamic processes accurring in the lake and its surrounding environment. The chief dynamic process occurring in the lake is lake breeze circulation. Day time breeze does not usually exceed 4 m s^{-1}. Beyond any doubt the lake breeze increases lake surface evaporation. Lake breeze and pressure variations generate free waves-seiches. The spectral analysis of lake level fluctuations from the limnograph shows that the period of free seiches is equal to 5.7 h, and its amplitude is equal to 7.5 mm. The seiches generate periodic currents which can reach 1 cm s^{-1} in the area of the nodal line.

The CTD measurements suggest a complex thermodynamic regime of lake Chapala, and a significant vertical stratification. The temperature differences between the surface and the bottom waters ranged from 0.5 to 1.0 °C at the lake's centre, and from 2.0 to 3.0 °C at the eastern part. In fact, the warmest water occurred at the eastern section where the lake is shallow and the water is turbid due to the Lerma River discharge (depth here ranges from 0.5 to 2.0 m). During calm weather, the water quickly heats up developing sharp thermicline gradients of 1–2 °C m^{-1}. The water of the central and southern parts of the lake are 1–2 °C cooler than the ones at the northern shore.

The CTD towing yielded particular features of surface temperature that has not been described before. For example, at the eastern and particularly at transects 12, 13 and 14, the instrument recorded spatial temperature variations of 1–2 °C. The largest variations recorded a value of 3 °C and occurred in the form of cold water patches 50–100 m in width. A great number of patches were recorded at transect 13 with spatial variation of 100–400 m. Larger variations of about 4.7 kms were also recorded, and they were oriented on wind direction.

A buoy station registered temperature fluctuations of 2 °C that belonged to the movements of the internal thermal front. A well-defined group of internal waves were also recorded at the leading edge of the front. They had the form of KdeV solitons and an amplitude of 1.5 m. It is implied that the front by the buoy station was generated by the movement of warm water travelling from the eastern part of the lake, where shallow water occur. The main force for warm water movement was morning breeze.

Dynamic processes in the lake and neighbouring areas are complex and variable in time and space, and they seem to be governed by local synoptic processes as well as orographic features. However, it is uncertain to fully understand the formation mechanisms of the observed currents and the thermal structure of the lake from the results of just one experiment. A full picture of the lake's thermodynamic would have been made possible by establishing a wide range of meteorological stations within the area to monitor the parameters at different seasons of the year. It is also important to study the character and the interaction of breeze circulation with valley winds in different parts of the lake. Particular attention should be given to breeze penetration depth when it is landward and then travels back to the lake. How lake breeze influence the water mass dynamics is the other phenomena to study. Finally it would be particularly interesting to study local peculiarities of the energy transfer mechanism from the main diurnal harmic to high-frequency oscillations, as well as the role of non-linear energy transformation.

Acknowledgements

This paper presents the results of research conducted as part of the project 'Integral Study of Ecological System of Lake Chapala, for its Preservation and Advantageous use'. I thank Dr Manuel Guzmán A. from the University of Guadalajara for his support in our investigations. The author is grateful to the CNA (Comision Nacional del Agua), particularly to engineers Carlos Alberto Hernandez Solís and Salvador Tínoco Gutiérrez for providing the data. Finally the author express gratitude to the MSc. Arturo Figueroa M. from the Department of Physics, University of Guadalajara for editorial help and help in translation of the article.

References

Bendat, J. S. & A. G. Piersol, 1967. Measurement and Analysis of Random Data. Wiley & Sons, New York: 409 pp.

Bergamasco, A. & M. Gacic, 1996. Baroclinic response of the Adriatic Sea to an Episode of Bora Wind. JPO, 26: 1354–1369.

Burman, E. A., 1969. Lacal Wind. Hydrometeorizdat, Leningrad: 344 pp.

Burman, E. A., G. P. Ivus & A. E. Filonov, 1983. The use of the spectral analysis for estimation of some characteristics of breeze structure. Izvestia. Atmosph. and Ocean Physics. 19: 376–381.

Filatov, N. N., 1983. Dynamics of Lakes. Hydrometeorizdat, Sant-Petersburg: 167 pp.

Filonov, A. E., C. O. Monzon & I. E. Tereshchenko, 1996. A technique for fast conductivity–temperature–depth Oceanographic surveys. Geofisica Internacional. 35: 415–420.

Filonov, A. E. & I. E. Tereshchenko, 1997. Preliminary results on the thermic regime of Lake Chapala, Mexico. Suppl. to EOS, Transact., AGU 78: 46.

Filonov, A. E., I. E. Tereshchenko & C. O. Monzon, 1998. On the Oscillations of the Hydrometeorological characteristics in the region of Lake Chapala in time frames of the days to tens of years. Geofisi. Int. 37: 293–307.

Filonov, A. E., 1998. Morira el lago de Chapala? Existe la pocibilidad de salvarlo. Teorema. 3: 16–18.

Filonov, A. E. & I. E. Tereshchenko, 1999a. Thermal Fronts and Nonlinear Internal Waves in Shallow Tropical Lake Chapala. Russ. Meteorol. Hydrol. 1: 94–102.

Filonov, A. E. & I. E. Tereshchenko, 1999b. Thermal lenses and internal solitons en Lake Chapala, Mexico. Chin. J. Oceanol. Limnol. 17(4): 308–314.

Gonella, J., 1972. A rotary-components method for analysis meteorological and oceanographic vector time series. Deep Sea Res. 19: 833–846.

Graf, W. H. & C. H. Mortimer (eds), 1978. Hydrodynamics of Lakes. Elsevier, Amsterdam: 360 pp.

Grimshaw, R., 1998. Internal solitary waves in shallow seas and lakes. Phys. Processes Lakes Oceans: coastal and estuerine studies. 54: 227–240.

Holloway, P. E., 1987. Internal hydraulic jumps and solitons at a shelf break region on the Australian North West Shelf. J. Geophys. Res. 92: 5405–5416.

Hunkins, K. & M. Fiegel, 1973. Internal undular surges in Seneca Lake: a natural occurrence of solitons. J. Geophys. Res. 78: 539–548.

Jaurégui, E., 1995. Rainfall fluctuations and tropical storm activity in Mexico. Erdrunde, Band 49: 39–48.

Jenkins, G. M. & D. G. Watts, 1969. Spectral Analysis and its Applications. Holden-Day, San Francisco: 672 pp.

Kakutani, T. & N. Yamasaki, 1978. Solitary waves on a two-layer fluid, J. Phys. Soc. Japon, 45: 674–679.

Konyaev, K. V., 1990. Spectral Analysis of Physical Oceanographic Data. A. A. Balkema, Rotterdam: 200 pp.

Konyaev, K. V. & K. D. Sabinin, 1992. Waves in the Interior of the Ocean. Hydrometeorizdat, Saint-Petersburg: 272 pp.

LeBlond, P. H. & L. A. Mysak, 1978. Waves in the Ocean. Elsevier, Amsterdam: 602 pp.

Mason, M., C. G. Guzkovska & F. A. Sreet-Perrot, 1994. The response of lake levels and areas to climate change. Climate change, 27: 124–136.

Mooers, C. N. K., 1973. A technique for the cross-spectrum analysis of pairs of complex valued time series with emphasis on properties of polarized components and rotational invariants. Deep Sea Res. 20: 1129–1141.

Ostrovsky, L. A. &, Yu. A. Stepanyants, 1989. Do internal solitons exist in the ocean? Rev. of Geophys. 27: 293–310.

Parsmar, R. & A. Stigebrandt, 1997. Observed damping of barotropic seiches throught baroclinic waves drag in the Gullmar Fjord. JPO 27: 849–857.

Pelinovsky, E. N., 1996. Tsunami waves hydrodynamics. Acad. Sci. Russia, Inst. Appl. Physics, Nigni Novgorod: 276 pp.

Riehl, H., 1979. Climate and Weather in the Tropics. Academic. Press. New York: 342 pp.

Sandoval, F., 1994. Pasado y futuro del lago de Chapala. UNED. Gudalajara, Mèxico: 94 pp.

Scorer, R. S., 1978. Environmental Aerodinamics. Elsevier, New York: 523 pp.

Simons, T. J., 1984. Effect of outflow diversion on calculation and water quality of Lake Chapala. Report Project MKX CWS-01: 23 pp.

Thorpe, S. A., 1971. Asymmetry of the internal seichesd in Loch Ness. Nature 231: 306–308.

Thorpe, S. A., 1992. The breakup of Langmuir circulation and the instability of an array of vortices. J. Phys. Oceanogr. 22 (4): 1–35.

Thorpe, S. A., 1998. Some dynamical effects of internal waves and the sloping sides of lakes. Phys. Processes Lakes Oceans: coast. estuar. Stud. 54: 441–460.

Thorpe, S. A., J. M. Keen; R. Jiang & U. Lemming, 1995. High frequency internal waves in Lake Genova. Phil. Trans. r. Soc. Lond. A 354: 237–257.

Turner, J. S., 1973. Buoyancy Effects in Fluids. Cambridge University Press Cambridge: 521 pp.

Volzinger, N. E., K. A. Klevanny & E. N. Pelinovsky, 1989. Long-Wave Dynamics of the Coastal Zone. Hydrometeoizdat, Leningrad: 272 pp..

Hydrobiologia **467**: 159–167, 2002.
J. Alcocer & S.S.S. Sarma (eds), Advances in Mexican Limnology: Basic and Applied Aspects.
© 2002 *Kluwer Academic Publishers.*

Interaction of water quantity with water quality: the Lake Chapala example

Owen T. Lind & L.O. Dávalos-Lind
*Limnology Laboratory, Biology Department, Baylor University and The Chapala Ecology Station,
Universidad Autónoma de Guadalajara and Baylor University, Waco, TX 76798, U.S.A.*

Key words: water quality, water quantity, pollution, turbidity, *Eichhornia*, phytoplankton

Abstract

Water quality may be significantly determined by water quantity. Lake Chapala, México is a large lake beset with numerous water quality problems. The decline in water volume over the past 20 years, a serious problem itself, is associated with causing or enhancing several problems of quality. Five such problems are explored herein. These are: extensive infestations of water hyacinth (*Eichhornia crassipes*), a declining native fishery, light limitation of phytoplankton production at the base of the food chain, shallow-water algal blooms resulting in water supply treatment problems, and the presence of toxic metals in the harvested and sold fishes.

Introduction

Water quality is a function of the volume of the receiving lake – dilution is the solution. But, the relationship goes beyond this concentration dependency. Many aquatic ecosystems worldwide – Aral Sea, Lake Chad, Lake Kinneret as examples – are experiencing declining volumes because of human over-exploitation of the resource (Hutchinson et al., 1992; Berman et al., 1997; Stone, 1999). In this paper, we provide examples of water quality effects brought on by declining water quantity of a large tropical lake – changes related primarily to decreased mixing depth and greater shoreline exposure. These examples are based on 15 years of investigations by our limnology group augmented with data from various Mexican government agencies.

Lake history and morphometry

Lake Chapala, México's largest lake, is located primarily in the state of Jalisco at approximately 20° north latitude and with a 1999 surface elevation of approximately 1520 m (Fig. 1). The lake's geologic history is not well known, Clements (1963) described it as a rift lake that is possibly the remnant of much larger 'inland sea' dating from the late Pliocene. If correct, it is one of the oldest lakes, if not the oldest, in North America. There is clear evidence of significant changes during and since the Pleistocene. The depths during the Illinoian and Wisconsin glacial periods are compared with the present in Figure 2. The lake is only a vestige of the original.

This lake receives drainage from a single river system – the Río Lerma. The drainage is one of México's largest. It arises in the east of the country near México City and flows generally from east to west into the lake. The original outflow was from the west end and into the Pacific Ocean. Tectonic uplifting during the mid-Pleistocene created a small mountain range damming the river and producing a second outflow, the Río Santiago, at the northeast corner of lake only 15 km from the Río Lerma inflow.

In modern times, the greatest lake depth reported was 14 m during the first part of the 19th century. No sediment studies have been completed, but the present morphometry, a flat basin with steep sides, suggests considerable loss of depth by sedimentation. With the decline in surface elevation, much of the lake bottom has been converted to recreational and agricultural uses. Table 1 with morphometric characteristics shows how this morphometry has resulted in a lake of greatly diminishing depth and volume, but little loss of surface area. A 3-m loss of surface elevation was accompanied by a 42% loss of volume, but only an 8% loss in

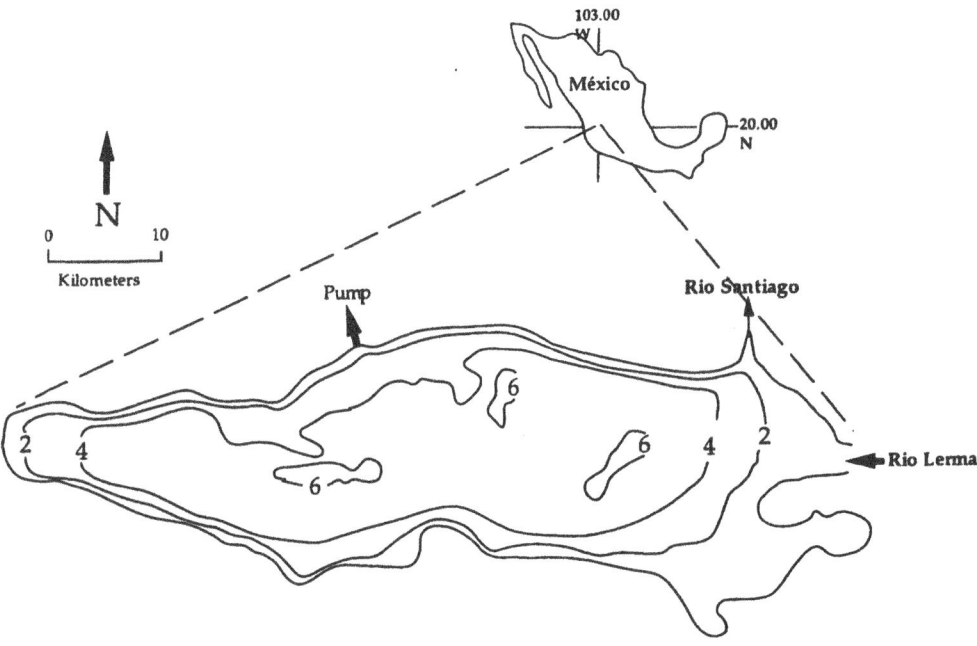

Figure 1. Lake Chapala, Jalisco, México. The depths shown are those when the surface elevation is 1520 m AMSL.

Table 1. Lake Chapala's morphometric variables for three periods of water quantity – the historical normal (pre-1977) 1524 m, declining (1982–1983) 1521 m and low (recent) 1519 m. (Data from Lind et al., 1997)

	Elevation (m amsl)		
	1524	1521	1519
Volume (10^6 m^3)	7962	4667	2700
Area (km^2)	1112	1039	930
Maximum length (km)	77	75	72
Maximum width (km)	22.5	22.5	19
Maximum depth (m)	10.5	7.5	5.5
Mean depth (m)	7.2	4.5	2.9
Relative depth	0.028	0.021	0.016

Note: Max. depth in 1816=14 m, & 1926=10 m (Clements, 1963).

surface area. As we discuss later, this volume to area relationship is about to change – with serious water quality consequences. The space-shuttle photograph (Fig. 3) allows for appreciation of the lake area when compared with the area of the city of Guadalajara, with population of approximately 5 million. With such a great area and shallow depth, this lake has one of the least relative depths (0.02%) reported.

Water balance

Being in the north tropics, this lake is subject to strong rainy-dry season cycles. A summer-fall rainy season of 4–5 months is followed by an extended winter–spring dry season. The 'saw-toothed' pattern in lake surface elevation seen in Figure 4 with an approximate 1-m annual change between seasons is typical. The consequence of consecutive years without a strong rainy season is evident in the 1981–1983 periods. Although such water-poor years produce dramatic changes, the consistent failure for the rainy season to make up for water extractions (which up to 1977 were in balance) has resulted in the lake's volume being reduced by one-half. A 50-year (1934–1984) water budget had the Río Lerma contributing 65% of the lake's water (Limon & Lind, 1990). Presently it contributes less than one-half. Economic and particularly agricultural development in the river's basin has significantly reduced the water available to the lake from the Río Lerma. At the same time, population growth of Guadalajara, which uses the lake as its primary water source, has greatly increased.

Water physics and chemistry

Water chemistry has been monitored at numerous lake sites by the Mexican Comisión Nacional del Agua and its predecessors for several decades. This monitoring has been sporadic primarily as a consequence of Mexico's economic conditions. Although valuable, much of these data were gathered without application

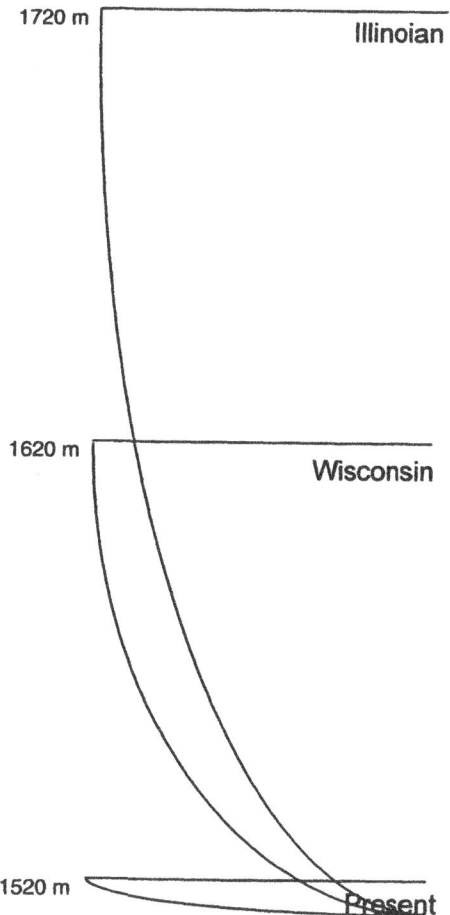

Figure 2. Illustration of the relative changes in Lake Chapala's depth during and since the Pleistocene.

Table 2. Mean±sd of physical and chemical variables and phaeophytin corrected chlorophyll *a* by region, June 1989 to July 1990, *N*=16. (Data from Lind et al., 1997)

	Eastern	Central	Western
Station depth (m)	1.0±0.2	4.9±0.4	3.9±0.4
Turbidity (NTU)	121±74	36±10	47±48
ISS (mg l^{-1})	114±71	25±15	38±40
Photic depth (cm)	51±25	92±7	88±21
NH$_3$-N (μg l^{-1})	113±126	68±41	65±35
NO$_3$-N (μg l^{-1})	235±187	188±138	235±136
POC (mg l^{-1})	19±9	12±3	12±5
DOC (mg l^{-1})	28±10	25±7	23±6
Chlor. *a* (μg l^{-1})	19±14	7±2	7±2

of modern quality control/quality assurance practices and should be so evaluated. Our investigations in 1982–1984 and 1988–1990 were to assess productivity – thus large scale water quality monitoring was not involved. Nevertheless, several variables thought to be associated with either phyto- or bacterioplankton production were measured with appropriate QA/QC and are summarized in Table 2. Because phytoplankton production in the lake is light limited (Lind et al., 1992), considerable attention was given to measurement of optical properties. Under bioassay conditions, phytoplankton production was nitrogen limited (Dávalos et al., 1989) thus the emphasis on this element rather than phosphorus which was always at high concentrations (up to 1 mg l^{-1}).

As described above, the lake is an elongate cul-de-sac with inflow and natural outflow in the same lake region. However, after the decline in water level starting in the late 1970s, the only outflow was by pumping for water supply to Guadalajara. The pumping station originally was near the mouth of the outflowing Río Santiago. In 1995, a new pumping station replaced it. The new station is located on the north shoreline near the town of Chapala, much nearer the west end of the lake. A hydrodynamic model was constructed for the lake based on the old outflow configuration (Simons, 1980, 1984). In this model, and with prevailing wind conditions (east to west), water and its contained dissolved materials circulates from east to west along the south lakeshore and returns to the east in the center of the lake. This circulation pattern is evident in the turbid gyre shown in Figure 3. A model of current circulation patterns is being made (deSentis, personal communication). We do not know the present distributions of river-borne nutrients and contaminants, but deSentis suggests that the strong pull by the pumps creates a mid-lake return current directed to the pumps ($\sim$7.5 m^3 s^{-1}).

Approximately one-half of the output from the lake is pumped by the new plant, thus, significant changes in current patterns could exist. Because of the relatively small volume and large surface area, the remaining output is to evaporation. The proportionately greater evaporation has resulted in an increase in the lake's concentration of total dissolved solids (Limón & Lind, 1990).

162

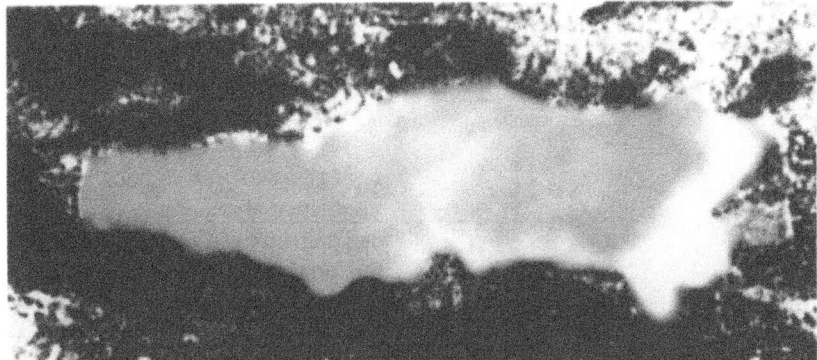

Figure 3. Lake Chapala showing turbid water circulation patterns along south shore as a return gyre at mid-lake. Computer enhancement of Ektachrome photograph (handheld camera) from NASA Space Shuttle.

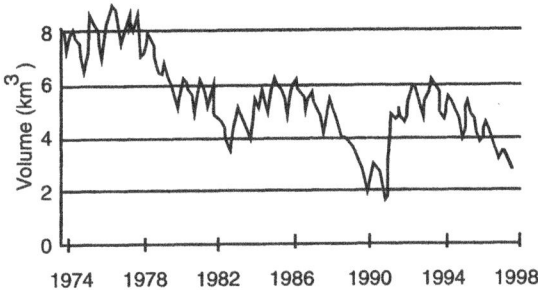

Figure 4. Long-term decline of lake volume. The annual wet–dry season cycle of approximately 1 m is evident as is the effect of water-poor years. (Modified from Limón et al., 1989.)

Problems associated with declining water quantity

With the above background, we will consider five problems that have received great, but differing degrees of attention by government officials and public alike. These are: extensive infestations of water hyacinth (*Eichhornia crassipes*), a declining native fishery, light limitation of phytoplankton production at the base of the food chain, shallow water algal blooms resulting in water supply treatment problems, and the presence of toxic substances in the harvested and sold fishes.

Eichhornia crassipes

Water hyacinth has been a generally minor but recurring problem for many years. The initial date of lake infestation is unknown, but two periods of extreme infestation are known – the first in the 1950s and the second in 1992–1993. What is the ecological factor causing a minor annoyance to become a major concern for the public, commercial fishing and tourism?

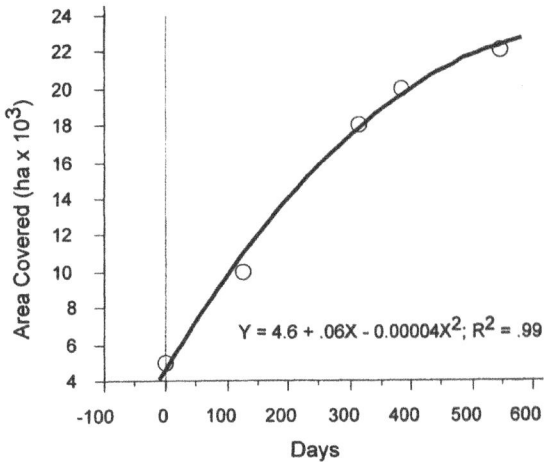

Figure 5. Expansion of the water hyacinth (*Eichhornia crassipes*) during the most recent problem period, 1992–1993.

The common factor was that these outbreaks each occurred after a period of extremely low lake volumes. Figure 5 shows the rapid expansion of lake surface coverage during the latter eruption. At its greatest extent, it covered more than 20% of the lake surface and moved as a great floating island that wreaked havoc with the fishermen's gill nets. The hyacinth was sufficiently stable as to permit secondary growth of cattails (*Typha angustifolia*) on its surface with colonization by some bird species. During the expansion, we (Lind & Dávalos-Lind, unpublished) fit a regression model with which we were able to predict the time of decline within 1 month and thus discouraged the proposed aerial application of glyphosate herbicide.

How was this hyacinth problem the product of declining water level? The answer lies in the reproductive biology of the plant. The typical annual dry–rainy season cycle is ideally tailored to its needs.

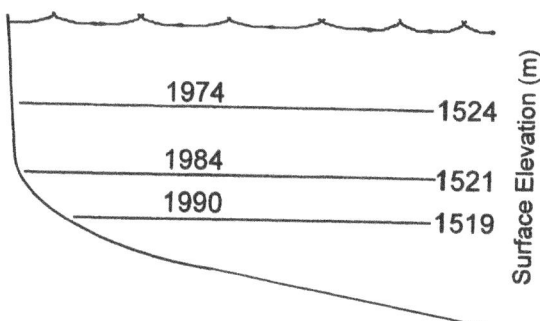

Figure 6. Illustrating the change in morphometry with change in surface elevation. Because of the general rectangular configuration, a decline in surface elevation from 1524 to 1521 m produces a large loss of volume, but little loss of surface area. Further declines in elevation produce a much greater loss of surface area (and beach exposure) relative to volume.

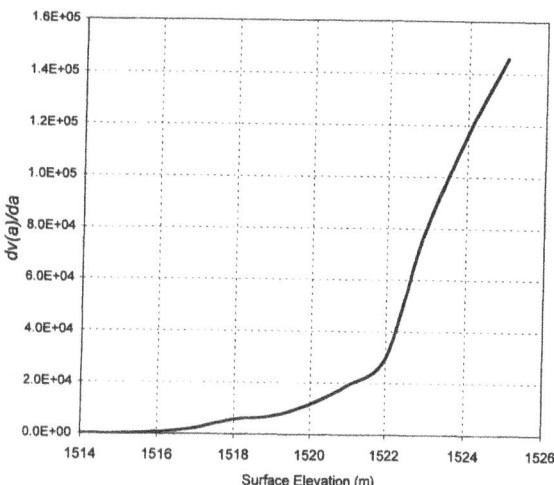

Figure 7. Determination of critical depth for significant increase in relative beach exposure as a function of surface elevation based on changes in area and in volume. The critical elevation is between 1521 and 1522 m.

Seed stocks are deposited into the water and then wash up on the beach. The tough seed coat requires desiccation to crack and/or scarification, which could be accomplished by abrasion with the beach sand and cobble during movement both by water and wind (Parija, 1934). The abraded seed is left on the beach as water levels continue to recede. Water, shallow water and high light intensities favor germination (Ted Center, USDA, personal communication). With the subsequent rainy season, the seed is wetted, imbibes, and germinates – often establishing weak roots in the loose beach soil. With continued rise in water level the plant breaks loose and becomes floating where it will complete its life cycle with further reproduction via stolons. The problem arises when there are suc-

cessive dry seasons without significant rainy seasons. Thus more than 1 year's production of seeds accumulates and remains awaiting the next time of wetting and germination. As a consequence, after a period of continued decline in water level, multiple years' accumulation of seeds germinate and produce the observed problem. This problem has been accentuated by the overall lake level decline. Because of its rectangular morphometry, loss of volume initially had little effect on surface area and beach exposure (Fig. 6). However, at recent low volumes, a small change in volume produces a proportionately greater change in area. An analysis of the area-volume curves suggests that the critical surface elevation at which this disproportionately greater increase occurs is between 1522 and 1521 m above sea level (Fig. 7).

Fish and fisheries

The fishery of Lake Chapala is important to a large segment of the peoples of the nearby communities – both as food and an economic resource. Pomeroy (1993, 1994) described the socioeconomic importance. This importance has produced studies of the abundance and distribution of harvested fish species. Nevertheless, most fisheries data exists only as 'gray literature'. Government reports showed a decline in the fishery during the 1980s (unpublished report, Delegación de la Secretaria de Pesca en Jalisco), but it is not clear whether that was due to fewer fishes and catch per unit effort or to less effort as fishers sought alternate means of income. Of those reported catches, 77% by biomass were sight-feeding planktivorous species (Lind et al., 1994). Lyons et al. (1998) concluded that the total fishery is not in decline with catch of exotic species (Tilapia and Carp) more than compensating for the loss of native species. Soto-Galera et al. (1998) reported a recent decline in the number of fish taxa present in the lower Río Lerma basin including Lake Chapala.

Morphoedaphic (MEI) models relate water fertility and depth to fish harvest (Ryder, 1965; Oglesby, 1977) in the form of nutrient concentration (usually as TDS or conductivity) divided by mean depth. Based on the MEI for the 1980s the fishery should have been approximately 10^7 kg year^{-1}. The average for the 1980s was 10^6 kg year^{-1}. On the other hand, given the levels of primary production, the fishery was large. Various fishery models (Melack, 1976; Oglesby, 1977; Jones & Hoyer, 1982) based on primary production or phytoplankton biomass predicted a smaller fishery than realized (Lind et al., 1994). According to MEI theory, as Lake Chapala has become shallower and saltier, the fishery should have increased. But in this case, it also became more light limited. Nevertheless, there must be undetermined limits to the nutrient concentrations and more likely to the minimum mean depth (as relective of total lake volume) beyond which the MEI fishery-yield model fails.

The unusual occurrence of the *Chirostoma* species flock has attracted evolutionary biologists for study of this taxon (Barbour, 1973). The functional role, or niche, of these fishes in the lake ecosystem has not been investigated, but study of niche partitioning in the *Chirostoma* flock is now in progress (J. Kelly, personal communication). The one piscivorous species, *C. luceius*, has been nearly exterpated from the lake. The reason for this is unknown, but over-fishing, introduction of exotic species, poor predator efficiency due to greater turbidity resulting from the declining surface elevation and lower primary production associated with quantity decline (see below) have been suggested (Lind et al., 1994).

Low primary production

The lake's phytoplankton production of approximately 100 g C m^{-2} year^{-1} places it on the borderline of oligo- and mesotrophy (Lind et al., 1992, 1997). Why is this so? Nutrient concentrations are high. Inorganic nitrogen concentrations – the algal growth-limiting element in the lake (Dávalos et al., 1989) – could support a hypereutrophic system. The water is warm (~20 °) and there is a year-round growing season. The reason for the low photosynthetic production is strong light limitation by suspended clay. Clay micelles in this lake are relatively small with a mean diameter of 0.5 μm (F. Schiebe, personal communication). The shallow depth, orientation to prevailing winds, holomixis, and small clay size assure this light limitation. More importantly, in determining phytoplankton production and biomass accumulation, is the proportion of time the phytoplankton has sufficient illumination to assure net production. Absolute turbidity is less important than the ratio of mixing depth to photic depth (Z_{mix} : Z_{eu}) which covaries with turbidity and, because of holomixis, with lake depth. As lake level declines, mixing brings increased clay resuspension. Seasonal patterns of phytoplankton photosynthesis show highest rates late in the rainy season

when the lake is at its seasonal maximum depth and turbidity is lowest. Thus, a falling water level alters both the photic depth and the mixing depth. Figure 8 shows this interaction and the different response for relatively deep versus relatively shallow lake regions. For the deeper region, a falling water level results in a less favorable light climate for phytoplankton due to increased turbidity. But, for a shallower region, the opposite occurs because of a shallower mixing depth. This is illustrated by comparing the ratio for the deeper central portion and shallower eastern portion of the lake in years of different water levels. For the shallow year, the light climate was less favorable to algal production in the central region, but more favorable in the eastern region where an algae bloom occurred. The critical Z_{mix} : Z_{eu} in Lake Chapala is apparently approximately two – the phytoplankton are exposed to adequate illumination for one-half time (or given the approximate equal day–night photoperiod in the tropical location, one-fourth of a day). We have only found high chlorophyll concentrations and, at times, algal blooms in areas shallow enough to provide this critical ratio.

Algal blooms

As shown above, declining water levels can limit phytoplankton production but only up to a critical ratio. When the water level declines sufficiently – albeit with increasing resuspension of clay – so that the algae have sufficient light energy, the excessively high nutrient concentrations support blooms. As described above, the lake's morphometry becomes an important factor. When the water level remains above 1521-m elevation, a decline in level results in only a small increase in shallow water area. When the level declines below this elevation, a proportionately greater area of the lake consists of shallow regions capable of reaching the critical Z_{mix} : Z_{eu} of two and blooms result. When the lake level was below 1520 m, the mean annual chlorophyll concentration of the shallow eastern region was 23.0 μg l^{-1} while the rest of the lake was 9.3 μg l^{-1}. Much of this greater mean is attributable to blooms. The coefficient of variation for chlorophyll concentrations in the shallow region was 65% while that for the rest of the lake was only 33%. This decline is of particular concern to the City of Guadalajara whose intake for supply is located in a shallow northshore area where algal and cyanobacterial blooms occur.

Central Region

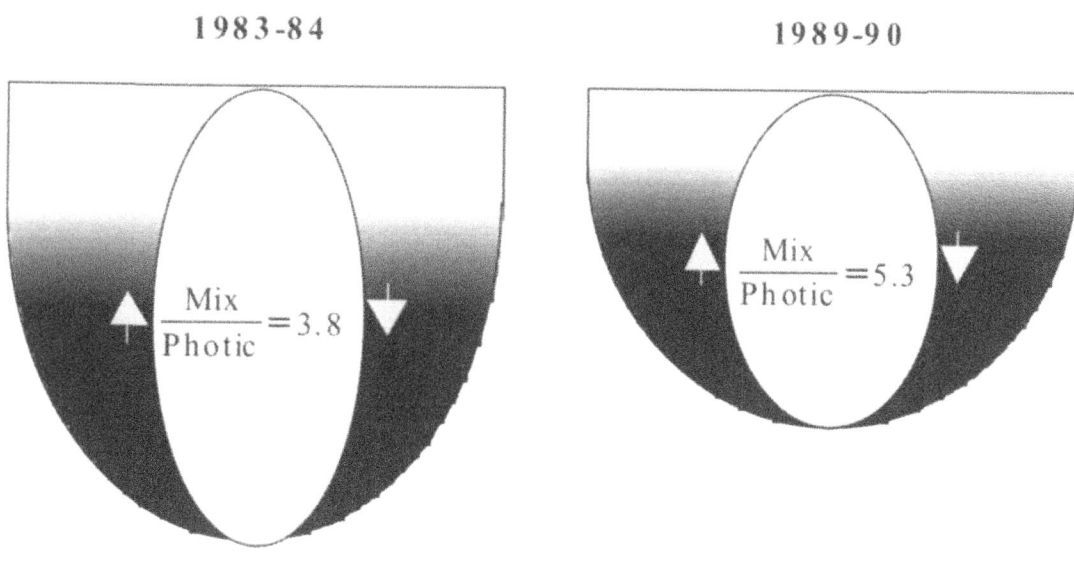

Eastern Region

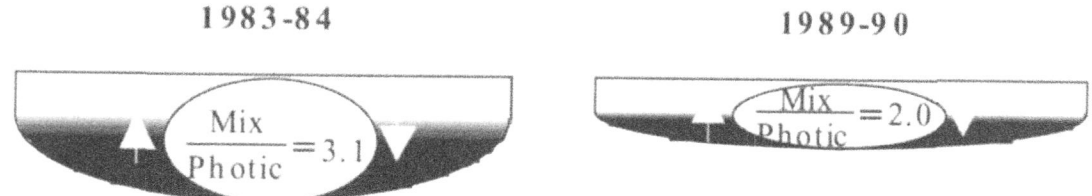

Figure 8. Illustration of how lake depth changes the ratio of lake mixing depth to lake photic depth ($Z_{mix} : Z_{eu}$) to produce light climates favorable and unfavorable to algal production. When the lake level declined in 1989–1990 the light climate in the deeper central region became less favorable while that in the shallow eastern region became favorable and algal blooms occurred. (From Lind et al., 1994.)

Table 3. Concentration of four metals in the tissues of fish from Lake Chapala. (Data from Shine et al., 1998)

	Metals (ppm dry wt)			
	Cr	Cu	Pb	Zn
Tilapia liver	6	2200	33	105
Tilapia muscle	0.5	0.3	0.6	36
Carp liver	0.8	13	2.5	1090
Carp muscle	0.2	2.2	3.1	180

Toxic substances and the fishery

The Río Lerma, which supplies most of Lake Chapala's water, drains a large industrial and agricultural region of Mexico. This leads one to suspect that the lake serves as a sink for multiple toxic substances. Of the possible toxicants, only certain heavy metals have been quantified in the water, biota or sediments (Hansen, 1992; Ford & Ryan, 1995; Shine et al., 1998). Shine et al. (1998) found high concentrations in the tissues of Tilapia and Carp (Table 3). Toxic properties of lake sediments were measured

166

by bacteria metabolism and algal growth elutriation bioassays (Dávalos-Lind, 1996). Toxicity of sediment elutriates varied with lake region. Sediments from the shallow eastern region (the region of the Río Lerma inflow) significantly reduced bacteria metabolism at a sediment concentration equal to that naturally in suspension ($\sim$100-mg l^{-1}). Sediments from the other lake regions were toxic to bacteria but at sediment concentrations approximately twice that normally found in the water. Assuming that declining water levels with increased suspension of sediments is analogous to elutriation, one can expect an increased exposure of the bacterio- and phytoplankton to the clay-associated toxicants. A possible pathway for transport of these metals into indigenous species of food fishes of the genus *Chirostoma* spp. (locally known as Charal) was proposed by Dávalos-Lind et al. (1992). They found that *Chirostoma* ingest clay-bacterial aggregates, presumably as a food source. Any toxic substance associated with the clay would directly (as opposed to food chain links) find its way into the fishes. *Chirostoma* may favor shallow lake regions. Becerra-Muñoz et al. (1997) reported the greatest densities in Lake Chapala occur in the shallow eastern region where sediments contain the greatest toxic load and resuspension is greatest.

Conclusions

Lake Chapala is a lake with serious water quality problems. Nevertheless, water quantity is the ultimate problem, both as a factor unto itself and as an interacting factor with quality. To protect the quality of the lake, water levels must be maintained above 1521-m elevation as a minimum. This will require significant changes in social–economic and political practices in the five-state catchment above the lake. It also will require dramatic conservation measures by Guadalajara to reduce the quantity extracted. Projections predict that Guadalajara will require 22 m^3 s^{-1} by 2010, which, if current proportions of sources is maintained, will require twice (15 m^3 s^{-1}) the present extraction from the lake. The low lake levels of the past two decades have resulted in almost no river outflow to permit flushing. Control of water quantity is a necessary step toward achieving a desirable water quality.

Acknowledgements

Contribution #16 from The Chapala Ecology Station, a joint facility of Universidad Autónoma de Guadalajara and Baylor University. The research upon which this paper is based was supported by National Science Foundation grants #BSR-8807341 and INT-8213149, and National Institutes of Health Grant #ES 07439-01TOX. We thank Centro Estudios Limnologicos, Escuela de Biología de Universidad Autónoma de Guadalajara, Club de Yates de La Floresta and Club de Yates de Chapala for facilities and field assistance.

References

Barbour, C., 1973. A biogeographical history of Chirostoma (Pisces: Atherinidae): A species flock from the Mexican plateau. Copeia 3: 533–556.

Becerra-Muñoz, S., H. Buelna-Osben, J., J. Catalan-Romero & A. Mejia-Muratalla, 1997. The atherinidae and goodeidae fish community as an indicator of ecosystem health. 5th International Conference on Aquatic Ecosystem Health: Linking Science, Education, Politics and Society. October 26–29, Ajijic, Jalisco, Mexico.

Berman, T., A. Nishri, A. Parparov, B. Kaplan, S. Chava, M. Schlicter & U. Pollingher, 1997. Relationships between water quality parameters and water levels in Lake Kinneret. Verh. Internat. Verein. Limnol. 26: 671–674.

Clements, T., 1963. Pleistocene history of Lake Chapala Jalisco, Mexico. Essays in Marine Geology in Honor of K. O. Emery. Univ. Calif. Press, Los Angeles.

Dávalos-Lind, L., 1996. Phytoplankton and bacterioplankton stress by sediment-borne pollutants. J. Aquat. Ecosyst. Health 5: 99–105.

Dávalos, L., O. Lind & R. Doyle, 1989. Evaluation of phytoplankton-limiting factors in Lake Chapala, Mexico: Turbidity and the spatial and temporal variation in algal assay response. Lake Reservoir Mgmt. 5: 99–104.

Dávalos-Lind, L., O. Lind, R. Sada, A. Guerra, G. Velarde, L. Orozco & T. Chrzanowski, 1992. La producción bacteriana y su importancia en la cadena trófica. Ingeniería Hidráulica en México 7: 30–26.

Ford, T. & D. Ryan, 1995. Toxic metals in aquatic ecosystems: a microbiological perspective. Environ. Health Perspect 103 (Suppl. 1): 25–28.

Hansen, A., 1992. Metales pesados en sl sistema Lerma-Chapala: distribución y migración. Ingeniería Hidráulica en México. Mayo-Diciembre: 92–98.

Hutchinson, C., P. Warshall, E. Arnould & J. Kindler, 1992. Development in arid lakes: lessons from Lake Chad. Environment 34: 16–20.

Jones, J. & M. Hoyer, 1982. Sportfish harvest predicted by summer chlorophyll-*a* concentration in midwestern lakes and reservoirs. Trans. am. Fish. Soc. 111: 176–179.

Limón, J. & O. Lind, 1990. The management of Lake Chapala (México): Considerations after significant changes in the water regime. Lake Reservoir Mgmt. 6: 61–70.

Limón, J. G., O. T. Lind, D. S. Vodopich, R. D. Doyle & B. G. Trotter, 1989. Long- and short-term variation in the physical and chemical limnology of a large, shallow, turbid tropical lake (Lake Chapala, Mexico). Arch. Hydrobiol. 83 (Monogr. Beit.) 1: 57–83.

Lind, O., T. Chrzanowski & L. Dávalos-Lind, 1997. Clay turbidity and the relative production of bacterioplankton and phytoplankton. Hydrobiologia 353: 1–18.

Lind, O., L. Dávalos-Lind, T. Chrzanowski & J. Limón, 1994. Inorganic turbidity and the failure of fishery models. Int. Revue ges. Hydrobiol. 79: 7–16.

Lind, O., R. Doyle, D. Vodopich, B. Trotter, J. Limón & L. Dávalos-Lind, 1992. Clay turbidity: governing of phytoplankton production in a large, nutrient-rich tropical lake (Lago de Chapala, México). Limnol. Oceanogr. 37: 549–565.

Lyons, J., G. Gonsalez, E. Soto & M. Guzman, 1998. Decline of freshwater fishes and fisheries in selected drainages of west-central México. Fisheries 23: 10–18.

Melack, J., 1976. Primary productivity and fish yields in tropical lakes. Oecologia 44: 1–7.

Oglesby, R., 1977. Relationships of fish yield to lake phytoplankton standing crop, production and morphoedaphic factors. J. Fish. Res. Bd Can. 34: 2271–2279.

Parija, P., 1934. Physiological investigations on water hyacinth (Eichhornia crassipes) in Orissa with notes on some other aquatic weeds. Indian J. Agricultural Sci. 4: 399–429.

Pomeroy, C., 1993. Organized fishers' responses to social dilemmas of common pool resource use PhD dissertation, Texas A & M Univ: 175 pp.

Pomeroy, C., 1994. Obstacles to institutional development in the fishery of Lake Chapala, Mexico. In Dyer, C. & J. McGoodwin (eds), Folk Management in the World's Fisheries. Lessons for Modern Fisheries Management. U. Colorado Press: 17–41.

Ryder, R., 1965. A method of estimating fish production in north-temperate lakes. Trans. am. Fish. Soc. 94: 214–218.

Shine, J., D. Ryan & T. Ford, 1998. Annual cycle of heavy metals in a tropical lake – Lake Chapala, Mexico. J. Environ. Sci. Health, Part A 33: 23–43.

Simons, T., 1980. Circulation models of lakes and inland seas. Can J. Fish. Aquatic. Sci. 203: 1–146.

Soto-Galera, E., E. Diaz-Pardo, E. Lopez-Lopez & J. Lyons, 1998. Fish as indicators of environmental quality in the Rio Lerma Basin, México. Aquat. Ecosyst. Health Mgt. 1: 267–276.

Stone, R., 1999. Coming to grips with Aral Sea's grim legacy. Science 284: 33–33.

Hydrobiologia **467**: 169–176, 2002.
J. Alcocer & S.S.S. Sarma (eds), Advances in Mexican Limnology: Basic and Applied Aspects.
© 2002 *Kluwer Academic Publishers.*

Sensitivity of Mexican water bodies to regional climatic change: three study alternatives applied to remote sensed data of Lake Patzcuaro

A. Gomez-Tagle Chavez[1], F.W. Bernal-Brooks[2] & J. Alcocer[3]

[1]*Instituto Nacional de Investigaciones Forestales y Agropecuarias, Campo Experimental en Morelia*
E-mail: yolalber@unimedia.net.mx   pichucho@lycos.com
[2]*Estacion Limnologica de Patzcuaro, Centro Regional de Investigacion Pesquera de Patzcuaro.*
Calzada Ibarra 28, Colonia Ibarra, Patzcuaro, Michoacan 61609
E-mail: bbrooks@jupiter.ccu.umich.mx
[3]*Limnology Lab, Environmental Conservation and Improvement Project, UIICSE, FES Iztacala, UNAM.*
Av. de los Barrios sin, Los Reyes Iztacala, Tlalnepantla, Edo. de Mex. 54090
E-mail: jalcocer@servidor.unam.mx

Key words: climatic change, lake levels, Central Mexico, remote sensing, closed lakes, Patzcuaro

Abstract

Historical aerial photography over closed basins enables the observation of lake surfaces at variable water levels and climatic conditions. Thus, the assessment at the landscape level of subtle variations in inundated areas depends on a suitable framework for processing historical imagery, frequently available at different scales and resolutions.

The present paper deals with Lake Patzcuaro as a case study and the application of three analytical procedures to remote sensed data of 1974, including two maps commercially available and nine aerial photographs. Computer-based processing of images and further incorporation into GIS led to three different outputs: topographic map (TM-GIS), photomosaic 1 (P_1) and photomosaic 2 (P_2). Aerial photographs assemblages were georeferenced by means of GCP's, 86 located by means of INEGI vectorial polygons (P_1), and 40 determined directly at the field by GPS (P_2).

In conclusion, TM-GIS lack of photointerpretation for the lake as an image derived from a topographic map; while P_2 involved an unacceptable RMS. P_1 thus became the best option for the calculation of areas not only because of the lower RMS associated, but the additional photointerpretation of limnological zones exclusive to aerial photographs. As a background picture, P_1 also facilitates the analysis of historical images on a comparative basis.

Introduction

Lakes in closed basins often undergo variations in water levels and area relative to changes in the regional water balance between precipitation (P) and evaporation (E). This magnitude of response to fluctuations in P–E depends on the relative contribution of groundwater inflow and outflow to the hydrologic budget; lake-level change is greatest in terminal basins which have neither surface nor groundwater outflow (Fritz et al., 1999). The analysis of climatic effects over lakes morphometry requires historical imagery as well as equipment for manual or computer-aid photointerpretation. Meaden & Kapetsky (1992) have provided a

view on the application of remote sensing techniques in inland waters. Johnson & Cage (1997) discuss current and future trends in technologies and tools used for catchment research and comment on their use as they has been applied to regional assessments.

Because of limited data availability about climatically sensitive lakes located in arid and semiarid regions, worldwide related programs have called for hydrological studies to be put into a more global perspective with the use of remote sensing techniques (Birkett & Mason, 1995). In consequence, the Mullard Space Science Laboratory (MSSL) is currently developing a Global Lake and Catchment Conservation Database (GLCCD) based on measurements of lake

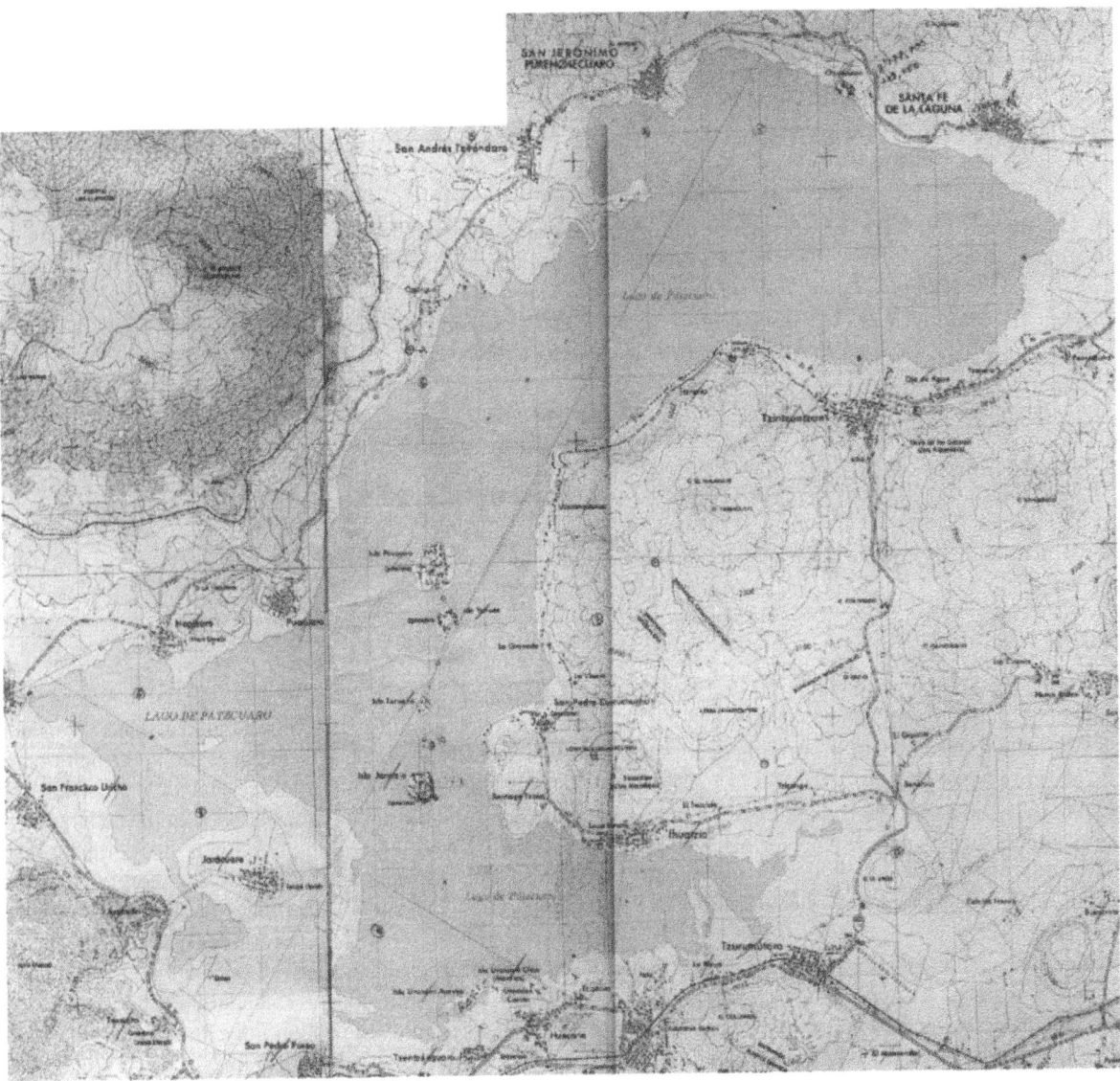

Figure 1. Lake Patzcuaro topographic map (source: CETENAL, 1987).

areas and levels by satellite imaging radiometers and radar altimeters respectively. An inherent limitation of 100 km^2 has been established as a minimum threshold area. We believe that studies related to smaller climatically sensitive lakes might also be informative when ground-based information and historical aerial photography are available. Moreover, the existence in Mexico of registered time-series on water levels for some closed lakes (Bernal-Brooks & MacCrimmon, 2000) encourage the possibility of retrospective studies relative to climate-based effects.

Since retrospective studies encompass decadal time scales, black and white aerial photographs rep-

resents the only remotely sensed data. Recent examples include studies of vegetation change in the South African Savanna (Hudak & Wessman, 1998), the Negrito Creek watershed in southwestern New Mexico (Miller, 1999), and the Mediterranean Maquis in Northern Israel (Kadmon & Harari-Kremer, 1999). The high-spatial resolution needed to discriminate shrubs and trees in landscape studies like those mentioned before, led us to consider simultaneously the suitability of aerial photos in old files of land use/cover as a useful material for the analysis of climatically sensitive lakes. Historical sequences available for the case study herewith presented, Lake

Figure 2. Lake Patzcuaro photomosaic.

Patzcuaro, demand a framework for both, photomosaics assemblages at different scales and resolutions as well as correction of geometric distortions caused by aircraft drifting. As a preamble to obtaining comparable results through time, we applied the procedures described below.

Study area

See Bernal-Brooks et al. (2002).

Materials and methods

Computer hardware included a personal computer with an Intel Pentium II processor, 32 MB in RAM, 16 MB in video memory, and a scanner (Scan Port SQ 4800) with flat bed and fluorescent lamp. Images management were made by means of Adobe Photo-Shop ver. 5.0 (Adobe), Cartalinx ver. 1.0 (Cartalinx), and Idrisi for Windows ver. 2.010 (GIS) over the following working references:

(1) Topographic maps Cheran E14A21 and Patzcuaro E14A22 (CETENAL, 1977), including Lake Patzcuaro at 1:50 000 scale.

(2) Nine sequential aerial photographs of the lake basin (1:75 000 scale) taken in 1974, which are also the basis for the elaboration of maps mentioned before.

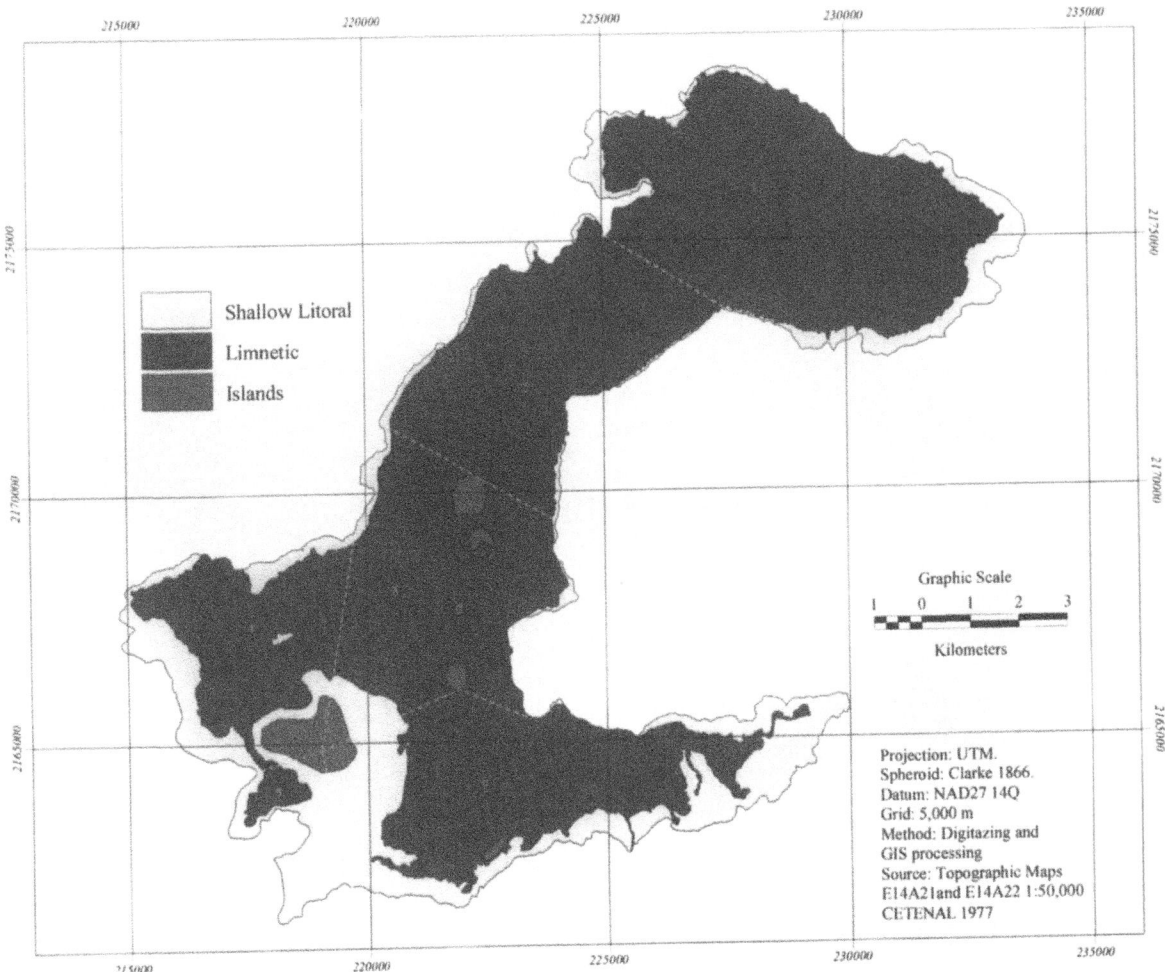

Figure 3. Output 1: Lake Patzcuaro (TM-GIS), including six sectors considered in the present research.

(3) Vectorial polygons of maps E14A21 and E14A22, 1:50 000, established by INEGI (1995) with an accuracy of 2 m.

General procedure

Six sectors were arbitrarily established inside the lake shoreline: (1) Quiroga, (2) Cuello, (3) Islands, (4) Erongaricuaro, (5) Jaracuaro, and (6) Ihuatzio. Also, a stereoscopic analysis of aerial photographs defined emergent and submersed vegetation on the basis of texture, as well as a limnetic zone (free of vegetation). Aerial photographs must be scanned in order to convert their information content into digital form (Kadmon & Harari-Kremer, 1999). Thus, maps and photos were transformed into digital data by means of graphic software (Adobe). Scanning of imagery at 600 dpi led to assemblage of entire Lake Patzcuaro

images and adjustment of stereoscopic pairs by means of entities present in overlapping areas. Files in a *.tif format allowed image importation to Cartalinx (resample module). Then, a georeferencing process relied on Ground Control Points (GCP's).

Digitizing followed contours of lakeshore, sectors, and limnological features over georeferenced images set up as backdrops. Vectorial polygons were incorporated into GIS and converted into raster format for the elaboration of output maps and areas calculation.

Specific outputs

Topographic map by GIS (TM-GIS)

The two parts of the lake image, as shown in CETENAL (1977) maps were carefully fitted together into the computer memory by means of vectorial polygons (INEGI, 1995), before proceeding to Cartalinx.

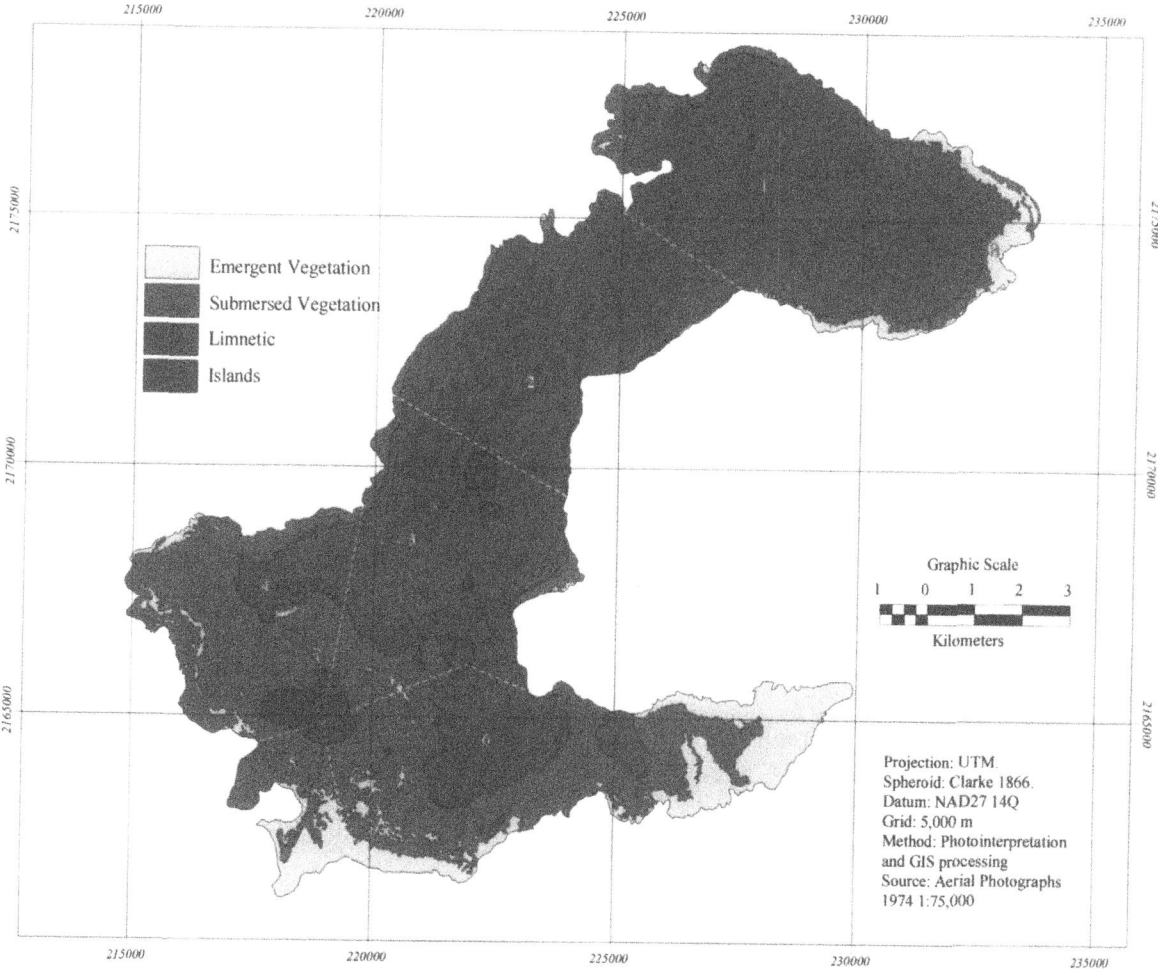

Figure 4. Output 2: photomosaic 1 (P₁), including six sectors considered in the present research and photointerpretation based on aerial photographs.

Photomosaic 1 (P₁)

Mosaicking a global lake image from aerial photographs involved an overlap and analogic adjustment of every single photo over a background file of the topographic map. The latter included an intricate matrix of digitized contours, including diverse terrestrial features in the lake nearby, such as watercourse junctions and road intersections. At Cartalinx level, 86 GCP's were included by means of an easy location of at both topographic maps and aerial photographs, and determined by the INEGI (1995) vectorial polygons. Thus, geopositioned references enabled the development of a correspondence file, which could be reprocessed into GIS resample module along with all digitized polygons.

Photomosaic 2 (P₂)

Procedures were basically the same as those for P₁. In this case, though, 40 GCP's were identified at working references and determined directly in the field by means of GPS (Garmin 45).

Results

Remote sensed data for the present investigation includes the assemblage of both topographic map (Fig. 1) and photomosaic (Fig. 2). Output maps (TM-GIS, P₁ and P₂) enabled the calculation of areas shown in Table 1. Estimations of the total lake area differed as follows: 11 839.93 ha (TM-GIS); 11 648.19 ha (P₁), and 11 912.04 ha (P₂), including islands. Figures 3

Table 1. Area estimation for three outputs (TM-GIS, P$_1$ and P$_2$) including A – aquatic surface; V – Vegetation; EV – Emergent vegetation; SV – submersed vegetation; L – lake; I – islands and S – sectors

Outputs/	Sectors	A	V	EV	SV	L	I	Total
CM	Quroga	27.70	4.45			32.15	0.00	32.15
	Cuello	18.00	0.81			18.82	0.00	18.82
	Islas	17.22	1.75			18.97	1.22	20.18
	Erongarícuaro	10.16	4.41			14.57	1.11	15.68
	Jarácuaro	0.85	3.41			4.26	0.14	4.40
	Ihuatzio	16.42	10.37			26.79	0.38	27.16
	Total	90.34	25.21			115.55	2.85	118.40
P1	Quroga	27.78		3.24	1.28	32.31	0.00	32.31
	Cuello	17.74		0.39	0.34	18.47	0.00	18.47
	Islas	14.43		0.23	3.02	17.68	1.07	18.75
	Erongarícuaro	2.87		1.33	10.03	14.24	1.13	15.37
	Jarácuaro	0.00		1.47	2.58	4.05	0.17	4.22
	Ihuatzio	7.77		8.09	10.77	26.63	0.72	27.35
	Total	70.59		14.76	28.01	113.38	3.10	116.48
P2	Quroga	28.42		1.31	3.31	33.04	0.00	33.04
	Cuello	18.14		0.34	0.40	18.89	0.00	18.89
	Islas	14.75		3.09	0.24	18.08	1.10	19.18
	Erongarícuaro	2.93		10.26	1.37	14.57	1.16	15.72
	Jarácuaro	0.00		2.63	0.00	2.63	1.68	4.31
	Ihuatzio	7.95		11.02	8.26	27.23	0.74	27.98
	Total	72.20		28.65	13.58	114.44	4.68	119.12

and 4 show TM-GIS and P$_1$, respectively, as the most relevant lake maps obtained herewith.

Discussion

Changes in the regional climatic regime induce responses of sensitive water bodies in terms of varying water levels and areas. The incorporation of aerial photographs into GIS enabled, in our study case of Lake Patzcuaro, an analysis of the situation at an unusually high water level (1974) different from what it currently is. As long as a generalized lake-level decline occurred in the western part of the Mesa Central of Mexico since 1979, lake surface shrinking and volume reductions indicate a progression to a dryer climate at the regional level (Bernal-Brooks & Mac-Crimmon, 2000). A graphic demonstration of climatic effects over sensitive lakes in Mexico is the aim of this and further research.

The analysis of aerial photographs on the basis of graphic software bear implicit different levels of vari-

ability, as a cumulative root mean square (RMS) rose in our case study to 24.5 m (TM-GIS), 19.5 m (P$_1$), and 197.6 m (P$_2$), with the following implications:

TM-GIS, despite the low RMS associated, includes a rough photointerpretation for the lake as it represents another view of the topographic chart (Fig. 2), focusing on terrestrial features at the water body nearby. Thus, the shallow lakeshore lack of vegetation-zone differentiation and, as such, overestimates of the limnetic area at the southern littoral.

P$_1$, in contrast, reaches a finer image adjustment reinforced by 86 GCP's determined on the basis of INEGI (1995) vectorial polygons. Also, the areas with emergent and submersed vegetation acquired definition compared to the diffuse wetland shown in TM-GIS at the same place. Therefore, P$_1$ incorporates photointerpretation exclusive to aerial photographs becoming a true limnologic map that surpasses TM-GIS limitations. Moreover, an intricate matrix of digitized layers set up a framework for mosaicing aerial photographs from any other time, a methodological frame-

Table 2. Lake Patzcuaro: reported estimations of total area

Author	Apparent source	Methodology	Result (km^2)
De Buen (1944)	Topographic survey	Grid enumeration analysis	111.00
Tellez & Motte (1976)	CM and aerial photographs (1974)	Photointerpretation	107.73
Herrera (1979)	CM	Planimetry	104.60
Velasco (1982)	CM	Planimetry	88.70
Chacon (1989)	CM	GIS	130.00
Present study	CM and aerial photographs (1974)	Photointerpretation-GIS	116.48

work that enables the retrospective studies mentioned above.

An unacceptable RMS associated with P_2 arose from data obtained directly at the field with a commercial GPS. Low points coverage (6.6 points/photo in average) accompanies such variability, particularly outstanding at the neighborhood of the Cuello and Islands sectors (2 and 3, respectively). This fact draws our attention as long as the steep slopes of mountains beside the lakeshore interfered with the reception quality of microwaves coming from satellites. A codification process is recommended in the future to tackle the high variation in the process of P_2 georeferencing, if that is the case.

An analysis of total area estimations for Lake Patzcuaro by other authors is shown in Table 2, including the apparent source of primary remote sensed data. Tellez & Motte (1980) described the existence of different kinds of vegetation at the littoral zone based on aerial photographs, although the purpose was merely descriptive. Nonetheless, the latter contribution remained irrelevant, as further research still relied on the topographic map of 1974 as the basic reference for lake description and areas calculation. At the shallowest southeastern shoreline, in particular, a westward displacement occurred during the 1980s and 1990s as the lake surface shrunk at lower water levels. To date, no assessment has been performed of area vs. water level position through time (and presumed changes in vegetation areas).

In our study, current advances in computer technology facilitated handling of geographical images on the basis of limited resources. Today, this type of analysis is within reach of those with the insight and desire to conduct the analysis (Johnson & Gage, 1999). In particular, the potentials of graphic software for incorporation of remotely sensed data into a GIS demonstrate an effective tool in our case study. Furthermore, the procedure is no less accurate than data

obtained from manual photointerpretation (Kadmon & Harari-Kremer, 1999). Our estimation, surprisingly, approximates the results obtained by a team of students and professors from the Escuela Superior de Ingeniería Mecánica y Eléctrica, Instituto Politécnico Nacional (supervision by Eng. Carlos Quintana and Eng. Jorge Mendez), which carried out a precise topographic survey in 1938–1939 (Zozaya, 1940 1a, b; De Buen, 1944). Both cases involved levels (Bernal-Brooks et al., 2002) with roughly similar areas, which enables comparisons. The analysis of Lake Patzcuaro, among other climatically sensitive lakes in Mexico, is on-going on the basis of the methodology applied here.

Acknowledgements

The authors appreciate the willingness of MC Araceli Orbe Mendoza (CRIP Patzcuaro) and Yolanda Chavez (INIFAP Campo Morelia) in providing us with facilities and equipment. Financial support was given, in part, by CONACyT project 3626P-B9608.

References

Bernal-Brooks, F. W. & H. R. MacCrimmon, 2000. Lake Zirahuen (Mexico): an assessment of morphometry change based on evidence of water level fluctuations and sediment inputs. In Munawar, M., S. Lawrence, I. F. Munawar & D. Malley (eds), Aquatic Ecosystems of Mexico: Status & Scope. Backhuys, The Netherlands: 61–76.

Bernal-Brooks, F. W., A. Gomez-Tagle Rojas & J. Alcocer, 2002. Lake Patzcuaro (Mexico): a controversy about the ecosystem water regime approached by field references, climatic variables, and GIS. Hydrobiologia 467 (Dev. Hydrobiol. 163): 187–197.

Birkett, C. M. & I. A. Mason, 1995. A new global lakes database for a remote sensing programme studying climatically sensitive large lakes. J. of Great Lakes Research 21: 307–318.

CETENAL (Centro de Estudios para el Territorio Nacional), 1977. Topographic charts E14A21 and E14A22 (1:50 000). 3rd edn.

176

Chacon, A., L. G. Ross & M. C. M. Beveridge, 1989. Lake Patzcuaro, Mexico: results of a new morphometric study and its implications for productivity assessments. Hydrobiologia 184: 125–132.

De Buen, F., 1944. Los lagos michoacanos. II. El Lago de Pátzcuaro. Rev. Soc. Mex. Hist. Nat. 5: 99–125.

Fritz, S. C., B. F. Cumming, F. Gasse & K. Laird, 1999. In Stoermer, E. F. & J. P. Smol (eds), The Diatoms. Applications for the Environmental and Earth Sciences. Cambridge University Press, Cambridge: 41–72.

García, E., 1988. Modificaciones al sistema climático de Köppen. Talleres de offset Larios, S.A., México, D.F: 217 pp.

Herrera, E., 1979. Características y manejo del Lago de Pátzcuaro, Michoacán. BSc thesis. National University of Mexico (UNAM). Mexico.

Hudak, A. T. & C. A. Wessman, 1998. Textural analysis of historical aerial photography to characterize woody plant encroachment in South African Savanna. Remote Sens. Environ. 66: 317–330.

Johnson, L. B. & S. H. Gage, 1997. Landscape approaches to the analysis of aquatic ecosystems. Fresh wat. Biol. 37: 113–132.

Kadman, R. & R. Harari-Kremar, 1999. Studying long-term vegetation dynamics using digital processing of historical aerial photographs. Remote Sens. Environ. 68: 164–176.

Meaden, G. J. & J. M. Kapetsky, 1992. Los sistemas de información geográfica y la telepercepción en la telepercepción en la pesca continental y la acuicultura. FAO technical Report no. 318: 266 pp.

Miller, M. E., 1999. Use of historic aerial photography to study vegetation change in the Negrito Creek Watershed, Southwestern New Mexico. The Southwestern Naturalist 44: 121–137.

SMN (Servicio Meteorológico Nacional), 1922–1986. Climatic variables registered at 16087 Meteorological Station.

Tellez, R. & O. Motte, 1980. Estudio planctonológico preliminar del Lago de Pátzcuaro, Michoacán, México. Mayo de 1976. 2nd. Latinamerican Symposium of Aquaculture: 1797–1836.

Velasco, A., 1982. Evaluación de la calidad del agua con base en algunos aspectos de la comunidad planctónica del Lago de Pátzcuaro, Michoacán. BSc thesis. National University of Mexico (UNAM), Mexico.

Zozaya, M., 1940a. Informes de la Estación Limnológica de Pátzcuaro. Mayo 1940: 14 pp.

Zozaya, M., 1940b. Informes de la Estación Limnológica de Pátzcuaro. Junio 1940: 14 pp.

Hydrobiologia **467**: 177–185, 2002.
J. Alcocer & S.S.S. Sarma (eds), Advances in Mexican Limnology: Basic and Applied Aspects.
© 2002 *Kluwer Academic Publishers.*

Characterization of small shallow ponds with color video imagery in Central Mexico

Jorge López-Blanco[1] & Luis Zambrano[2]

[1]*Instituto de Geografía UNAM, Circuito Exterior, Ciudad Universitaria, A.P. 20-850, C.P. 04510, México D.F.*
Tel.: (52)-55-6224335. Fax (52)-55-6162145. E-mail: jlblanco@servidor.unam.mx
[2]*Instituto de Biología UNAM, Departamento de Zoología, Circuito Exterior, Ciudad Universitaria, A.P. 70-153,*
C.P. 04510, México D.F.
E-mail: zambrano@ibiologia.unam.mx

Key words: shallow lakes, video remote sensing, macrophytes, zooplankton, phytoplankton, suspended solids, water turbidity, Acambay, Central Mexico

Abstract

We present a method to evaluate ecological characteristics of small shallow ponds in Central Mexico based on video remote sensing and image processing techniques in a GIS environment. We used a set of color video imagery obtained from heights lower than 700 m above ground. Our analysis established statistical correlations between the average reflectance values contained in video imagery (digital numbers DN per Blue-Green-Red band) and the average values of limnetic variables: (1) water suspended solids concentration; (2) water turbidity; (3) total macrophytes coverage; (4) free floaters and emergent macrophytes coverage; (5) zooplankton abundance; and (6) chlorophyll-*a* concentration in water. We found strong correlation between DN values and vegetation presence, suspended solids concentration and water turbidity ($R = 0.85$ to 0.98), but weak correlations with phytoplankton and zooplankton abundance. This cheap and fast method can be used to describe general conditions of ponds related with vegetation abundance, turbidity and suspended solids.

Introduction

Recent studies have established the ecological importance of shallow artificial ponds for maintaining regional aquatic diversity, as well as providing ecosystem services (Margalef, 1983: 37; Hernández-Aviléz, et al., 1995; Scheffer, 1997; Zambrano & Macias García, 1999). The importance of such ponds increases in areas where the existence of natural water bodies is not possible due to relief, or rainfall and stream flow conditions.

Small ponds account for more than 67% of the 14 000 lentic stationary water bodies (non-flowing water) the whole country of Mexico (Athié, 1987). Together, these ponds cover almost 189 000 ha; which is around 15% of the total area of inland water bodies (Hernández-Aviléz, et al., 1995: 291). Ponds are broadly distributed in the central-western region of Mexico, which is semi-arid. In this region, water bodies are used to supply early season irrigation for the

surrounding agricultural lands (primarly maize), during February and March. Occasionally, ponds are also used for carp culture (*Cyprinus Carpio* L., Hernández Aviléz, et al., 1995) as an alternative food source for domestic consumption (Zambrano et al., 1999; Lopez-Blanco & Zambrano, 2001).

Pond general attributes (high evaporation during dry season, big temperature changes during the day, and practically without stratification) are similar to the conditions of natural lakes in the region, such as Pátzcuaro (Rosas et al., 1993). The presence of regional endemic organism in these ponds (Zambrano et al., 1999) suggests that aquatic organisms such as native fish, amphibians and reptiles find these bodies as a place of refuge and dispersion. Because of their ecological and agricultural importance, a fast and cheap method to characterize the most relevant characteristics of these ponds would be useful for studies in biogeography and freshwater management.

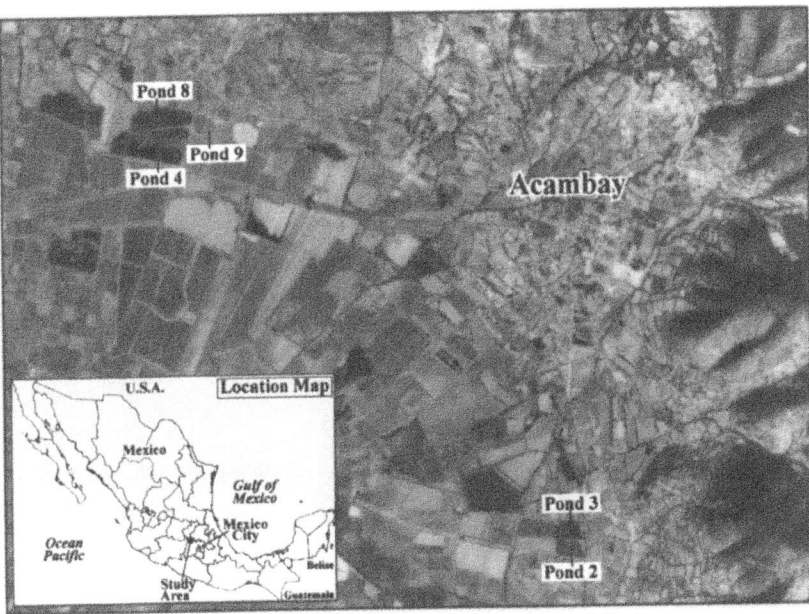

Figure 1. Location of the study area in Acambay, Mexico. Pond names are shown in each fresh water system. Pond numbers do not follow a particular meaning.

In the last few years, aerial photographs and video imagery have been widely used to evaluate aquatic plant abundance, and to obtain a better understanding of environmental variables of aquatic ecosystems, just as with oceanic as inland aquatic ecosystems (Lefevre et al., 1984; Schloesser et al., 1988; Welch et al., 1988; Nohara, 1991; Marshall & Lee, 1994; Pasqualini & Pergent-Martini, 1996; Malthus & George, 1997; Ramírez-García et al., 1998; López-Blanco et al., 1998). Digital image processing (DIP) by means of automatic techniques in a GIS environment has been a common procedure used to obtain statistical groups of image-elements (pixels) that correspond to ecological variables, such as the pond area covered by aquatic plants (particularly of macrophytes), at both the local and regional levels (see Lehmann & Lachavanne, 1997; Ferguson & Korfmacher, 1997; Norris et al., 1997; Robbins, 1997). Both aerial photographs and color videoimages taken in infrared and natural color are commonly used in aquatic evaluations (Harris et al., 1996; López-Blanco et al., 1998).

Here we evaluate video imagery as a method to characterize small shallow ponds in Central Mexico by correlating mean digital number values (DN, magnitude of radiance of a image's pixels set) from the image with respect some visual variables of the ponds (water turbidity and macrophytes coverage) and other values that affects indirectly water transparency (sus-

pended solids, chlorophyll-*a* and zooplankton abundance). Preliminary results suggest that colour images are better related with those parameters linked with water turbidity (López-Blanco & Zambrano 2001), which is a keystone in the shallow lakes dynamics (Scheffer, 1997).

Study site

Field data and images were collected from small (0.9 to 9.6 ha) shallow (mean ± 1SE = 0.84 ± 0.21 m) ponds in Acambay located at Northwest of Toluca, near to Mexico City, in the Trans-Mexican Volcanic Belt, at 2550 m above sea level (19° 57′ N, 99° 51′ W, Fig. 1). The site occupies an East–West oriented tectonic depression filled of alluvial and lacustrine sediments. This depression is bounded to the North by the Acambay–Tixmadeje Fault and to the South by the Venta de Bravo Fault. Average annual rainfall in the study area is 903.8 mm (García, 1988). Ponds tend to fill from June to October, coinciding with the heaviest rains. At first, the filling of ponds is provided directly by rainfall and, after that, mainly by the overland and stream flow through channels. The average annual temperature is of 14.2 °C (maximum 16.8 °C, minimum 10.6 °C; García, 1988). Acambay is in a gently sloping alluvial-lacustrine plain, in the uppermost watershed portion of the Lerma River. Full of endemic species, this river is one of the most important biodiversity spots for fresh water organisms in Mexico (Espinoza et al., 1993; Contreras-Balderas, 1999; Zambrano & Macias-García, 1999).

Method

We collected data from the ponds at onset of the dry season in February and March of 1995. We quantified limnetic variables and obtained video images within a very few days of each other to decrease the possibility of false non-correspondence between both methods.

Limnetic variables

For each water body the total area of the pond, the area covered by water and the mean depth was measured. Mean Secchi depth values were obtained crossing the ponds with two transects (each consisting of 20 points). Two 1 liter composite samples, from three subsamples were collected around midday in each pond to measure suspended solids and chlorophyll-*a* concentration, the late to provide an index of phytoplankton biomass. Suspended solids concentration (mg l^{-1}) was obtained by the gravimetric method after ignition to 105 °C according to the standard methods described by APHA (1985). Chlorophyll-*a* concentration was determined using the fluorometric method (Lorenzen, 1966) from 50–500 ml of filtered water. As an index of macrophyte abundance, we estimated the coverage of submerged , emergent and floating macrophytes coverage within a 40 cm diameter circular frame (Necchi et al., 1995) at five sample points along each of twelve 10 m length transects (60 samples per pond in total). As ponds were generally square, we made three transects at equal distances along each of the pond sides, originating from, and perpendicular to, the littoral margin. We collected zooplankton from five samples of the entire water column with a PVC tube (7 cm diameter). The resulting pooled sample was filtered through 200-μ-mesh plankton net and stored in alcohol to be quantified in the laboratory.

Image acquisition

Pond imagery was obtained with a standard videocamera Sony Handycam CCD-TR55, with focal length between 11 to 66 mm respectively, and wave-length limits of visible light sensitivity between 0.4 to 0.7 μm using an analog signal format with an 8 mm tape format. The recording device of videoimages is the CCD (Charge Coupled Device). This solid-state videosensor is formed by a large number of active photodetectors (270 000 in this case) installed in a microchip a half-inch in diagonal length. The camera was suspended from a domestic-manufactured tethered balloon, which was built with plastic bags inflated with Helium and manually controlled from the ground by means of two plastic threads.

Depending on the flight height of the balloon used, the ground area covered was as minimum as 0.17 ha (when flying at 100 m above ground level -AGL-), to a maximum as 8.3 ha (when flying at 700 m AGL). During fieldwork several topographic surveys were carried out using compass and tape measurements. This helped to established control points with reference marks in the video imagery, used for georeferencing to an orthogonal coordinate system, in order to integrate such images in a GIS environment.

The image acqusition procedure obtained average radiance values (digital numbers or DN) for the three 8-bit visible bands (Blue-Green-Red, BGR) which video images could be decomposed or separated using a 256-gray scale. Those video images had to be captured in digital format of natural color (at least in a 24-bit per pixel format).

Selected video images were digitized using a Video Reveal TV 500 frame grabber card, with a chosen spatial resolution of 320 × 240 pixels using a 24-bit TIFF format file ('real' or 'natural' color; see López-Blanco & Arias-Chalico, 1998). Such selected images were processed in the ILWIS Geographic Information System (Integrated Land and Water Information System, Version 1.41 MS-DOS and Versions 2.1 and 2.2 for Windows; ITC, 1993, 1994, and 1997), by means of applying conventional procedures of digital image processing (Lillesand & Kiefer, 1994).

Image processing and its correlation with limnetic variables

Nineteen images were selected for processing; three images to build a digital mosaic of Ponds 2 and 3 (Fig. 2) and sixteen to build the digital mosaic of Ponds 4, 8 and 9 (Fig. 3). A GIS procedure known as 'mosaicking' was applied to integrate a set of individual videoimages (see López-Blanco & Arias-Chalico, 1998).

A geometric correction procedure was necessary to create the two mosaics of videoimages. This procedure produced a north-oriented videomap with a pixel size of 68 cm for the Videomap of Ponds 2 and 3, and a

180

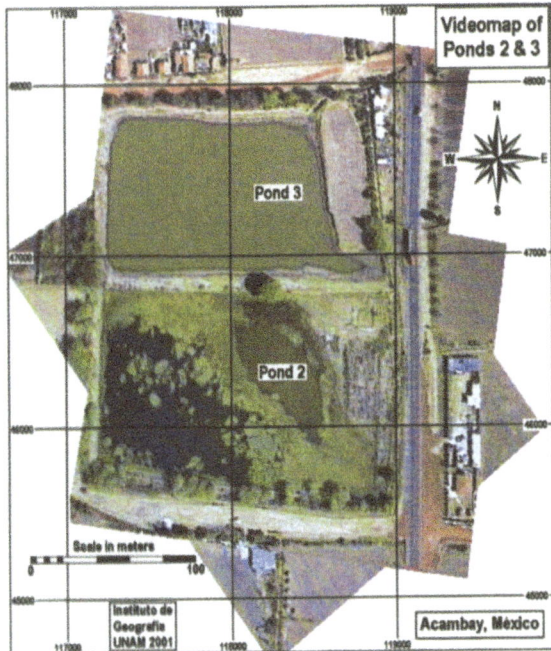

Figure 2. Videomap of Ponds 2 and 3 made with a digital mosaic from the three separate images.

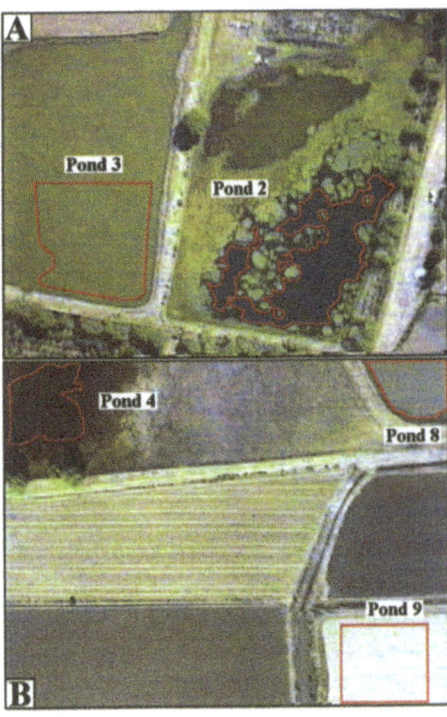

Figure 4. 'Sampled' areas of DN values of pixels within the water area of ponds to determine the average radiance per band: (A) Ponds 2 and 3; (B) Ponds 4, 8 and 9.

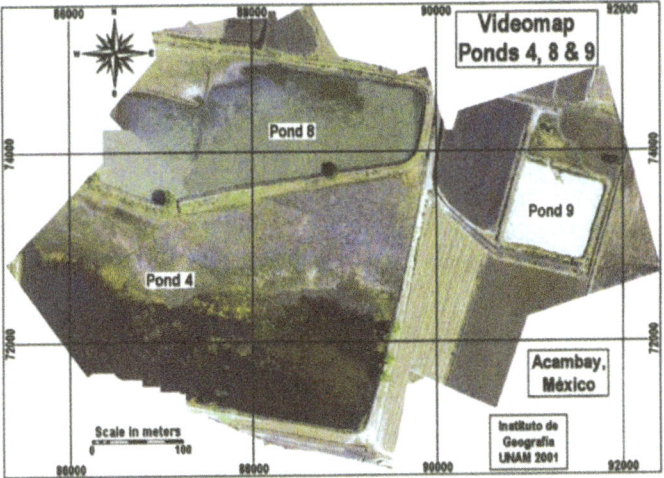

Figure 3. Videomap of Ponds 4, 8 and 9 made with a digital mosaic from the sixteen separate images.

60 cm pixel size in the case of the Videomap for Ponds 4, 8, and 9 (Figs 2 and 3 respectively). This allowed us to cover all the study area with only two mosaics.

In order to characterize digital numbers of pixels of the five ponds evaluated, we chose two of the nineteen selected videoimages. Figure 4 (A and B) shows the 'sampled' areas in which values of the three corresponding bands (Blue, Green and Red) were used to determine radiance average values. In Figure 4 (A and B) is possible to distinguish very remarkable differences of color and tone characteristics between the ponds. This also demonstrates that 'sampled' areas which were considered to have average radiance values per pixel have sufficient homogeneity that they can be taken as representative of the water surface of each pond.

We used a linear correlation analysis to establish the proximity and association of DN average values (light intensity transformed to integer values that rank between 0 to 255) for the three BGR bands to the average values of the six limnetic variables, which could be considered as representative of the ecological conditions of any water body. We established as an alternative hypothesis that the differences in ecological variables could be sufficient to introduce visible differences in the videoimages.

Results

Morphologic characteristics are quite variable between the study ponds, particularly in extent and volume. Largest pond extents in 9.59 ha (5 ha full of water at the moment which the video images were obtained), while smallest pond occupied only 0.86 ha (0.76 ha full of water). Although differences in depth are less extreme, 1.02 m on average in the deepest, and 0.49 m in the shallowest, the area difference produce a difference on volume of 700% between the biggest and the smallest pond (Table 1).

It is possible to distinguish very remarkable differences between the average values of the limnetic variables for the five evaluated ponds (Table 2). The main differences are apparent in: Total Suspended-Solids (TSS), Secchi depth (SEC) and total macrophytes coverage (TMAC) (Columns 1 to 3, Table 2). From these contrasting variables, we divided the studied ponds into three groups: the first group (Ponds 2 and 4) are ponds with low values of TSS (≤ 7 mg l^{-1}) but high values of SEC (≥ 0.5 m) and TMAC ($\cong 95\%$). The second group (Ponds 3 and 8) with middle values (TSS $\cong 40$ mg l^{-1}, SEC $\cong 0.25$ m and TMAC $\cong 25\%$). Finally the third group only with Pond 9 with high in TSS $\cong 190$ mg l^{-1}, and low values in SEC $\cong 5$ cm, and TMAC $\cong 10\%$ (Table 2). However, the rest of the variables: Surface Macrophyte Coverage (SMAC), Abundance of Zooplankton (ZOO) and chlorophyll-a Concentration in water (CLOA) (Table 2, Columns 4 to 6) do not seem to have the same relationships.

Descriptive statistical parameters of the DN values of pixels from each pond have been simplified for each of the three analyzed band-image BGR in Table 3. Analyzing such values, it is possible to point out that there exists an important relative closeness between DN values for Ponds 2 and 4. On the other hand, there exists dissimilarity between values for Pond 9 with

regard to the others. Highest values of standard error (SE) are presented in Ponds 8, 4, and 9.

As in some of the limnetic variables, the DN value for the average of the three bands BGR separates the studied ponds in three groups (Table 3 and Fig. 4): The first one (Ponds 2 and 4) is with a 'darker' tone of water than the others, a second group (Ponds 3 and 8) having medium tones, and the third having the 'whitest' tone of water (Pond 9).

Despite their DN similarities, there are differences between them once each color is analyzed. Ponds 2 and 4 (with low DN average) are relatively darker than the other ponds. Pond 2 shows a color-domain from blue and green bands (76, 68) with respect to the red one (45). On the other hand, the average values of the three bands of Pond 4 have similar magnitudes between them (55, 52 and 50). With these values, Pond 2 can be described as blue-greenish dark, and Pond 4 can be described as brownish dark (Fig. 4).

Contrasting with the internal differences in the first group, the second group of ponds (3 and 8) present similarities in color-domain. They form a group in that they generally have medium DN average values (from 105 to 139). However, there are color-domain differences between the water of these ponds that could be described as greenish (for Pond 3) and blue-greenish color (for Pond 8), both with a medium-light tone (Fig. 4). Finally DN average values for Pond 9 are the highest of all the ponds (221, 226, and 228) and its color-domain confirms it as the least dark of all (Table 3 and Fig. 4).

Analyzing values and tendencies shown by the determination coefficients R in Table 4, it is possible to establish the following observations: There are several levels of significant correlation between DN values of images and the six limnetic variables. R values were high for variables as TSS, SEC and TMAC (0.98 to $- 0.77$). Regressions that considered the percentage of area covered with surface macrophytes SMAC (emergent and free-floating) had intermediate levels of correlation. The coefficients for ZOO and CLOA were low. The probability levels were significant at the $P < 0.05$ level, for the three bands with regard to TSS and of the same manner for two bands in relation to SEC.

Correlation analysis suggests a strong direct relationship between the TSS levels and the DN values ($P < 0.05$). Not surprisingly, the whitest of ponds, has more concentration of total suspended-solids. Macrophytes and turbidity also had high correlation values with DN. In both cases, the relation is inverse, sug-

Table 1. Ponds morphometric characteristics

Pond no.	Total area (ha)	Water area (ha)	water area/ total area (%)	Depth (m) and (stand. dev.)	Volume (m³)
2	2.36	1.68	71.3	0.73 (0.18)	12219.3
3	1.32	1.13	86.3	1.02 (0.45)	11624.8
4	9.59	5.00	52.1	0.53 (0.20)	26360.0
8	5.42	4.04	74.6	0.72 (0.37)	29044.3
9	0.86	0.76	87.9	0.49 (0.18)	3691.7

Table 2. Average values of limnetic variables measured in field and laboratory from samples taken in the five ponds studied

Pond no.	Susp. solids total (mg l⁻¹)	Secchi disc depth (m)	Total macro-phytes(%)	Surface macrophytes (%)	Zooplankton (org l⁻¹)	Chlorophil-*a* (mg m⁻³)
2	2.33	0.73	97.78	79.43	19.14	10.08
3	32.33	0.25	12.06	1.67	16.60	3.62
4	7.00	0.54	93.02	57.53	23.30	1.77
8	64.00	0.29	29.78	29.45	6.80	17.48
9	188.00	0.05	11.05	4.73	16.30	0.70

gesting that darker DN values are given by pristine and highly vegetated ponds (Table 4).

Discussion

Correlation values suggest that the pond evaluation technique based on GIS processing of aerial video images is useful for some obvious visible variables such as water turbidity and suspended solids concentration. In other variables such macrophytes coverage seems to be also useful although correlation values were not significant in this variable, but it has high correlation values. It is useful to use a particular band-color of the three image-band to describe one of the variables. Differences in its color-domain per pond gave particular correlation value at different limnetic variables for each color-band. Therefore, different color band describes better changes in some variables than the other two (Table 4). For instance, the blue band is the best to estimate the total suspended solids concentration in ponds. The red band is the most useful to estimate the dimension of water turbidity; and finally, the green band is the best to estimate the percentage of area covered of total macrophytes.

It seems that these correlations also may help to elucidate which of the factors are more related with water turbidity in these shallow ponds. For ex-

ample, food web relationships, such as top-down control dynamics, could suggest a relationship between zooplankton and turbidity. High zooplankton abundance may produce a reduction on phytoplankton community, and therefore, in water turbidity (Carpenter et al., 1985). However, it does not seem to be a close relationship between phytoplankton and zooplankton in these ponds (Zambrano et al., 1999). Food chain between zooplankton and phytoplankton seems to be broken by different factors in these ponds.

The relatively high values obtained for chlorophyll-*a* concentration in water suggest that these ponds are in a permanent eutrophic condition that decreases the strength of the link between these factors (Scheffer, 1997). Also, we note the low values of abundance of zooplankton in ponds, in comparison with values for ponds outside of the intertropical zones (with more temperate climate; Zambrano et al., 1999) are not helping for the top-down control. Although ponds have a high chlorophyll-*a* concentration, their values are not reflected in the DN values. Therefore, phytoplankton as a turbidity agent in these ponds seems to be different of temperate lakes. It is necessarily to have further research to start the understanding of pathways and trends on the trophic web of this type of ponds.

On the opposite, high correlation values between suspended solids and DN, that also is related with Secchi values, suggest that most of the turbidity of

Table 3. Statistical description of DN average values of pixels of 'sampled' areas in videoimages of ponds (see areas in Fig. 4)

Band-Pond	No. of pixels	DN average	Stand. dev.	Median	Minimum	Maximum	Stand. error
Blue-Pond 2	4942	76.3	8.3	76	30	104	0.117
Green-Pond 2	4942	67.6	6.6	68	27	92	0.094
Red-Pond 2	4942	44.6	8.4	44	8	76	0.120
Blue-Pond 3	5775	105.1	8.3	106	75	137	0.110
Green-Pond 3	5775	134.5	5.4	134	110	156	0.071
Red-Pond 3	5775	124.0	8.3	124	82	152	0.109
Blue-Pond 4	2377	55.2	8.4	55	26	86	0.172
Green-Pond 4	2377	51.6	8.3	50	19	85	0.171
Red-Pond 4	2377	50.3	9.8	50	10	85	0.202
Blue-Pond 8	1875	138.8	11.0	141	77	175	0.254
Green-Pond 8	1875	138.7	9.4	139	83	179	0.217
Red-Pond 8	1875	129.4	11.9	129	65	175	0.275
Blue-Pond 9	3286	220.7	10.1	221	167	255	0.176
Green-Pond 9	3286	225.7	8.0	225	193	255	0.139
Red-Pond 9	3286	227.8	9.9	229	185	255	0.172

Table 4. Parameters correlation values between DN average values of pixels (per band) regarding to limnetic variables

Image band-limnetic variable	Interception (standard error)	Slope (standard error)	Standard deviation	R	P
Blue-TSS	70.498 (09.5)	0.8298 (0.1)	16.164	0.97642	0.00433
Green-TSS	73.483 (15.7)	0.8546 (0.2)	26.692	0.94241	0.01645
Red-TSS	60.431 (14.7)	0.9330 (0.2)	24.945	0.95692	0.01066
Blue-SEC	196.574 (32.5)	−208.621 (74.0)	39.181	−0.85218	0.06669
Green-SEC	212.860 (25.6)	−240.737 (58.1)	30.769	−0.92269	0.02550
Red-SEC	213.584 (23.4)	−265.314 (53.2)	28.190	−0.94463	0.01551
Blue-TMAC	175.311 (34.5)	−1.151 (0.6)	47.978	−0.76777	0.12956
Green-TMAC	191.758 (27.7)	−1.398 (0.4)	38.576	−0.87542	0.05179
Red-TMAC	187.897 (30.8)	−1.491 (0.5)	42.776	−0.86723	0.05691
Blue-SMAC	165.384 (36.7)	−1.335 (0.8)	53.943	−0.69358	0.19398
Green-SMAC	181.882 (31.0)	−1.686 (0.7)	45.488	−0.82166	0.08795
Red-SMAC	179.016 (32.1)	−1.846 (0.7)	47.194	−0.83560	0.07802
Blue-ZOO	203.987 (93.5)	−5.159 (5.4)	65.576	−0.48280	0.41007
Green-ZOO	213.180 (99.9)	−5.452 (5.8)	70.067	−0.47873	0.41462
Red-ZOO	203.000 (110.2)	−5.343 (6.4)	77.323	−0.43581	0.46321
Blue-CLOA	125.531 (48.8)	−0.935 (5.3)	74.496	−0.10134	0.87119
Green-CLOA	134.483 (51.6)	−1.615 (5.6)	78.724	−0.16413	0.79196
Red-CLOA	131.191 (54.9)	−2.372 (6.0)	83.728	−0.22396	0.71725

Notes: TSS=Total Suspended-Solids, SEC=Disk of Secchi depth, TMAC=Total macrophytes cover, SMAC=Surface Macrophytes cover, ZOO=Zooplankton Abundance, CLOA=Chlorophyll-*a* concentration. N=5. Single underlined values of P are significant in $P<0.05$. Double underlined values of P are significant in $P<0.10$.

the ponds is based on suspended solids concentration. Solids can be re-suspended by wind action and benthivorous effects, such as carp presence (Zambrano & Hinojosa, 1999). Suspended solids concentration is related with submerged macrophytes coverage. Macrophytes form a barrier between bottom and the water, and therefore they are a factor that decreases the amount of solids that can be re-suspended (Scheffer, 1997; Zambrano & Hinojosa, 1999). Therefore, it is possible that high correlation values of DN with macrophytes coverage are not only related with macrophytes own color but with the lack of suspended solids in the water (that make them whiter), which produce images with darker values in the selected areas.

Freshwater communities respond to more detailed temporal and spatial scales than terrestrial communities. Changes in limnetic variables occur rapidly, thus it is difficult to obtain them in a direct manner at the speed necessarily to understand pond dynamic. With video images, we had a detailed record, at the necessary moment, and with enough spatial resolution, to characterize limnetic variables of the typical shallow ponds in Central Mexico. This method can be repeated as many times as is necessarily in many ponds at relatively short period of time. The possibility of predicting limnetic variables by means of processing DN values of videoimages bands has given promising results with adequate statistical significance, even if the method used here has restrictions for application in other study areas where sediments, food webs dynamics, and general environment are different. Although the approach presented in this paper still in a pilot project stage, we found that the method to be effective and inexpensive, and therefore practical. The main potential use of this approach is to identify rapidly ponds with high TSS, SEC or TMAC, and is simple enough that can be used efficiently to describe the characteristics of a large number of ponds making slight changes in such method, such as the use of an aircraft cabin door mount for photo and video cameras is possible to record and obtain images of a large area.

Acknowledgements

We thank Enrique Elizarrarás, Leopoldo Galicia, Eduardo Pérez, Víctor Aguirre and Demián Hinojosa for their help and support during field work. We also thank Maclovio Ruiz and his brothers, owners of the ponds for their assistance and cooperation. We thank Dave Montgomery and anonymous reviewers for helpful critiques of the manuscript. This work would not have been possible without the economic and logistic support of the Institutes of Ecology and Geography (UNAM) and the Biodiversity Support Program (USAID Fundend Consortium World Wildlife Fund, the Nature Conservancy and World Resources Institute).

References

APHA, 1985. Standard Methods for the Examination of Water and Wastewater. Am. Publish Health Assoc., 16th edn. Washington D.C.

Athié, L., 1987. Calidad y cantidad de agua en México. Universo Veintiuno, México.

Carpenter, S. R., J. F. Kitchell & J. R. Hodgson, 1985. Cascading trophic interactions in a lake productivity. Bioscience 35: 634–639.

Contreras-Balderas, S., 1999. Annotated checklist of introduced invasive fishes in Mexico, with examples of some recent introductions. In Claudi, R. & J. Leach (eds), Non-indigenous Freshwater Organisms in North America: Vectors of Introduction, Biology and Impacts. Lewis Publishers, 500 pp.

Espinoza Pérez, H., M. T. Gaspar Dillanes & P. Fuentes Mata, 1993. Listados faunísticos de México. III. Los peces dulceacuícolas mexicanos. Depto. de Zoología, Instituto de Biología, UNAM, México, 100 pp.

Ferguson, R. L. & K. Korfmacher, 1997. Remote sensing and GIS analysis of seagrass meadows in North Carolina, USA. Aquat. Bot. 58: 241–258.

García, E., 1988. Modificaciones al Sistema de Clasificación Climática de Köppen, Instituto de Geografía UNAM, México, 246 pp.

Harris, N. R., D. E. Johnson, T. L. Righetti, & M. R. Barrington, 1996. A blimp Borne Camera System for Monitoring Rangelands, Riparian Zones, or Critical Areas, Geocarto International, Vol.11, No.3: 99–104.

Hernández-Aviléz, J. S., M. C. Galindo de Santiago & J. Loera Pérez, 1995. Bordos o microembalses. In De la Lanza Espino & J. L. García Calderón (Compiladores). Lagos y Presas de México. Centro de Ecología y Desarrollo, Mexico: 291–308.

ITC, 1993. The Integrated Land and Water Information System ILWIS Version 1.4, User's Manual, 1st edn. International Institute for Aerospace Survey and Earth Sciences, Enschede, The Netherlands, 412 pp.

ITC, 1994. The Integrated Land and Water Information System ILWIS Version 1.41, Supplement to 1.4 User's Manual, International Institute for Aerospace Survey and Earth Sciences, Enschede, The Netherlands, 87 pp.

ITC, 1997. The Integrated Land and Water Information System ILWIS Version 2.1 for Windows, User's Guide, 1st edn. International Institute for Aerospace Survey and Earth Sciences, Enschede, The Netherlands, 511 pp.

Lefevre, J.R., C. Valerio, & A. Meinesz. 1984. Optimisation de la technique de la Photographie aerienne pour la cartographie des Herbiers de Posidonies. International Workshop Posidonia Oceanica Beds. In Boudouresque C. F., J. de Grissac A. & J. Olivier (eds), GIS Posidonie Publ., Fr., 1984:1: 49–55.

Lehmann, A. & J.-B. Lachavanne. 1997. Geographic Information Systems and Remote Sensing in Aquatic Botany. Aquat. Bot. 58: 195–207.

Lillesand, T. M. & R. W. Kiefer. 1994. Remote Sensing and Image Interpretation, 3rd edn. Wiley & Sons, New York, 750 pp.

López-Blanco, J. & T. Arias-Chalico. 1998. Elaboración de video-mapas mediante la corrección fotogramétrica de imágenes de video en color: La región de La Montaña de Guerrero, Investigaciones Geográficas Boletín del Inst. de Geogr. UNAM 37: 21–35.

López-Blanco, J., P. Ramírez-García & A. Lot. 1998. Local distribution of seagrass (Phyllospadix spp) by means of processing infrared and color imagery obtained with a tethered blimp to the North of Ensenada, Mexico. Proceedings of The Fifth International Conference on Remote Sensing for Marine and Coastal Management, San Diego, USA, 5–7 October 1998, Vol. I: 97–104.

López-Blanco J. & L. Zambrano, 2001 Propiedades limnéticas de sistemas dulceacuícolas pequeños en Acambay, México: Correlación de datos de campo con imágenes de video en color. Investigaciones Geográficas Boletín del Instituto de Geografía, UNAM 44: 94–84.

Lorenzen, C. J. 1966. A method for the continuous measurement of in vivo chlorophyll concentrations. Deep-Sea Research 13: 223–227.

Malthus, T. J. & D. G. George. 1997. Airborne remote sensing of macrophytes in Cefni Reservoir, Anglesey, UK. Aquat. Bot. 58: 317–332.

Margalef, R. 1983. Limnología, Editorial Omega, Barcelona, 1010 pp.

Marshall, T. & P. F. Lee. 1994. Mapping aquatic macrophytes through digital image analysis of aerial photographs: an assessment, J. Aquat. Plant. Manage. 32: 61–66.

Necchi, O. Jr., L. H. Z. Branco & C. C. Z. Branco. 1995. Comparison of three techniques for estimating periphyton abundance in bedrock streams, Arch. Hydrobiol. 134: 393–402.

Nohara, S. 1991. A study on annual changes in surface cover of floating-leaved plants in a lake using aerial photography, Vegetatio 97: 125–136.

Norris, J. G., S. Wyllie-Echeverria, T. Mumford, A. Bailey & T. Turner. 1997. Estimating basal area coverage of subtidal seagrass bed using underwater videography, Aquat. Bot. 58: 269–287.

Pasqualini, V. & C. Pergent-Martini, 1996. Monitoring of Posidonia oceanica Meadows Using Image Processing. In Kuo, J., R. C. Phillips, D. I. Walker & H. Kirkman (eds), Seagrass Biology. Proceedings of an International Workshop. Faculty of Sciences, The University of Western Australia, Nedlands, Western Australia: 351–358.

Ramírez-García, P., J. López-Blanco & Daniel Ocaña, 1998. Mangrove vegetation assessment in the Santiago River Mouth, Mexico, by means of supervised classification using Landsat TM Imagery, Forest Ecology and Management 105: 217–229.

Robbins, B. D., 1997. Quantifying Temporal Change in Seagrass Areal Coverage The Use of GIS and Low Resolution Aerial Photography, Aquat. Bot. 58: 259–267.

Rosas, I. A., R. Velasco, A. Belmont, A. Baez, & A. Martinez, 1993. The algal community as an indicator of the trophic status of Lake Patzcuaro, Mexico, Envir. Pollut. 80: 255–264.

Scheffer, M., 1997. Ecology of Shallow Lakes, Kluwer Academic Publishers, Dordrecht: 384 pp.

Schloesser, D., C. Brown & B. A. Manny, 1988. Use of Aerial Photography to Inventory Aquatic Vegetation, J. Aerosp. Eng. 1(3): 142–150.

Welch, M. Madden Remillard & R. B. Slack, 1988. Remote Sensing and Geographic Information System Techniques for Aquatic Resource Evaluation, PE&RS, Vol. 54, No. 2: 177–185.

Zambrano, L. & D. Hinojosa, 1999. Direct and indirect effects of carp (Cyprinus carpio L.) on macrophytes and benthic communities in experimental shallow ponds in Central Mexico. Hydrobiologia 408/409: 131–138.

Zambrano, L. & C. Macias García, 1999. Impact of intentional fish introduction in Mexican freshwater systems. In Claudi, R. & J. Leach (eds), Non-indigenous Freshwater Organisms in North America: Vectors of Introduction, Biology and Impacts. Lewis Publishers: 113- -127

Zambrano, L., M. Perrow, C. Macias Garcia & V. Aguirre-Hidalgo, 1999. Impact of introduced carp (Cyprinus carpio) in subtropical ponds of Central Mexico. J. Aquat. Ecosystem Stress and Recovery 6: 281–288

Hydrobiologia **467**: 187–197, 2002.
J. Alcocer & S.S.S. Sarma (eds), Advances in Mexican Limnology: Basic and Applied Aspects.
© 2002 *Kluwer Academic Publishers.*

Lake Patzcuaro (Mexico): a controversy about the ecosystem water regime approached by field references, climatic variables, and GIS

F. W. Bernal-Brooks[1], A. Gomez-Tagle Rojas[2] & J. Alcocer[3]
[1]*Estacion Limnologica de Patzcuaro, Centro Regional de Investigacion Pesquera de Patzcuaro.*
Calzada Ibarra 28, Colonia Ibarra, Patzcuaro, Michoacan. 61609
E-mail: bbrooks@jupiter.ccu.umich.mx
[2]*Instituto Nacional de Investigaciones Forestales y Agropecuarias, Campo Experimental en Morelia*
E-mail: yolalber@mail.giga.com
[3]*Limnology Laboratory, Environmental Conservation and Improvement Project, UIICSE, UNAM Campus Iztacala.*
Av. de los Barrios s/n, Los Reyes Iztacala, Tlalnepantla, Edo. de Mex. 54090
E-mail: jalcocer@servidor.unam.mx

Key words: climatic change, lake-levels, Central Mexico, closed lakes, Patzcuaro

Abstract

In this paper, a basic question is asked about a well-documented case study in Mexico: how well do we know Lake Patzcuaro? We address water balance as fundamental to the question. Past studies provide ambiguous explanations about the role of either underground infiltration and/or runoff, relative to the lake-level fluctuation. Thus, our suspicions over the database reliability led us to inspect historic records on water levels and climatic variables; check out the altitude of ground references, and analyze traces of runoff watercourses over the terrestrial basin by means of GIS. By making data re-arrangement and corrections, it became evident that the lake is subject to long-term cycles with ca. 40-year peaks, including short-term seasonal cycles within. Sensitivity to climatic conditions was determined, as well as the active influence of runoff as an important hydrologic component that contributes to cause serious damage to the land surface by erosion. Rearrangement of raw data highlights the occurrence of past misinterpretations founded on biased information.

Introduction

Deevey (1957) considered Lake Patzcuaro as 'the only lake in Middle America that has been studied at all comprehensively' and also 'one of the best known lakes in the continent'. These conclusions acknowledge the pioneer limnological work made in Mexico by the Estación Limnológica de Patzcuaro (summarized in De Buen, 1941b, 1944a, 1944b) and the Instituto de Biología (Ancona et al., 1940) along with the analysis of supplementary data obtained by himself on July 9, 1941. Thereafter, Limnology remained dormant for many years in this country before colleagues recommended studies some 25 years ago, and today, after so many years, a question is put forth relative to this case study: to what extent the lake is really known, at the beginning of a new century? This paper focuses on one fundamental topic:

the lake-water balance, including viewpoints without unanimous conclusions.

Planas & Moreau (1990) developed a theory on natural eutrophication, assuming the importance of nutrient loading released from the terrestrial substrate to the lake. To these authors, the postulation of internal springs in the Erongarícuaro basin (De Buen, 1941a, 1944b) explains 'that the lake's phosphorus levels could be naturally high by way of underground water inflow'. A carriage of nutrients produced by infiltration of rainwater presumes a direct relationship to chlorophyll *a* peaks in the water body. Besides, a reduction in Secchi disc transparency since the mid 1930s was identified in the sediment transport via runoff rather than a continuous progress of eutrophication signs.

Chacon (1989, 1992, 1993a, 1993b) independently stated that the negative balance between precipitation and evaporation at the lake surface has to be necessarily offset by the influx of underground waters to the lake basin. Registered data show a slight increase in water level (2035.4 m.a.s.l. for 1939–1942; 2036.45 m.a.s.l. for 1971–1989), coupled with a reduction in maximum depth from 15 m (De Buen, 1944b) to 12 m (Chacon et al., 1989). Thus, an explanation arose that extensive sediment loads has been incorporated into the water body as a consequence of deforestation, land erosion, and downhill transportation of particles. Strangely, runoff became a negligible component in the corresponding model of water balance to demerit the importance of water as sediment carryover.

In contrast, Barrera (1992) estimated a significant runoff component not ever considered: 'the whole watershed received a water input of some 1000 million m^3... three fourths of it return to the atmosphere as evaporation... [and] 300 million m^3 drain at the watershed grounds as surface runoff (200) or infiltration (100)'. Additionally, out of the scientific arena, photographic and pictorial evidences, as well as personal communications of people living in the area nearby Lake Patzcuaro for more than 60 years, reveal an exceptionally high water level in the '30s which was never reached again during the past century. These images in particular discredit the available information on water levels and leave suspicions about the veracity of available databases. Hence, this research targets a revision of basic information and the analysis of the watershed hydrology by means of GIS.

Background

Lake Patzcuaro water level has been registered through time on the basis of different ground references:

1st time-series (De Buen, 1944b). Measurements made approximately every 15-days from April 15, 1939, to August 31, 1943; on a reference altitude of 2044 m.a.s.l. at the railroad tracks of the train station in Patzcuaro.

2nd time-series (SARH, 1949–1970). Monthly data from October 1949 to December 1970, based on 2034.08 m.a.s.l. as the zero of a metal scale located at the Santa Fe de la Laguna lakeshore.

3rd time-series (SARH, 1950–1987). Information obtained on a once-every-15-days basis from January 1950 to December 1987, without a ground reference associated.

Figure 1. Geographical location of Lake Patzcuaro in Central Mexico.

4th time-series (SCT, 1986–1990). Monthly data from January 1986 to December 1990, overlapping the third and fifth time-series. This information was recalculated by the Coordinación de Dragado (SDAF) to meet the position of the current level bank.

5th time-series (SDAF, 1989). Water levels determined almost daily since 1989 at the last reference point established in the far extreme of Lake Patzcuaro main dock.

SMN (1922–1986), based on the 16087 Meteorological Station, reported climatic information for the area of Lake Patzcuaro. Annual records of temperature, precipitation, and evaporation, among others, include frequently incomplete time-series (also available in a CD entitled ERIC by CNA, 1995). Mean values are as follows: temperature, 15.8 °C (1922–1930), 16.7 °C (1931–1940), and 16.3 °C (1973–1979); precipitation, 1060 mm (1922–1930), 1049 mm (1931–1940), and 922 mm (1971–1979); evaporation, 1810 mm (1939–1940), 1351 mm (1941–1943), 1438 mm (1973–1975), and 1556 mm (1984–1985). Zozaya (1941) registered data simultaneously for 1922–1940, presumably obtained from the same Station.

CETENAL (1987) charts include the Lake Patzcuaro region (1:50 000), as well as traces of runoff watercourses (channels) as they appear in aerial photographs of March 1974. The Instituto Nacional de Estadística, Geografía e Informática, developed a Digital Elevation Model from remote sensed information.

Study area

Lake Patzcuaro (Fig. 1) lies within an endorheic basin of the East–West Volcanic Axis at 19° 32′ N; 19° 42′ N; 101° 32′ W and 101° 42′ W approximately, in the highlands of Central Mexico. The water body varies in surface altitude between 2035 and 2041 m.a.s.l., with a maximum area detected of 116 km² (Gomez-Tagle Chavez et al., 2002). No evident outlets are present.

The lake origin dates back to the Tertiary and Quaternary (Tamayo and West, 1964; Demant, 1975). Tectonic and volcanic events cut off an ancestral tributary to the Lerma river, which, in turn, created a closed basin and hence a lake (De Buen, 1944a; Barbour, 1973). De Buen (1941c, 1944b), Osorio-Tafall (1941, 1944) and Deevey (1957) described the shallow water body ($Z_{max}=10$ m) as tropical since pioneer limnological studies in Mexico. However, except for the thermal regime, Lake Patzcuaro and the temperate zone of North America show a close similarity of biotic elements (Osorio-Tafall, 1944; Hutchinson et al., 1956).

Method

A geodesic point was established in Patzcuaro downtown by CETENAL (1974) and signalized at the floor next to the Shrine front door by means of a mortised aluminum plate (Ref. a). The corresponding file indicates 19° 30′ 52″ N; 101° 36′ 23″ W; and 2154.285 m.a.s.l (DGG, 1991). From here, the difference in altitude relative to other terrestrial references used in the past for the registration of the time-series on water levels (Refs. b, c and d) was determined by means of a level (Carl Zeiss Jena NI 050). Measurements with ±1 mm accuracy were carried out on November 25, 1997, at Refs. b and c in the same town of Patzcuaro. Later, on May 25, 1998, we checked out the altitude at Ref. d located 20 km to the north relative to the other three references, by previous knowledge of the water level at Ref. c on calm meteorological conditions.

The fact of unstable water-level conditions before 1971 and after 1979 precludes the analysis of proportional and consistent lake responses to climatic variables. Therefore, we focused on data available for 1971–1979, as the lake level, during this particular time, maintained a fairly steady-state condition defined by $A_{LE} = R/E_L - P_L$ in the case of closed lakes (Harris, 1994). Monthly data of precipitation (P), evaporation (E), and changes in water level were

Table 1. Altitude (m.a.s.l.) of ground references used for the registration of Lake Patzcuaro water levels

References	Reported m.a.s.l.	Measured m.a.s.l.	Difference (m)
R. a	2154.29		
R. b	2044	2049.026	5.026
R. c	2034.08	2038.471	4.391
R. d		2038.385	0

converted to volumes. Next, we determined $E_L + V/P_W$ as the rate of watershed precipitation (P_W) able to offset the lake evaporation (E_L) and provide a surplus amount of water equivalent to the volume change (V) into the lake, determined by the fluctuation in water level. In a steady state, the water balance ($P_L - E_L)/R = 0$ (P_L = lake precipitation; E_L = lake evaporation; R = runoff), therefore $R = P_L - R$ and the total water supply to the lake $W_S = 2E_L + V = P_L + R$. Thus, the relationship between W_S and V were determined.

An incorporation of topographic charts (CETENAL, 1987) into a Geographic Information System (ILWIS Ver. 1.3) enabled digitizing of watercourses (channels) in order to assess number, length, and density.

Results

Water levels

The altitude of ground-based references before and after verification in the field is shown in Table 1. Here, Refs. b and d display the highest bias in contrast with c, which has no difference relative to the geodesic point.

Climatic variables

The seasonal alternation of wet and dry seasons results in lake-level fluctuations following the predominance of one or the other climatic extreme situations (Fig. 2). When steady water additions begun with early rains (1972–1973), the lake level shows a lag-time response in its recovery of higher positions. In the case that an initial water supply by early rains stops, water additions offset losses by evaporation at some 23 million m³, reaching an equilibrium lake level. Thereon, new additions of water cause a proportional change in the lake level, following a close linear relationship

190

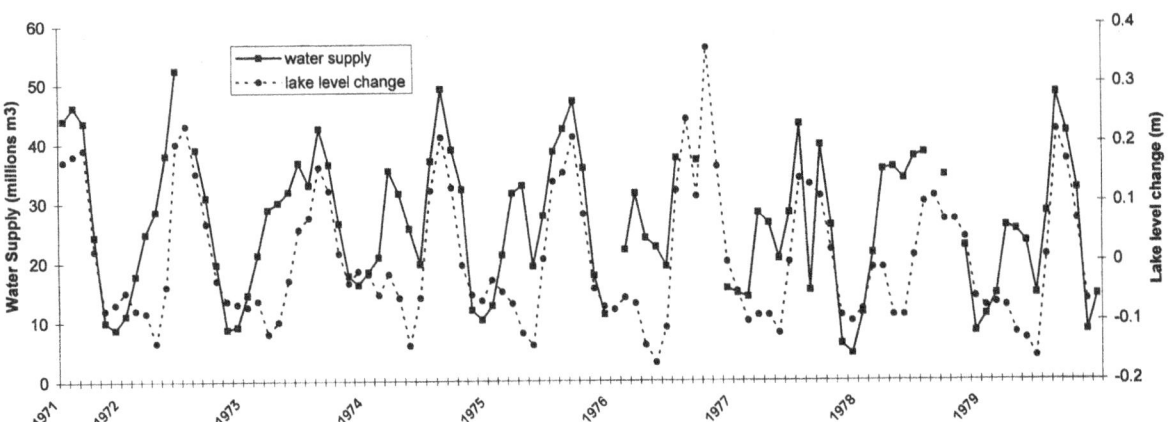

Figure 2. Sequences of water supply to the lake (W$_S$) and corresponding lake volume changes (1971–1979).

Table 2. Lake Patzcuaro water balance variables (1971–1979) including *P*=precipitation (mm); *E*=evaporation (mm); *P$_W$*=watershed precipitation (10^6 m^3); *E$_L$*=lake evaporation (10^6 m^3); *V*=lake volume fluctuation (10^6 m^3); *P$_L$*=precipitation over the lake surface (10^6 m^3); *R*=runoff (10^6 m^3); and *R$_c$*=runoff coefficient.

		P	E	P_W	E_L	V	W_S	P_L	R	R_c
1971	J	0.2487	0.1038	232.4	12.1	19.8	44.0	29.0	15.0	0.07
	A	0.1942	0.1083	181.4	12.6	21.0	46.2	22.6	23.6	0.15
	S	0.2208	0.092	206.3	10.7	22.1	43.6	25.7	17.8	0.10
	O	0.1091	0.0946	101.9	11.0	2.3	24.4	12.7	11.7	0.13
1972	A	0.1768	0.0926	165.2	10.8	17.5	39.1	20.6	18.5	0.13
	S	0.1713	0.1001	160.0	11.7	7.6	30.9	20.0	10.9	0.08
1973	J	0.1389	0.1042	129.8	12.1	8.7	33.0	16.2	16.8	0.15
	A	0.2407	0.1026	224.9	12.0	18.6	42.5	28.0	14.5	0.07
	S	0.1617	0.0966	151.1	11.3	14.0	36.5	18.8	17.6	0.13
	O	0.1322	0.1062	123.5	12.4	17.5	42.2	15.4	26.8	0.25
1974	J	0.1567	0.0989	146.4	11.5	14.0	37.0	18.3	18.8	0.15
	A	0.2093	0.1058	195.5	12.3	24.5	49.1	24.4	24.7	0.14
	S	0.1078	0.1045	100.7	12.2	14.6	38.9	12.6	26.4	0.30
1975	J	0.1663	0.098	155.4	11.4	15.7	38.6	19.4	19.2	0.14
	A	0.2679	0.1069	250.3	12.5	17.5	42.4	31.2	11.2	0.05
	S	0.143	0.097	133.6	11.3	24.5	47.1	16.7	30.4	0.26
	O	0.0244	0.1135	22.8	13.2	9.3	35.8	2.8	32.9	1.65
1976	J	0.2695	0.1007	251.8	11.7	14.0	37.4	31.4	6.0	0.03
	S	0.2235	0.1043	208.8	12.2	12.8	37.1	26.0	11.1	0.06
1977	J	0.2941	0.1152	274.8	13.4	16.3	43.2	34.3	8.9	0.04
	S	0.1052	0.115	98.3	13.4	12.8	39.6	12.3	27.4	0.32
	O	0.0585	0.1016	54.7	11.8	2.3	26.0	6.8	19.2	0.40
1978	J	0.183	0.1144	171.0	13.3	11.7	38.3	21.3	17.0	0.11
	S	0.0837	0.1131	78.2	13.2	8.2	34.5	9.8	24.8	0.36
1979	J	0.2036	0.1165	190.2	13.6	11.7	38.8	23.7	15.1	0.09
	A	0.1821	0.0971	170.1	11.3	25.6	48.3	21.2	27.0	0.18
	S	0.1926	0.0943	179.9	11.0	19.8	41.8	22.4	19.3	0.12

191

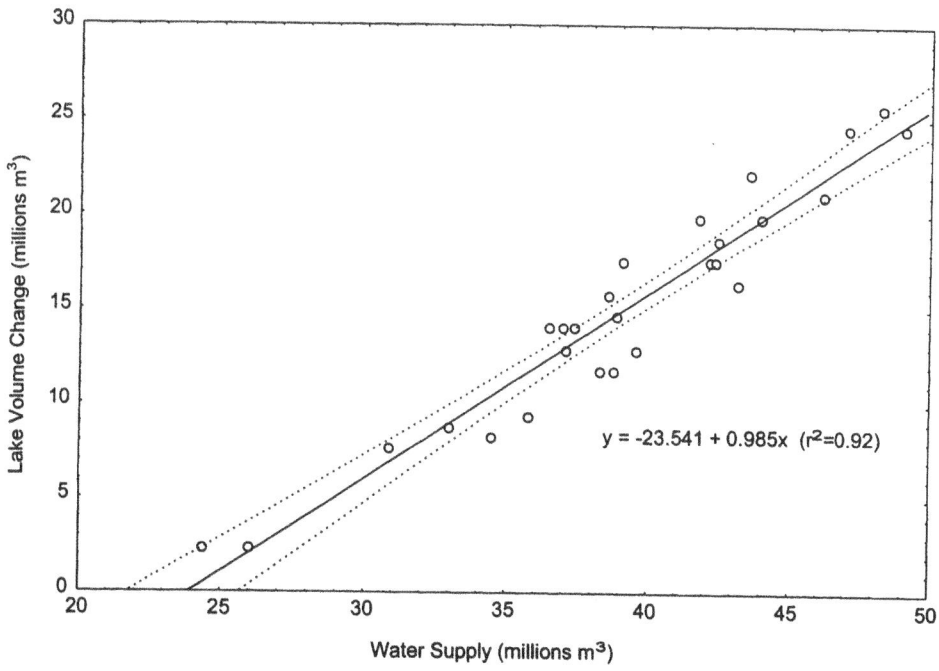

Figure 3. Water supply for the rainy seasons of 1971–1979 and corresponding changes in lake volume.

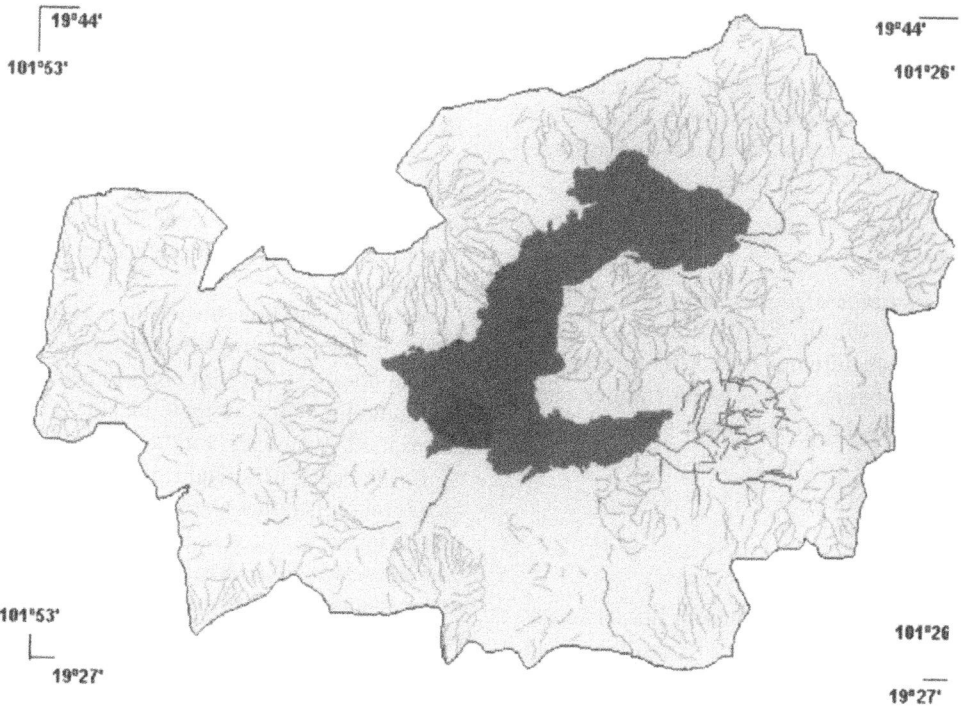

Figure 4. Lake Patzcuaro channels network.

Table 3. Morphometric and hydrologic characteristics of Lake Patzcuaro watershed.

	I Zurumutaro	II Patzcuaro	III Ajuno	IV Erongaricuaro	V Napizaro	VI San Andres	VII Quiroga	VIII Ihuatzio
Area, ha	17406	9015	12131	19296	1332	7350	15897	2221
Mean slope, %	11.27	14.34	13.06	14	91.22	15.43	12.34	88.29
Channels	132	47	101	166	18	52	178	20
Length of main channel, km.	53	49.6	37.05	19.75	11.5	30.9	56.85	5.95
Channels density, number/ha	0.76	0.52	0.83	0.86	1.35	0.71	1.12	0.9
Channels density, km/ha	1.25	0.55	0.97	1.19	2.31	0.96	1.57	0.9

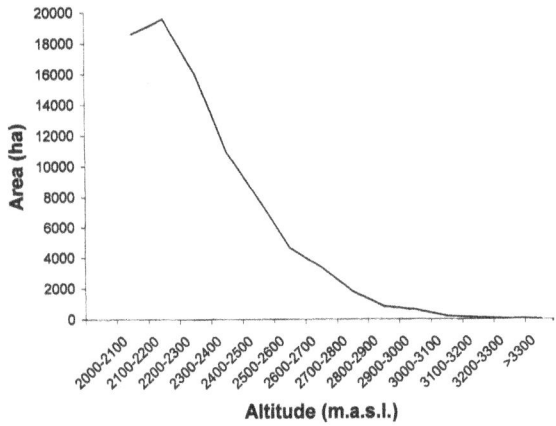

Figure 5. Lake Patzcuaro watershed area relative to the altitude.

(Fig. 3). Conversely, the volume loss during the dry season impacts the water level in a parallel decreasing sequence. Runoff involves 49% of W_s on average (Table 2).

GIS output

Lake Patzcuaro watershed (Fig. 4) measures 93 430 ha between 19° 27′ N, 19° 44′ N, 101° 26′ N, and 101° 53′ N, in an altitude differential of 1265 m (Fig. 5). Eight internal sub-basins (Fig. 6) bear implicit specific morphological and hydrological characteristics (Table 3).

Discussion

Goldman (1988) states that 'long-term data sets made it possible to detect eutrophication in Lake Tahoe and to sort out the climatic effects of rainfall, solar radiation, depth of winter mixing, and possibly in future the role of grazing by zooplankton on the annual primary productivity. To improve our understanding of both the structure and function of lakes, we need sustained data collection over many years in order to interpret lake processes that are often obscured in short time by interannual variation'. In Mexico, unfortunately, part of the limnological tradition that begun with Cuesta-Terrón, De Buen, Osorio-Tafall, Rioja, and Alvarez, has been lost and there are now only isolated efforts...' (Arredondo, 1987) mostly analogous to photographs taken in particular moments by different methods. Our case study, as well as case studies of some other lakes in Michoacan, demands a conceptual framework of regional applicability (Bernal-Brooks, 1998) based on reliable long-term databases. Here, we highlight the need to profoundly revise any past information available in regard to Mexican water bodies, especially that coming from anonymous or government sources, before proceeding to interpretations. The following drawbacks were detected for Patzcuaro:

Ground-references for water levels seem to be established by means of altimeter measurements, and the high variance associated with this method imposes limitations when comparing data between the different time-series.

De Buen (1944b:108–9, Fig. 4) data on water levels include an error between September 1, 1941, and February 28, 1942. In order to be consistent with the figure, the typist forgot to include 2035 before the corresponding numbers for those dates.

In the case of the 2nd time-series, the scales for lake-level registration were changed several times and moved to other places in the nearby, trying to maintain the zero at 2034.08 m.a.s.l. Since the first wooden scale attached to a tree (October 19, 1949), scale re-locations proceeded as the lake level went down (December 23, 1949; March 24, 1952) and up (May 22, 1962), respectively. Sometime during the 1960s, a metal scale was established permanently, although it currently lies over dry land some 100 m away from

Figure 6. Lake Patzcuaro basin and sub-basins in a perspective view from Cerro 'El Zirate'.

the shoreline. An unknown ground reference at the town of Quiroga (2075 m.a.s.l.) was not indicated in the reference paper.

Data contained in the third time-series appear with a difference of +2.43 m relative to the 2nd time-series, and also was moved one position (the number for February should be in the place of January, as it is in the 2nd time series) for unknown reasons. However, this sequence by 1987 connects smoothly with the 4th time-series, pointing out a convenient anonymous intervention to change Ref. d for Ref. b.

The time-series on climatic variables frequently includes missing data or some of them doubtful in the case of evaporation.

CNA (1995) reissued the climatic information provided by the SMM (1922–1986) for all the meteorological stations in Mexico but, in the case of the one numbered 16087, the evaporation data appear different and significantly lower than they are in the original file.

The pertinent data corrections applied reveal cyclic movements (<6 m) at the lake surface in the long-term of approximately every 40 years, including also short-term seasonal fluctuations (<1 m). Figure 7 show the whole assembly of water-level positions over time be-

fore and after verification in the field and corrections. The maximum depth of 15 m determined at Station VII on February 17, 1941 (De Buen, 1941c, 1944b), approximates the current maximum depth (10-11 m) following the rearrangement made on the water levels time-series, which also served to double-check the recently developed database.

Chacon (1989, 1992, 1993a, b) model of water balance bears implicit high evaporation rates at the terrestrial watershed able to prevent the development of surface runoff over the topsoil. In his own words (Chacon, 1993a, p. 32): 'On an annual basis, if the average monthly precipitation (with the exception of the rainy season) is less than the average monthly evaporation, then any runoff and precipitation over the lake surface would have negligible effects in the lake water levels'. However, the shape of seasonal cycles highlights an alternated influence in the lake water level of either precipitation or evaporation predominance (Osorio-Tafall, 1944; De Buen, 1944b) rather than the arrival of seepage flows. The speed of rain infiltration underground and then, towards the lake, requires further investigation to accept or reject the hypothesis of water movement in terms of one season. However, such a complex commitment involving land use, soil

194

Figure 7a.

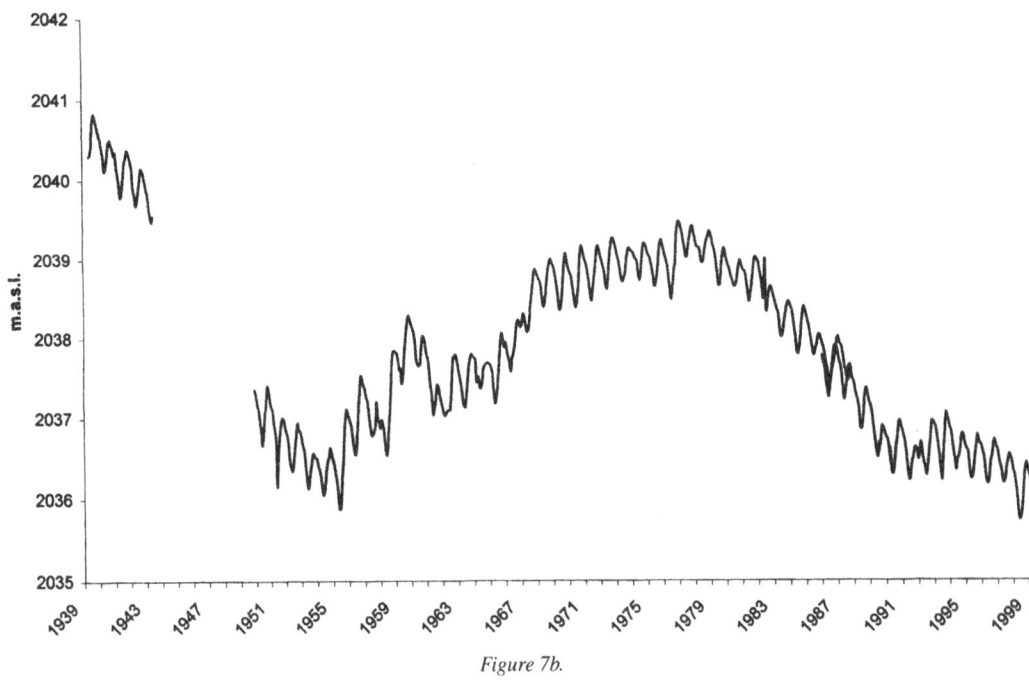

Figure 7b.

Figure 7. Lake Patzcuaro water levels, (a) before and (b) after rearrangement of data and corrections applied. (Time series 1: 1939–1943; Time series 2: 1949–1970; Time series 3: 1950–1987; Time series 4: 1986–1990; Time series 5: 1989–).

type, and erosion degrees and slopes, led us to consider a shortcut by the analysis of lake water levels versus climatic variables. Herewith, a close relationship ($r^2 = 0.92$) between the lake level fluctuations and water additions (Fig. 3) addresses the significant role of runoff as a hydrological component in combination with direct precipitation over the lake surface.

GIS output shows the existence of abundant runoff channels in intimate interaction with the land features mentioned above. Along the watershed, sub-basins I–III include more channels, main water courses at maximum length, and low to intermediate channel density running over smoother slopes of null to slight soil erosion. In contrast, sub-basins V and VIII with less area than the other sub-basins, have steeper slopes, shorter water channels and high channel density in deteriorated grounds where the hydrologic erosion has caused strong to severe impacts. Sub-basins IV, VI and especially VII, maintain a hybrid situation between the two extremes mentioned above. Furthermore, land use following deforestation, agriculture, and extensive cattle management, catalyzes an hydrological network deterioration due to vegetation cover reduction and the stimulation of damages caused to the grounds by increasing water erosion. Spatial and temporal variations in water additions seem to induce variable R values between years as detected in our analysis. A parallel study (Gomez-Tagle, 1997) shows the ultimate consequence of land mismanagement in terms of erosion.

Other arguments in support to the presumed climatic sensitivity of the closed basins located in Central Mexico as a response to variations in rainfall amounts through time are as follows:

Hutchinson et al. (1956, p. 1493) considered that 'the high level of Nahua time, which permitted Cortes to conquer Tenochtitlan (Mexico City today), is also established for Patzcuaro by the evidence of the Beaumont map. Thus, while the archaeological data from Michoacan are scanty, they provide some independent evidence for the view that moist and dry phases were synchronous in Michoacan and in the Valley, and must have been controlled by climate'.

Wetzel (1983, p. 38) states that:

"Much of the intercepted water and surface runoff (up to 80%) is returned to the atmosphere by evaporation. Infiltrated water may be temporarily stored (average renewal time approximately 280 days) as soil moisture prior to being evapotranspired. Some of the water percolates to deeper zones to be stored as groundwater (average re-

newal of 300 years). Groundwater is actively exchanged and may be used by plants, flow out as springs, or seep as runoff. The runoff phase is exceedingly complex and variable because of the extensive involvement with biota, the extreme heterogeneity of soil structure and composition, and variation in climate. . . . When the rate of rainfall or melt water influx exceeds the adsorptive capacity of the soil, the excess water flow over the surface as overland flow. Overland flow is most common in arid and semiarid regions. . .".

The same author considers groundwater seepage below the surface for lakes in rock basins and lake basins in glacial till that extend below the water table. This is not the case for Lake Patzcuaro.

Watts and Bradbury (1982) and O'Hara (1991), quoted by Metcalfe et al. (1994, p. 139), pointed out that 'the level of Lake Patzcuaro has fluctuated considerably over both the long and the short term. There is no evidence that the lake has ever desiccated completely or had a clear tendency to dry up in the 44 000 years of its known existence'.

Harris (1994, p. 83) considers that 'measurements of lake-volume changes. . . provide a climate record. In particular, closed lakes. . . by integrating the precipitation over their catchment basins, provide a fuller picture of precipitation changes than the more localized *in situ* measurements, particularly in arid and semiarid regions, where such monitoring is often sparse'.

Grimm et al. (1927, 1997, Fig. 1) shows a map of western North America, including the Basin and Range, Southwest and Mexico region, with runoff ratios (runoff/precipitation) in the range of 0.1–0.2 for the area of Lake Patzcuaro.

Bernal-Brooks and MacCrimmon (2000) observed parallel movements in the water-level fluctuations of some lakes in the region of Michoacan (Zirahuen, Patzcuaro, Cuitzeo and Chapala), which support the argument of climatic sensitivity for them all.

Gomez-Tagle Chavez et al. (2002) analyzed aerial photographs from Lake Patzcuaro in order to set up a methodological basis for the graphic demonstration of climatic sensitivity through detection of changes in area over time.

A Global Lake and Catchment Conservation Database (GLCCD) under development includes some of the closed basins in Central Mexico as potential water bodies sensitive to climate change. Lake Cuitzeo and Lake Chapala appear on the basis of a threshold area of detection (>100 km^2) inherent to the satel-

196

lite technology used (Birkett & Mason, 1995). Long- and short-term water cycles seem to be an expression of the lake and its catchment functioning relative to climate change.

A decline in lake water level since 1979 in Lake Patzcuaro, along with the observations made by Williams (1993) in saline lakes, suggests an impairment of rainfall for the last 20 years, as well as that Lake Patzcuaro is a valuable study case for monitoring the regional climatic change. The answer to how much of the tendency to lower water levels responds to a natural cycle or an anthropogenic influence remains as a challenge. The next decade will demonstrate if a natural progression to wetter conditions will appear as it occurred during the 1970s or a global warming effect irremediably announces our lakes' disappearance and human life constraints in the area. A monitoring of lake levels in this region became an imperative for the Comisión de Pesca, Residencia en el Lago de Pátzcuaro, at the Government of Michoacan.

Finally, an ancillary observation made on another example based on altimeter measurements. Cruz (1995), on the basis of data obtained by De Buen (1943), reached the conclusion of a 'transgression stage' (*sensu* De Buen) for Lake Zirahuen, a water body located 17 km from Lake Patzcuaro. Again, data verification must be carefully undertaken before interpretation in order to avoid misuse of data to support prescribed theories. Certainly, in the study of Mexican waters, frequent and detailed observations based on standard methodology and systematized in long-term databases should become a priority, as Goldman (1988) shows in the case of Lake Tahoe.

Acknowledgements

This research would not have been possible without the participation of the topography team at the Comisión de Pesca del Gobierno del Estado de Michoacan, and the kindness of its Director, Biól. Pedro Tamayo Díaz. The authors appreciate the willingness of MC Araceli Orbe Mendoza (CRIP Patzcuaro) and Yolanda Chavez (INIFAP Campo Morelia) to provide us with facilities and equipment. Financial support was given, in part, by CONACyT project 3626P-B9608. We thank H. V. Grey for her valuable linguistic revision.

References

Ancona, I., M. A. Batalla, E. Caballero, C. C. Hoffman, R. Llamas, R. Martín del Campo, I. Ochoterena, E. Rioja, J. Roca, A. Sámano, C. Vega & F. Villagrán, 1940. Prospecto biológico del lago de Pátzcuaro. An. Inst. Biól. (Méx.). 11: 415–503

Arredondo, J. L., (1983) 1987. Bosquejo histórico de las investigaciones limnológicas realizadas en los lagos mexicanos, con especial enfasis en su ictiofauna. In Gomez-Aguirre, S. & V. Arenas-Fuentes (eds), Contribuciones en Hidrobiología. Proceedings of the meeting "Alejandro Villalobos" University of Mexico, 24–26 October 1983.

Barbour, C. D., 1973. A biogeographical history of *Chirostoma* (Pisces: Atherinidae): a species flock from the Mexican plateau. COPEIA 3: 553–556.

Barrera, N., 1992. Ecogeografia. In Toledo, V. M., P. Alvarez-Icaza & P. Avila (eds), Patzcuaro 2000. Fundación Friedrich Ebert Stifung, Mexico, D.F.: 11–35.

Bernal-Brooks, F. W., 1998. The lakes of Michoacan (Mexico): a brief history and alternative points of view. Freshwater Forum 10: 20–34.

Bernal-Brooks, F. W. & H. R. MacCrimmon, 2000. Lake Zirahuen (Mexico): an assessment of the morphometry change based on evidence of water level fluctuations and sediment inputs. In Munawar, M., S. Lawrence, I. Munawar & D. Malley (eds), Aquatic Ecosystems of Mexico. in press.

Birkett, C. M. & I. A. Mason, 1995. A new global lakes database for a remote sensing programme studying climatically sensitive large lakes. J. Great Lakes Res. 21: 307–318.

CETENAL (Centro de Estudios sobre el Territorio Nacional), 1987. Topographic charts E14A21, E14A22, E14A31 AND E14A32. 3rd edn.

Chacon, T. A., 1989. A limnological study of Lake Patzcuaro, Mexico, with a consideration of the applicability of remote sensing techniques. PhD thesis, Stirling Univ. Scotland: 340 pp.

Chacon, T. A., 1992. El Ecosistema Lacustre. In Toledo, V. M., P. Alvarez-Icaza & P. Avila (eds), Patzcuaro 2000. Fundación Friedrich Ebert Stifung, Mexico, D.F.: 37–70.

Chacon, T. A., 1993a. El Lago de Pátzcuaro, Michoacán, México. Bosquejo de un Lago Amenazado. Ed. Universitaria. Universidad Michoacana de San Nicolás de Hidalgo.

Chacon, T. A., 1993b. Lake Patzcuaro, Mexico: watershed and water quality deterioration in a tropical high-altitude Latin American Lake. Lake Reserv. Mgmt. 8: 37–47.

CNA (Comisión Nacional del Agua), 1995. A database of climatic variables registered by SSM (1969–1986).

Cruz, O., 1995. Balance hídrico en la cuenca del lago de Zirahuén. BSc thesis, University of Michoacan: 78 pp.

De Buen, F., 1941a. Dos cortas campañas limnológicas en el lago de Patzcuaro (febrero y julio de 1941). Investigaciones de la Estación Limnológica de Pátzcuaro 10: 1–16

De Buen, F., 1941b. El lago de Pátzcuaro. Rev. Geogr. 1: 20–44.

De Buen, F., 1941c. Las variaciones físicas y químicas de las aguas del lago de Pátzcuaro (st. X) desde octubre de 1939 a marzo de 1941. Investigaciones de la Estación Limnológica de Pátzcuaro 7: 1–25.

De Buen, F., 1943. Los lagos michoacanos. I. Caracteres generales. El lago de Zirahuén. Rev. Soc. Mex. Hist. Nat. 4: 211–232.

De Buen, F., 1944a. Limnobiología de Patzcuaro. Anales del Instituto de Biología (Mexico) 15: 261–312.

De Buen, F., 1944b. Los lagos michoacanos. II. El lago de Patzcuaro. Rev. Soc. Mex. Hist. Nat. 5: 99–125.

Deevey, E. S., 1957. Limnological studies in Middle America with a chapter on Aztec limnology. Trans. Connect. Acad. Arts Sci. 39: 213–328.

Demant, A., 1975. Les quatre provinces volcaniques du Mexique, relations avec l'evolution geodynamique, depuis le Cretace, II. Les deux provinces occidentales. C.r. Acad. Sci. 280: 1437–1440.

DGG (Dirección General de Geografía), 1991. Data file for the geodesic point BNT-601 established by CETENAL in 1974.

Goldman, C. R., 1988. Primary productivity, nutrients, and transparency during the early onset of eutrophication in utra-oligotrophic Lake Tahoe, California–Nevada. Limnol. Oceanogr. 33: 1321–1333.

Gomez-Tagle, A., 1997. Levantamiento Agrológico Forestal de la Cuenca de Patzcuaro, Mich., y Diagnóstico de Posibilidades de Recuperación y Desarrollo mediante Sistemas de Informacion Geográfica. PhD thesis. UNAM: 131 pp.

Gomez-Tagle, A., F. W. Bernal-Brooks & J. Alcocer, 2002. Sensitivity of Mexican water bodies to regional climatic change: three study alternatives applied to remote sensed data of Lake Patzcuaro. Hydrobiologia 467 (Dev. Hydrobiol. 163): 169–176.

Grimm, N. B., A. Chacon, C. N. Dahm, S. W. Hostetler, O. T. Lind, P. L. Starkweather & W. W. Wurtsbaugh, 1997. Sensistivity of aquatic ecosystems to climatic and anthropogenic changes: the Basin and Range, American Southwest and Mexico. Hydrological Processes 11: 1023–1041

Harris, A. R., 1994. Time series remote sensing of a climatically sensitive lake. Remote Sens. Envir. 50: 83–94

Hutchinson, G. E., Patrick, R. & E. S. Deevey, 1956. Sediments of Lake Patzcuaro, Michoacan, Mexico. Bull. geol. Soc. am. 67: 1491–1504

Metcalfe, S., F. A. Street-Perrot, S. O'Hara & P. E. Hales, 1994. The paleolimnological record of environmental change: examples from the arid frontier in Mesoamerica: In Millington, C. & K. Pye (eds), Environmental Change in Drylands: Biogeographical and Geomorphological Perspectives. Wiley & Sons: 131–145.

O'Hara, S. L., 1991. Late Holocene environmental change in the basin of lake Patzcuaro, Michoacan, Mexico. PhD thesis, Oxford University.

Osorio-Tafall, B. F., 1941. Materiales para el estudio del microplancton del lago de Pátzcuaro (Mexico). Anales de la Escuela Nacional de Ciencias Biológicas (Mexico) 2: 331–384.

Osorio-Tafall, B. F., 1944. Biodinámica del lago de Pátzcuaro. I. Ensayo de interpretación de sus relaciones tróficas. Rev. Soc. Mex. Hist. Nat. 5: 197–227.

Planas, D. & G. Moreau, 1990. Natural eutrophication in a warm volcanic lake. Verh. int. Ver. Limnol. 24: 554–559.

SARH (Secretaría de Agricultura y Recursos Hidráulicos), 1949–1970. Lake Patzcuaro water levels. (2nd time-series).

SARH (Secretaría de Agricultura y Recursos Hidráulicos), 1950–1987. Lake Patzcuaro water levels. (3rd time-series).

SCT (Secretaría de Comunicaciones y Transportes), 1986–1990. Lake Patzcuaro water levels. (4th time-series).

SDAF (Secretaría de Desarrollo Agropecuario y Forestal), 1989–. Lake Patzcuaro water levels (5th time-series).

SMN (Servicio Meteorológico Nacional), 1922–1986. Climatic variables registered at 16087 Meteorological Station.

Tamayo, J. L. & R. C. West, 1964. The hydrology of Middle America. In Wauchope, R. & R. C. West (eds), Handbook of Middle America Indians, I. University of Texas, Austin.

Watts, W. & J. P. Bradbury, 1982. Paleoecological studies at Lake Patzcuaro on the west-central Mexican plateau and at Chalco in the basin of Mexico. Quat. Res. 17: 56–70.

Wetzel, R., 1983. Limnology. CBS Colege Publishing Co.: 767 pp.

Williams, W. D., 1993. The worldwide occurrence amd limnological significance of falling water-levels in large, permanent saline lakes. Verh. int. Ver. Limnol. 25: 980–983.

Zozaya, M., 1941. Observaciones termopluviométricas en Pátzcuaro, Mich. Investigaciones de la Estación Limnológica de Pátzcuaro 2: 1–14.

Hydrobiologia **467**: 199–213, 2002.
J. Alcocer & S.S.S. Sarma (eds), Advances in Mexican Limnology: Basic and Applied Aspects.
© 2002 *Kluwer Academic Publishers.*

Developing diatom-based transfer functions for Central Mexican lakes

S.J. Davies[1], S.E. Metcalfe[1], M.E. Caballero[2] & S. Juggins[3]
[1]*Department of Geography, University of Edinburgh, Drummond St, Edinburgh EH8 9XP, U.K.*
[2]*Instituto de Geofisica, Universidad Nacional Autonoma de Mexico, Av. Universidad, Copilco, Mexico DF, Mexico*
[3]*Department of Geography, University of Newcastle-upon-Tyne, Daysh Building, Newcastle NE1 7RU, U.K.*

Key words: Mexico, diatoms, palaeolimnology, transfer function, CCA

Abstract

This paper is the first attempt to produce diatom-based transfer functions for the northern tropical Americas. A dataset of 53 modern diatom samples and associated hydrochemical variables from 31 sites in the volcanic highlands of central Mexico is presented. The relationship between diatom species distribution and water chemistry is explored using canonical correspondence analysis (CCA) and partial CCA. Variance partitioning indicates that ionic strength and ion type both account for significant and independent portions of this variation. Transfer functions are developed for electrical conductivity ($r^2 = 0.91$) and alkalinity (as a percentage of total anions) ($r^2 = 0.90$), reflecting ionic strength and ionic composition respectively. Prediction errors, estimated using jack-knifing, are low for the conductivity model, but the carbonate transfer function performs less well. This study highlights the potential for diatom-based quantitative palaeoenvironmental reconstructions in central Mexico. However, a number of key diatom species found in fossil material are not represented in the modern flora. Sampling of additional sites may resolve this, but it is thought that the lack of modern analogues may reflect the high degree of anthropogenic disturbance in many of the catchments. This highlights the problem of trying to reconstruct pre-disturbance environmental changes in highly modified ecosystems. One possible solution is to merge the central Mexican data with the African dataset, which includes sites of similar chemical composition, but which have not suffered the same degree of disturbance.

Introduction

Lakes in closed basins are sensitive to climatic fluctuations (Street-Perrott & Harrison, 1985; Fritz et al., 1993; Gasse et al., 1997). Variations in effective moisture, and hence water balance, are reflected in changes in ionic strength and composition of lake waters through dilution and evaporative concentration (Eugster & Hardie, 1978; Fritz et al., 1999). As a result, the physical, chemical and biological properties of the lake and its sediments change through time. Sediment cores from such basins therefore provide excellent opportunities for the reconstruction of past water chemistries (De Deckker & Forester, 1988). These records can then be used to infer periods of increased or decreased effective moisture, providing information on climatic variability through time. In the last 20 years, the application of diatoms in palaeoenvironmental research has increased dramatically. More

information is now available on the ecological requirements of diatoms in different parts of the world (e.g. Gasse et al., 1983; Servant-Vildary & Roux, 1990; Jones et al., 1993). This work has largely been driven by the need for ecological information in order to interpret palaeolimnological records accurately.

Significant progress in diatom-based palaeoenvironmental reconstructions has been made during the last decade due to the development of quantitative methods (Birks, 1998). Clear correlations have been established between species composition and pH (e.g. Gasse & Tekaia, 1983; Birks et al., 1990), salinity (e.g. Cumming & Smol, 1993), nutrient content (Bennion, 1994) and ionic composition (e.g. Gasse et al., 1983; Fritz et al., 1993). Modern calibration datasets have been established for a number of regions. For example, a dataset of 282 modern diatom samples from Africa has been used to create transfer functions for conductivity, pH and ionic ratios (Gasse et al., 1995). In the northern Great Plains region of North America, a salinity transfer function has been developed from a 66 lake dataset (Fritz, 1990; Fritz et al., 1991).

Diatom-based transfer functions have not been developed in the northern American tropics. Indeed, there is very little ecological information on diatoms from this region at all. Haberyan et al., (1997) report that cation concentration and related variables, such as hardness and pH have a significant influence on diatom species composition in a 25 lake dataset from Costa Rica. In the Bolivian altiplano Servant-Vildary & Roux (1990) found that diatom assemblages from saline lakes were strongly linked to ionic composition. Such studies highlight the potential for quantitative diatom-based reconstructions in the tropical Americas.

Mexico lies in a sensitive transition zone between temperate and tropical climatic regimes and is a key area for palaeoclimatic research. Closed-basin lakes in the highlands of central Mexico have been the subject of a number of diatom-based palaeolimnological studies (Bradbury, 1989; Metcalfe et al., 1991; Metcalfe & Hales, 1994; Metcalfe, 1995; Caballero & Ortega, 1998; Caballero et al., 1999). To date, interpretation of these records has been qualitative, based on ecological data from other regions and the limited information on the ecology of diatoms in Mexico. Previous studies on diatoms in central Mexico have identified large ecological gradients between different sampling sites, although the relationships between water chemistry and diatom assemblages have not been quantified (Bradbury, 1971, 1989; Metcalfe, 1988; Caballero, 1995). Metcalfe (1988) used ordination techniques to examine species–environment relationships but found that habitat differences between the samples tended to mask the effects of water chemistry variations. Caballero (1995), using a CCA on a relatively small dataset of 9 different water bodies, found that ionic composition along a gradient between Ca–Mg–HCO_3 to Na–Cl water chemistries was the main factor determining the distribution of diatom species.

The development of diatom-based transfer functions for central Mexican lakes is important for a number of reasons. It will allow quantitative interpretation of previous palaeolimnological studies and provide a basis for future investigations. The information can be used to test whether ecological requirements are comparable between regions by comparison with datasets from, for example, Africa or North America. If this is the case, then a long-term goal is to merge the Central Mexican dataset with those of other regions. This would increase the accuracy of transfer func-

tions as the coverage of chemical gradients would be greater (Gasse et al., 1995). It would also make it more likely that modern analogues for fossil data would be encountered (Gasse et al., 1997).

In this paper, a diatom calibration dataset, based on 54 surface sediment samples from 31 sites within the Trans-Mexican Volcanic Belt, is presented. Transfer functions have been developed for electrical conductivity (E.C.) and alkalinity (expressed as a percentage of $HCO_3^- + CO_3^{2-}$), reflecting ionic strength and ionic composition respectively. The dataset includes samples collected during earlier investigations (Metcalfe, 1985, 1988; Caballero, 1995) and new additions hitherto unreported.

Study area

The Trans-Mexican Volcanic Belt (TMVB) cuts across the country in a west-east orientation between 18 and 22° N, representing the southern limit of the Central Plateau. Primarily a product of Quaternary volcanism, much of the terrain within the TMVB is mountainous, lying above 2000 m a.s.l. A number of large stratovolcanoes also lie within these highlands, such as the Pico de Orizaba, Popocatepétl, Iztaccihuatl, and the Nevado de Toluca, all of which attain an altitude greater than 4000 m above sea level.

The climate of the TMVB is largely influenced by the Sub Tropical High Pressure Belt and the easterly trade winds, although altitude and relief are important in causing significant local variability (Cavazos & Hastenrath, 1990). The average temperature is ca. 15 °C, with frosts often occurring during the winter, due to the high altitude. The precipitation regime is highly seasonal with 80% of the annual rainfall occurring between June and September when a more northerly location of the inter-tropical convergence zone brings a deep easterly flow over the region (Mosiño-Aleman & García, 1974). Average annual precipitation is ca. 1000 mm yr^{-1} in the southern, humid temperate, part of the TMVB, decreasing northwards towards the semi-arid central plateau to around 500 mm yr^{-1}. The northern fringes of the volcanic highlands experience greater potential evaporation than southern parts, leading to an overall moisture deficit of as much as 1000 mm yr^{-1}. The effects of altitude greatly modify this climatic gradient. For example, the central part of the Basin of Mexico (2240 m a.s.l.) has a moisture deficit of approximately 1000 mm yr^{-1}, whilst the Nevado de Toluca (4600 m a.s.l.), approximately 80 km to the west, has a surplus of 318 mm yr^{-1}(Caballero, 1995). Given that water balance is one of the major factors affecting the chemical composition of lake waters (Eugster & Hardie, 1978), lakes along the TMVB should reflect these large variations in effective moisture.

A large number of closed lacustrine basins of tectonic and volcanic origin are found within the TMVB (Fig. 1). The large, hydrologically complex basins are The Basin of Mexico, on which Mexico City is built and the Oriental Basin, lying to the east. Other closed systems occur in the state of Michoacán (e.g. Cuitzeo, Pátzcuaro, Zirahuén and Zacapu) and in the state of Jalisco (e.g. Cajititlán, Atotonilco and Juanacatlán). Numerous crater lakes also lie along the TMVB, for example in the Valle de Santiago, in the Oriental Basin and on the Nevado de Toluca.

The large variety of aquatic habitats found within the volcanic highlands, along with a clear climatic gradient, mean that there is a large range in terms of physical and chemical properties of these water bodies (Metcalfe, 1988). Sampling sites have been chosen in order to reflect this diversity. The location of sites in the Central Mexican Diatom Dataset is illustrated in Figure 1. Further site information is found in Table 1.

Methods

Sample collection, preparation and analysis

Collection of modern diatom samples has been carried out during three separate periods: 1982 (SEM), 1991–1992 (MCM) and 1997–1998 (SJD). In all cases, the majority of sampling was undertaken during the dry season. For most lakes in the dataset, a number of different habitat types have been sampled. However, for the purposes of this study, only surface sediment samples have been included in the calibration dataset as these are felt to be most analogous to core assemblages, representing an average composition of species within the lake. This also avoids the problem of habitat preferences masking water chemistry variations as reported by Metcalfe (1988). Surface sediment samples were extracted using a grab from the deepest part of the lake wherever possible.

The choice of sampling sites within the TMVB reflects the area of focus of palaeolimnological investigations, with sampling clustered around the Basin of Mexico and the Michoacán lake district. Additional sites, such as the crater lakes of the Oriental Basin

(2, 15, 18) and the Nevado de Toluca (11, 12) have also been included to ensure a good range of physical and chemical properties. Most recently, sampling has focused on perceived gaps within the dataset. The dataset currently includes 54 diatom samples from 31 different sites within the TMVB (Fig. 1, Table 1), although addition of new samples is ongoing.

Electrical conductivity (E.C.) and pH were measured in the field and a note was made of any anthropogenic influences in the basin. Water samples were analysed for major anions and cations in the laboratories of the Instituto de Geografía and the Instituto de Geofísica, Universidad Nacional Autónoma de México, following standard methods described in Metcalfe (1985) and Armienta et al. (1987). Samples collected by MCM were not analysed directly for sulphate, these values were inferred by the cation-anion balance. For some lakes, where two or three surface sediment samples were taken at the same time, only one set of water chemistry data is available.

Diatom samples were treated with HCl and H_2O_2 as described in Battarbee (1986) to remove carbonate and organic material, then cleaned by washing in distilled water and centrifuging several times. A small amount of the clean sample was evaporated onto coverslips and then mounted onto slides using Naphrax® resin. Counts were carried out at 1000× magnification using an Olympus BX 50 microscope (SJD) and an Olympus VANOX with Normarski illumination (SEM). A Zeiss photomicroscope at 1250× magnification was also used (MCM). At least 400 valves were counted in each sample. The main diatom floras used for taxonomic identification were Krammer & Lange-Bertalot (1986, 1988, 1991a, 1991b), Patrick & Reimer (1966, 1975), Germain (1981), Gasse (1980, 1986) and Hustedt (1930–1966). The taxonomic nomenclature used in this study is based on that prior to the new generic names proposed by Round et al. (1990). Where appropriate, the new generic names have been placed in brackets in the text. Every effort has been made to ensure that the taxonomy is consistent between the three researchers involved in the identification of diatom species.

Data analysis

The final dataset of 53 diatom samples consisted of counts (expressed as relative abundances) for 186 taxa that were present in at least two lakes and reached a maximum abundance of at least 2.0%. A sample from Lago Atotonilco (4) was removed from the data-

Table 1. Selected geographical information, sample codes and water chemistry data for samples included in the dataset

Lake	Type	Lat.	Long	Alt. (m.a.s.l.)	Sample code	Date sampled	pH	E.C.	K meq	Na meq	Mg meq	Ca meq	SO4 meq	Cl meq	CO3 meq
Hoya La Alberca, Zacapu Basin	Crater Lake	19 52	101 50	1980	1i, 1ii	23/06/82	9.00	3600	0.51	6.09	6.25	2.15	0.34	5.64	6.00
Lago Alchichica	Crater Lake	19 24	97 24	2345	2a	08/04/97	9.31	10140	5.91	103.31	36.13	0.51	17.71	84.63	41.13
					2b	01/08/82	8.80	7521	5.90	108.10	33.70	0.25	24.90	85.10	28.75
El Angulo, Zacapu Basin	Irrigation channel	19 54	101 45	1980	3	21/06/82	7.30	305	0.06	3.04	0.08	1.10	1.37	0.21	1.80
Lago Atotonilco	Lake	20 34	103 70	1600	4	29/03/97	9.69	11260	3.90	139.01	1.76	2.59	19.28	44.43	78.04
Lago Cajititlan	Lake	20 42	103 35	1570	5	30/03/97	8.65	756	0.87	4.49	1.99	1.68	0.15	0.85	6.94
Lago Chapala	Lake	20 15	103 50	1523	6	29/03/97	8.59	954	0.70	5.39	10.18	2.99	1.77	1.13	7.43
Lago de Chalco	Lake	19 15	98 55	2240	7i, 7ii	16/01/91	9.20	2420	1.25	23.70	8.81	4.03	124.31	4.79	18.71
Lago de Juanacatlan	Lake	20 37	104 44	1995	8i, 8ii	25/03/97	8.78	148	0.06	0.30	0.35	0.86	0.11	0.06	1.39
Lagunilla de San Gregorio	Lake	19 25	101 30	3100	9	15/09/98	6.45	30	0.03	0.06	0.10	0.16	0.08	0.03	0.24
Lago de Yuriria	Lake	20 15	101 10	1730	10	31/03/97	8.69	1885	1.32	14.80	3.11	4.15	0.21	4.01	16.84
Lago de la Luna	Crater Lake	19 10	99 45	4550	11i, 11ii	03/03/91	4.90	17	0.00	0.00	0.89	0.00	0.79	0.00	0.10
Lago del Sol	Crater Lake	19 10	99 45	4550	12i, 12ii, 12iii	03/03/91	5.80	24	0.00	0.00	0.32	0.00	0.12	0.00	0.20
Lago de Patzcuaro	Lake	19 35	101 39	2044	13a	28/03/82	7.50	800	0.69	6.05	3.04	0.67	1.03	0.56	6.70
					13b	03/04/97	8.86	1021	1.07	6.64	3.31	0.75	1.12	0.91	10.48
Lago Potrerillo	Pool	20 34	104 15	1250	14	27/03/97	7.45	158	0.27	0.35	0.44	0.68	0.02	0.08	1.41
Lago Preciosa	Crater Lake	19 21	97 22	2365	15i, 15ii	08/04/97	8.90	2090	0.44	8.11	16.29	0.76	2.74	7.84	13.28
La Piscina de Yuriria	Crater Lake	20 13	101 08	1730	16a	08/04/82	11.00	26000	30.68	478.50	0.02	0.00	29.34	174.25	300.00
					16b	31/03/97	10.18	8130	6.99	95.79	1.86	5.91	9.19	27.50	68.25
Querendaro	Pool	19 49	100 57	2070	17	19/06/82	6.60	180	0.16	0.78	0.64	0.72	0.51	0.85	1.80
Laguna Quechulac	Crater Lake	19 21	97 21	2395	18	08/04/97	8.82	840	0.19	3.17	4.42	1.01	0.35	2.06	5.17
Tarajero, Zacapu Basin	Pool	19 49	101 43	1980	19	21/06/82	7.80	195	0.07	0.67	0.93	0.73	0.51	0.14	2.20
Lago Tecocomulco	Lake	19 50	98 21	2500	20	02/04/92	6.80	250	0.49	1.57	0.70	0.74	1.26	0.24	2.00
Lago Recreativo (Texcoco)	Lake	19 30	99 00	2236	21i, 21ii	16/04/92	9.60	7500	2.30	54.80	12.45	2.20	0.00	67.69	2.07
Jalapango marsh (Texcoco)	Marsh	19 30	99 00	2236	22i, 22ii	16/04/92	9.10	2360	2.62	36.10	1.92	1.00	21.38	7.61	12.63
Artemia Pond (Texcoco)	Pool	19 30	99 00	2236	23	16/04/92	9.50	44100	25.58	434.95	2.06	4.13	2.04	259.47	207.00
Xochimilco	Irrigation channel	19 17	99 10	2240	24	04/08/82	8.00	638	0.30	4.40	3.45	2.14	2.20	1.47	5.70
Lago Zacapu	Lake	19 49	101 48	1980	25i, 25ii, 25iii	23/04/82	8.80	146	0.01	0.58	0.50	0.43	0.69	0.14	1.80
Zacapu Celanese	Drainage channel	19 50	101 48	1980	26i, 26ii, 26iii	19/04/82	8.80	1000	0.06	9.57	0.44	0.60	10.47	0.42	1.40
Zempoala stream	Stream	19 03	99 19	2730	27	02/02/91	7.40	105	0.05	0.20	0.42	0.45	0.17	0.04	0.91
Lago Zempoala	Lake	19 03	99 19	2730	28i, 28ii, 28iii	02/02/91	7.70	91	0.03	0.13	0.33	0.45	0.04	0.04	0.84
Lago Compila (Zempoala)	Lake	19 03	99 19	2730	29i, 29ii	02/02/91	6.50	102	0.05	0.17	0.51	0.45	0.17	0.03	0.96
Lago de Zirahuen	Lake	19 21	101 46	2075	30ai, 30aii	02/04/97	8.40	119	0.09	0.25	0.49	0.44	0.01	0.10	0.10
					30b	20/06/82	8.60	102	0.06	0.19	0.43	0.47	0.34	0.00	1.70
Lago Zumpango	Lake	19 46	99 07	2242	31ai, 31aii	13/02/91	7.20	507	0.28	2.20	0.89	1.85	0.45	0.71	4.02
					31b	13/02/91	7.30	508	0.29	2.33	0.84	1.80	0.47	0.72	4.03

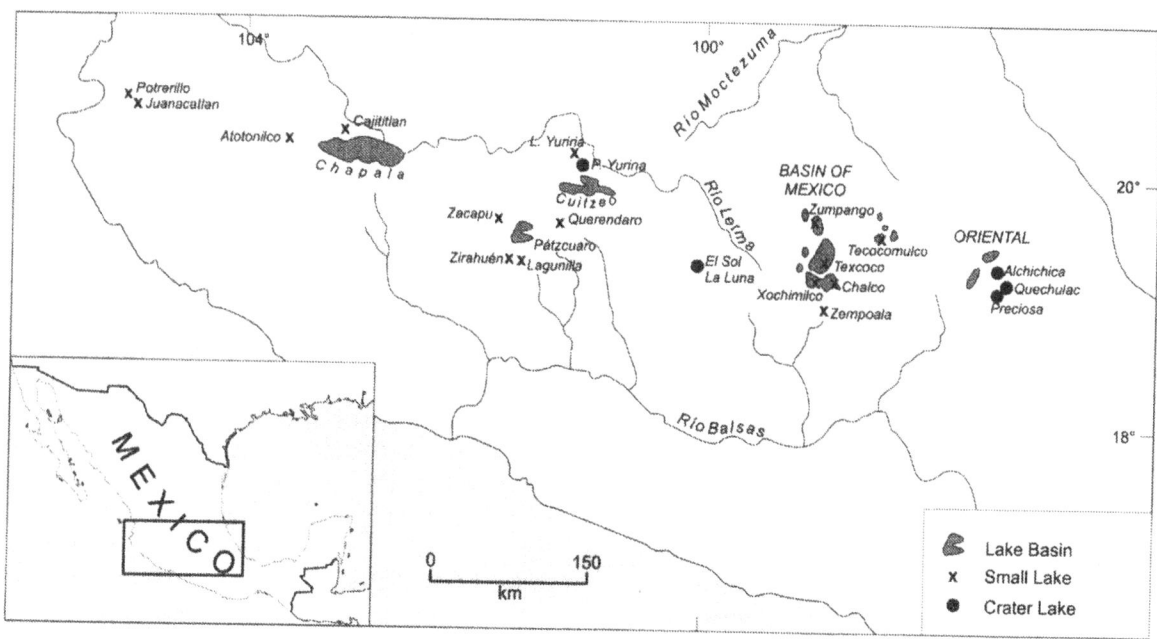

Figure 1. Location of sampling sites in the Central Mexican Diatom Dataset.

set as it appears to contain reworked diatoms from a nearby diatomite deposit (Bradbury, pers. comm.), and the flora may not accurately represent the present water chemistry of the lake. Relationships between diatom assemblage composition and water chemistry were explored using two-way indicator species analysis (TWINSPAN; Hill 1979), detrended correspondence analysis (DCA) and canonical correspondence analysis (CCA; ter Braak, 1986). The main chemical gradients in the dataset are represented by a number of highly inter-correlated variables. To reduce this redundancy we divided the variables into three groups representing gradients of electrical conductivity (ionic strength), ionic composition, and pH respectively, and used CCA with forward selection to select a subset of variables to represent each gradient (Gasse et al., 1995).

For a transfer function to be developed for a particular variable we require that the variable should explain a significant part of the total variation in the diatom data, independent of other variables (Birks, 1995). The relative strength and independence of each of the main chemical gradients was estimated using a series of CCAs and partial CCAs to partition the variance in the diatom data into components representing (1) the unique contribution of individual chemical gradients, (2) the contribution due to interactions between gradients, and (3) unexplained variance

(Gasse et al., 1995). The statistical significance of individual variables in forward-selection CCA, and of CCA and partial CCA ordination axes was determined using a Monte Carlo permutation test (999 permutations; $P = 0.01$), with significance levels in forward selection adjusted for the Bonferroni inequality (Manly, 1992). All ordinations were performed using the program CANOCO (ter Braak & Šmilauer, 1998). Conductivity and all variables expressed as concentrations showed strongly right-skewed distributions and were $\log_{10}$-transformed prior to all statistical analyses.

Transfer functions were developed using the method of weighted-averaging (WA) regression and calibration with inverse deshrinking (Birks et al., 1990; Birks, 1995). The performance of the transfer functions is reported in terms of the root mean square of the error (RMSE) (observed – inferred) and the squared correlation (r^2) between observed and inferred values for the modern dataset. These two measures are useful for comparison with other published transfer functions, but because they use the same data to both develop and evaluate the transfer function they inevitably underestimate the true prediction errors. We therefore also report the jack-knife, or 'leave-one-out' RMSE (RMSE of prediction, RMSEP; ter Braak & Juggins, 1993), as this measure is less biased by sample re-substitution and is thus a more reliable indicator of true predictive ability (Dixon, 1993). WA calculations were performed using the program CALIBRATE (Juggins & ter Braak, 1999).

Results and discussion

Water chemistry

The 36 samples analysed for hydrochemical variables span a large environmental gradient. Results are summarised in Table 1. Conductivity ranges from 17 μS cm^{-1} to 44 100 μS cm^{-1}. A wide range in pH is also observed, with values ranging from 4.9 to 11, although the majority of sites have a pH between 7 and 9. Figure 2 shows the distribution of the lakes according to their ionic composition, pH and conductivity. Most of the lakes are of bicarbonate-carbonate type, although those lakes with the highest conductivity values tend towards chloride dominance. Only three sites are dominated by sulphate anions: the two crater lakes (La Luna and El Sol) of the Nevado de Toluca (11, 12) and Zacapu Celanese (26). The cations of freshwater lakes comprise approximately equal amounts of calcium and magnesium. As conductivity increases, Mg^{2+} and Na$^+$+ K$^+$ begin to dominate. The dataset represents a clear evolutionary trend from fresh water systems of calcium–magnesium bicarbonate–carbonate type, through magnesium-sodium carbonate to the most highly evolved concentrated sodium–chloride systems. This corresponds to pathway IIIA of Eugster and Hardie's (1978) model of brine evolution with evaporative concentration. Figure 2 illustrates that, with the exception of sites 11 and 12, the crater lakes included in the dataset appear to have well-evolved brines, being of either Mg^{2+}+ Na$^+$+ CO$_3$$^{2-}$ or Na$^+$+ Cl$^-$ composition, ranging in conductivity from 840 μS cm^{-1} to 26 000 μS cm^{-1}. Water samples taken from the same lake but at different times show a remarkable degree of correspondence. Alchichica (2), Pátzcuaro (13), La Piscina de Yuriria (16) and Zirahuén (30) were all sampled in both 1981 and 1997, producing very similar results in terms of the proportions of major anions and cations.

The two crater lakes of the Nevado de Toluca (La Luna and El Sol) (11, 12) are clear outliers in the dataset. They have the lowest pH and conductivity values (Caballero, 1996). Their chemical composition is clearly different from the general trend in the dataset and may be a reflection of local geological and / or hydrological factors. These lakes are situated at 4550 m

a. s. l., which is c. 2000 m higher than most other lakes in the dataset. The other site, which does not follow the identified evolutionary trend, is Zacapu Celanese (26). This is immediately downstream from a synthetic fibre factory and it is thought that the high sulphate concentrations reflect human impact (Metcalfe, 1988).

TWINSPAN groupings

TWINSPAN classification was used to partition the sites into 11 main groups on the basis of their diatom assemblages. The diatom composition and chemical environment of each group is summarised in Figure 3. Group 1 represents the low pH and low conductivity crater lakes La Luna (11) and El Sol (12) and is characterised by species such as *Navicula pseudoscutiformis* and *Cymbella perpusilla*. The dominant taxa within this group do not occur in any other samples. Groups 2–6 all have broadly similar chemical composition dominated by bicarbonate/carbonate anions, with a range in conductivity from 119 μS cm^{-1} to 399 μS cm^{-1} and a pH of 7–8.8. A number of these samples have very distinctive species compositions, which accounts for the separation into different groups. For example, the Zirahuén samples 30ai and 30aii have been classified as a single group (Group 3). This is due to the high abundance of *Cyclotella ocellata*, which is not recorded at other sites. Species occurring across these five groups include *Fragilaria* (=*Staurosira*) *construens* var. *venter*, *Fragilaria* (= *Staurosirella*) *pinnata* var. *lancettula*, *Cocconeis placentula* and *Nitzschia amphibia*. Group 7 is characterised by *Aulacoseira granulata*, its variety, *angustissima* and *Aulacoseira ambigua*. Sites in this group are of low to medium conductivity and a medium pH, but display a considerable range in carbonate-bicarbonate content, from 11 to 93%. Group 8 includes the large, shallow basins of Chapala (6), Chalco (7) and Pátzcuaro (13), which are all of medium to high conductivity, but also includes the 1982 Zirahuén sample (30b) and that of Lagunilla (9). These two are low conductivity sites and are included in this group due to the common occurrence of species such as *Gomphonema gracile*, *Cocconeis placentula* and *Navicula* (=*Sellaphora*) *pupula*. Common taxa in Group 8 are *Aulacoseira granulata* var. *angustissima*, *Cocconeis placentula* and *Amphora veneta*. Group 9 represents the two crater lakes of Hoya La Alberca (1) and Lago Quechulac (18), the most common species being *Gomphonema parvulum*, *Gomphonema angustatum* and *Achnanthes minutissima*. The high

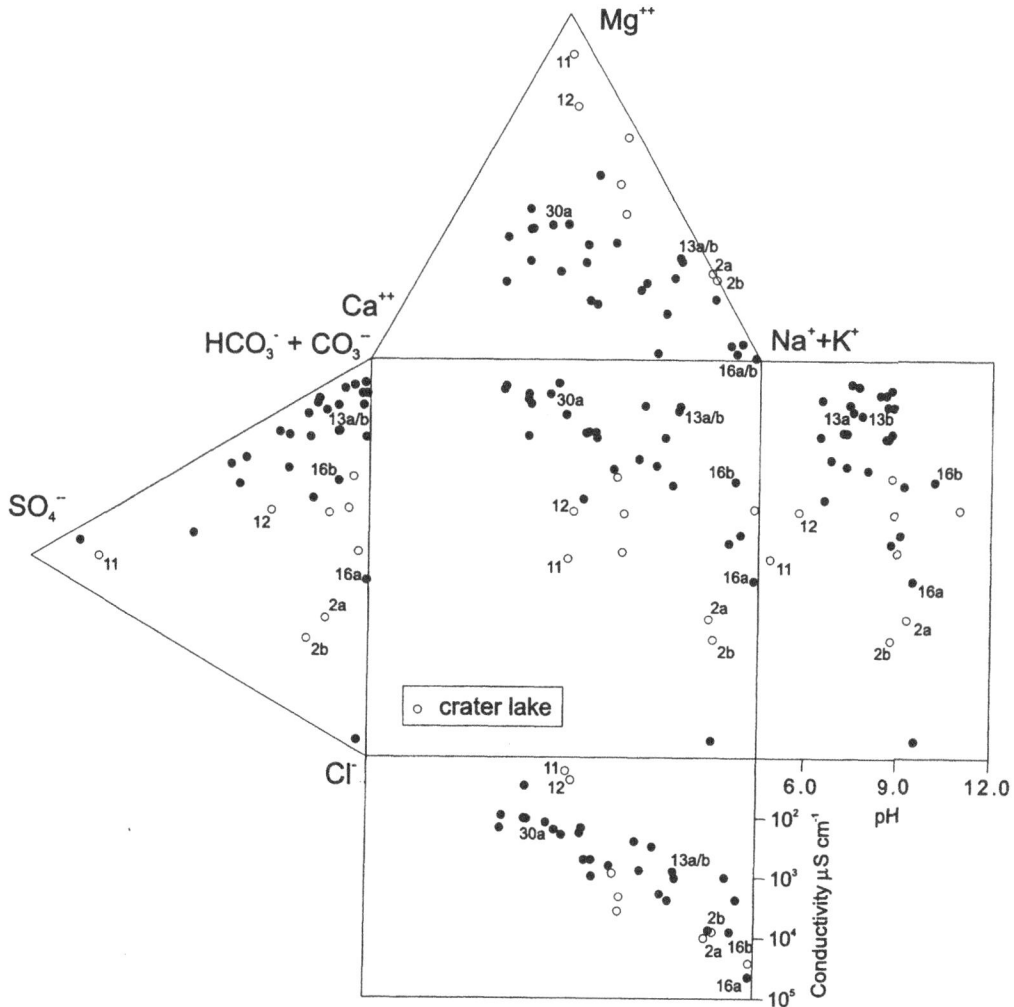

Figure 2. Durov diagram showing ionic composition, pH and conductivity for the Central Mexican Dataset. See Table 1 for key to lake numbers.

conductivity samples from the Texcoco sites (21–23) and sample 7i from Chalco, all in the Basin of Mexico, are placed in Group 10. Common diatoms in this group include *Cyclotella meneghiniana, Nitzschia frustulum* and *Amphora veneta.* Group 11 includes the two crater lakes of Lago Alchichica (2b) and La Piscina de Yuriria (16a and 16b). Dominant species in this group do not occur elsewhere in the dataset and include *Nitzschia inconspicua,Navicula elkab, Anomoeoneis sphaerophora,* and *Nitzschia romana.*

Canonical Correspondence Analysis

Results of the CCA analysis are shown in Figure 4 as a species- and sample-environment biplot. CCA Axes 1 (λ_1=0.71) and 2 (λ_2=0.59) account for 9.9% of the total variation in the diatom data. The low percentage

variance explained is typical of such noisy datasets with large numbers of taxa and many zero values in the species matrix. Despite the relatively low variance captured, the first two ordination axes are both highly significant ($P = 0.01$). Furthermore, eigenvalues for the CCA are similar to those obtained for DCA (λ_1=0.88) and 2 (λ_2=0.74). The similar configuration of samples and taxa in both ordinations suggests that the gradients of ionic strength, composition and pH included in the CCA account for the major patterns of variation in the diatom data.

The relative importance of individual variables in explaining diatom distributions is indicated by the length of each arrow, whilst their direction illustrates the relationship to the CCA axes. Axis 1 clearly reflects the conductivity gradient, from fresh through

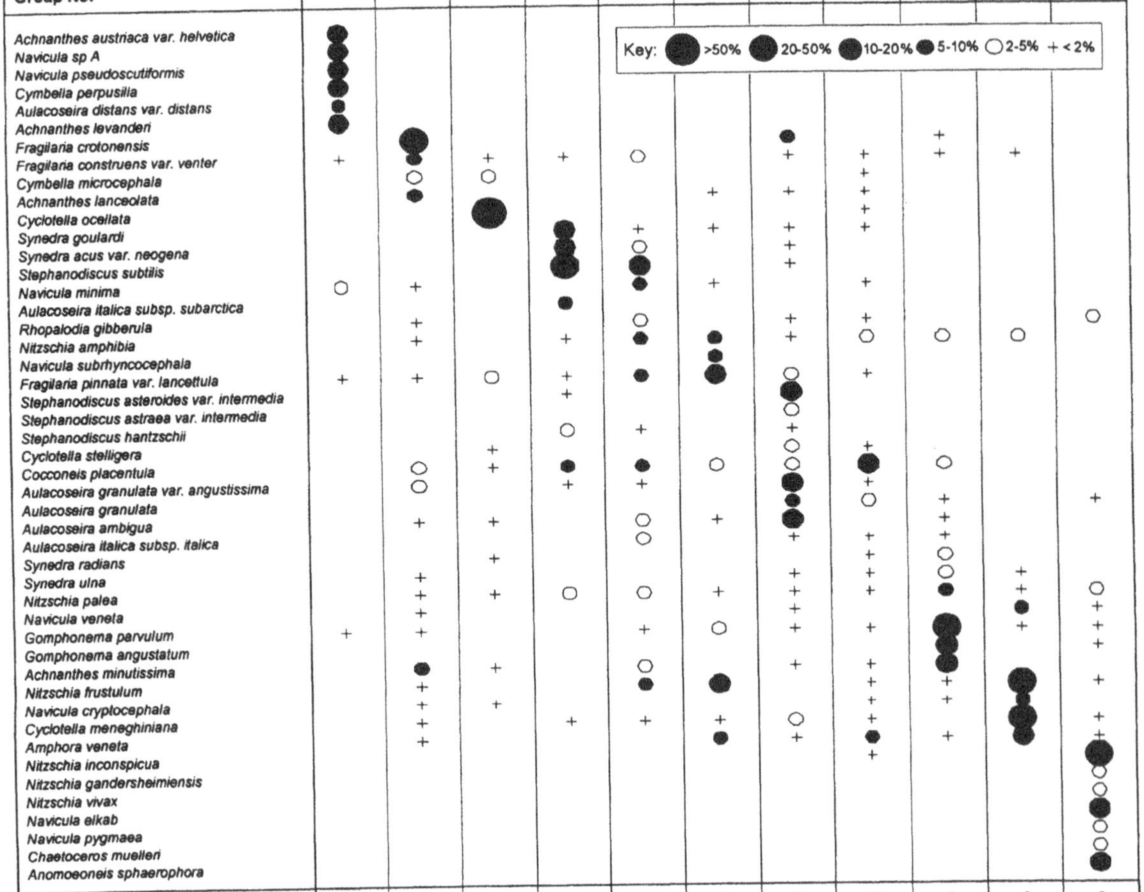

	1	2	3	4	5	6	7	8	9	10	11
Group No.											
Achnanthes austriaca var. helvetica	●										
Navicula sp A	●										
Navicula pseudoscutiformis	●										
Cymbella perpusilla	●										
Aulacoseira distans var. distans	●										
Achnanthes levanderi		●					●		+		
Fragilaria crotonensis			+	+	○		+	+	+	+	
Fragilaria construens var. venter	+							+			
Cymbella microcephala		○	○			+	+	+			
Achnanthes lanceolata		●				+	+	+			
Cyclotella ocellata			●				+	+			
Synedra goulardi				●	+		+	+			
Synedra acus var. neogena				●	○		+				
Stephanodiscus subtilis				●	●			+			
Navicula minima	○	+		●		+					
Aulacoseira italica subsp. subarctica					○		+	+			○
Rhopalodia gibberula		+			●		+	○	○	○	
Nitzschia amphibia		+		+	●	●	+	+			
Navicula subrhyncocephala					●	●	+				
Fragilaria pinnata var. lancettula	+	+	○	+	●	●	●				
Stephanodiscus asteroides var. intermedia				+			●				
Stephanodiscus astraea var. intermedia				○	+		○				
Stephanodiscus hantzschii			+				○				
Cyclotella stelligera		○	+		●	○	○	●	○		
Cocconeis placentula		○	+	●	●		○	+			
Aulacoseira granulata var. angustissima				+			●	+	○		
Aulacoseira granulata		+	+				●	+	+		+
Aulacoseira ambigua					○		+	+	+		
Aulacoseira italica subsp. italica			+					+	○		
Synedra radians							+	+	○		
Synedra ulna		+	+				+	+		+	+
Nitzschia palea		+	+	○	○	+	+	+	●	+	+
Navicula veneta		+					+			●	+
Gomphonema parvulum	+	+			+	○	+	+	●		+
Gomphonema angustatum							+		●		
Achnanthes minutissima		●	+		○	●		+	+	●	+
Nitzschia frustulum		+				●		+	+	●	
Navicula cryptocephala		+	+					+		●	+
Cyclotella meneghiniana		+		+	+	○		+	+	●	+
Amphora veneta		+				●	+	+			●
Nitzschia inconspicua								+			○
Nitzschia gandersheimiensis											○
Nitzschia vivax											○
Navicula elkab											●
Navicula pygmaea											○
Chaetoceros muelleri											○
Anomoeoneis sphaerophora											●
N	6	8	2	3	2	4	7	10	3	6	3
pH	5.4 (4.9-5.8)	7.8 (6.5-8.8)	8.4 (8.4)	8.8 (8.8)	7 (6.6-7.3)	7.4 (7.2-7.8)	8.2 (6.5-8.8)	8.3 (6.5-9.3)	8.9 (8.8-9.0)	9.4 (9.1-9.6)	10 (8.8-11.0)
EC (μS cm^{-1})	21 (17-24)	136 (91-638)	119 (119)	146 (146)	234 (180-305)	399 (195-508)	583 (102-1885)	760 (30-10140)	2218 (840-3600)	5675 (2360-4.4x10^4)	11668 (7521-2.6x10^5)
Salinity (g l^{-1})	1.2 (0.9-2.0)	1.1 (1.9-19.7)	2.5 (2.5)	4.1 (4.1)	6.5 (5.5-7.7)	8.8 (5.3-10.4)	13.2 (2.3-44.4)	19.7 (0.7-289)	22.9 (16.4-27.0)	147 (75.6-935.0)	398 (216-1020)
Carbonate (%)	33 (10-48)	85 (61-91)	91 (91)	68 (68)	55 (53-57)	77 (77-78)	54 (11-93)	64 (29-90)	56 (50-68)	67 (2.9-45.9)	48 (21-65)
Members	11i, 11ii, 12i, 12ii, 12iii	8i, 8ii, 24, 27, 28i, 28ii, 28iii, 29ii	30ai, 30aii	25i, 25ii, 25iii	3, 17	19, 31ai, 31aii, 31b	5, 10, 14, 26i, 26ii, 26iii, 29i	2a, 6, 7ii, 9, 13a, 13b, 15i, 15ii, 20, 30b	1i, 1ii, 18	7i, 21i, 21ii, 22i, 22ii, 23	2b, 16a, 16b

Key: ● >50% ● 20-50% ● 10-20% ● 5-10% ○ 2-5% + < 2%

Figure 3. Twinspan classification of the surface sediment diatom assemblages, showing for each group the number of sites (N), mean relative abundance of selected taxa, means and ranges for selected chemical variables, and site membership.

to hypersaline waters. TWINSPAN groups 1–5, representing fresh water sites plot out on the left-hand side of the sample biplot, whilst the more concentrated systems (TWINSPAN groups 8–11) are plotted on the right. Conductivity is closely related to the ionic composition of the samples, which is reflected in the biplot of environmental variables. Those sites with high percentages of Ca^{2+}, Mg^{2+} and HCO_3^- + CO_3^{2-} also have low conductivity, whilst increasing proportions of Na^+ and Cl^- are indicated on the right hand side. Axis 2 is closely related to the percentage sulphate content. The Zacapu Celanese samples (26, TWINSPAN group 7) all have high scores for this axis. This gradient is somewhat exaggerated due to

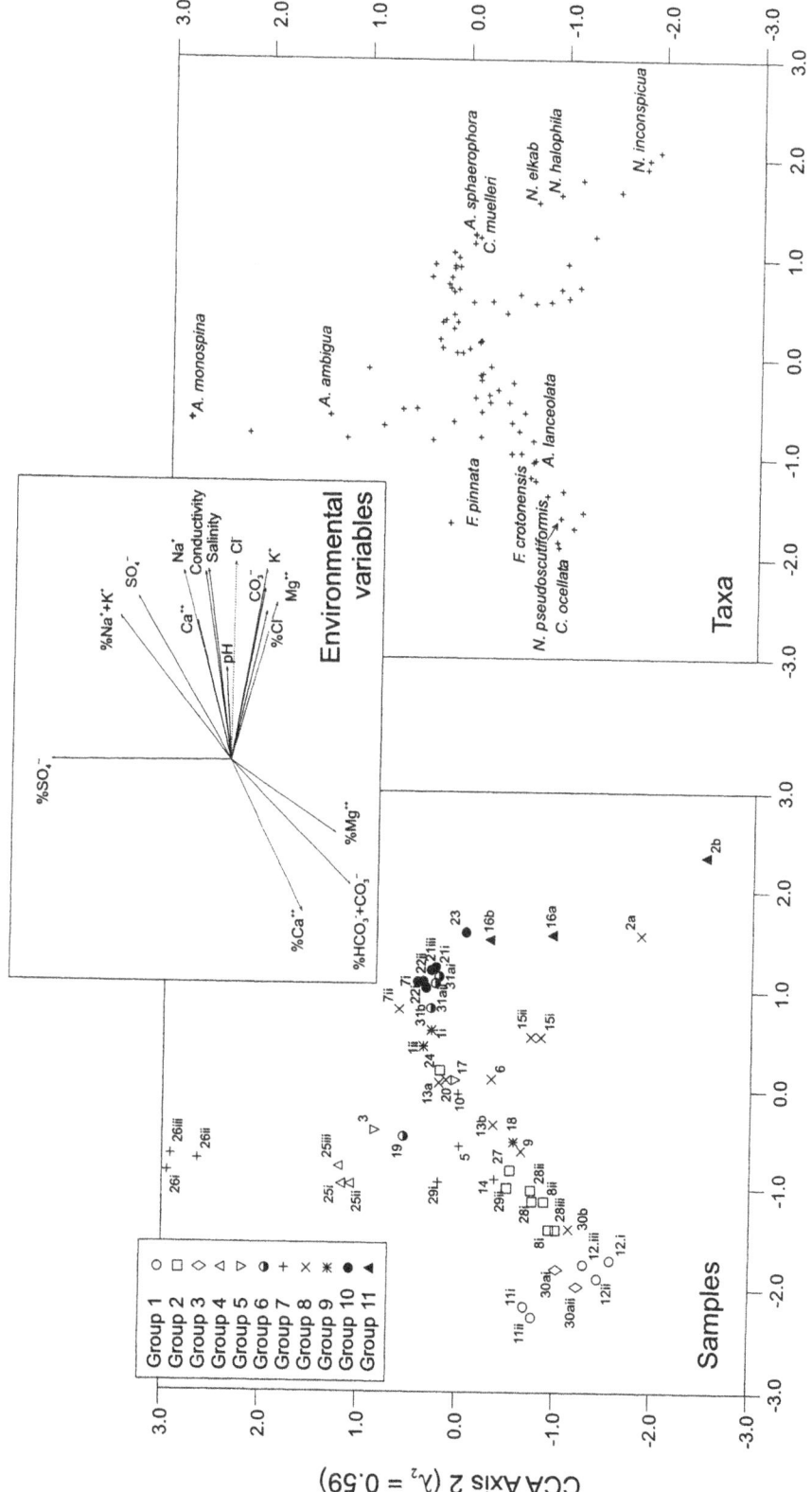

Figure 4. CCA ordination biplot, showing configuration of samples, taxa and environmental variables on axes 1 and 2. Solid arrows represent variables used in the CCA while dotted arrows represent additional variables added to the biplot *post-hoc*. See Table 1 for key to site numbers.

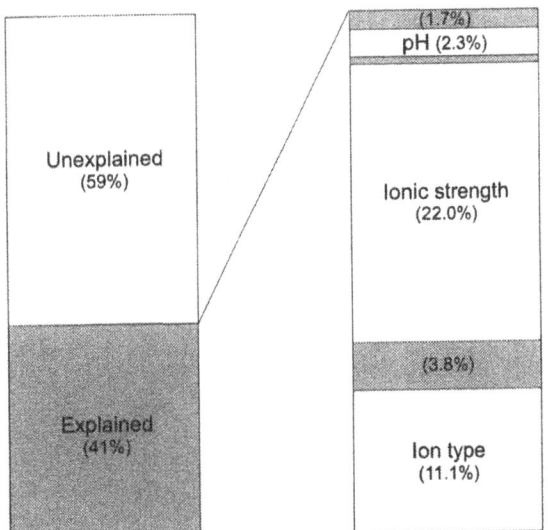

Figure 5. Results of the variance partitioning, showing the percentage variance accounted for by unique contributions of pH, ionic strength and ion type, and interactions or conditional effects between gradients (shaded areas).

the small number of samples with high sulphate content and their distinctive species composition. These samples are dominated by *Melosira* (= *Aulacoseira*) *monospina*, a taxon not recorded elsewhere in the dataset. Species, which plot on the left-hand portion of the biplot representing fresh water environments, include *Navicula pseudoscutiformis*, *Fragilaria crotonensis*, *Cyclotella ocellata* and *Achnanthes lanceolata*. Those species identified with more chemically evolved, saline systems include *Anomoeoneis sphaerophora*, *Navicula halophila* (= *Craticula halophila*), *Nitzschia inconspicua* and *Chaetoceros muelleri*.

Variance partitioning

Results of the variance partitioning are shown in Figure 5. The 12 environmental variables chosen by forward selection account for a total of 41% of the variance in the diatom data. The total explained variance is predominantly composed of unique contributions from variables representing ionic strength (22.0%) and ionic composition or brine type (11.1%). pH makes a statistically significant but small contribution (2.3%). Similarly, the variance due to interactions or conditional effects between pairs of gradients is low, indicating that gradients of ionic strength and ion type make large and unique contributions to the explained variance. These results indicate that statistically significant and

independent transfer functions can be developed for variables representing ionic strength and ion type.

Transfer functions

Transfer functions were developed for ionic strength and ion type using conductivity and alkalinity (percentage $HCO_3 + CO_3$). These variables were chosen because (1) they have clear hydrochemical interpretations, and (2) they accounted for the highest proportion of variance in the diatom data of the environmental variables tested. Scatterplots of observed against diatom–inferred values, and summary performance measures are shown in Figure 6. The transfer functions for both variables perform well and the squared correlations between observed and inferred values are high (0.91 and 0.90 for conductivity and alkalinity respectively). Jackknife estimates of the RMSEP are higher than corresponding RMSE, indicating the importance of cross-validation in estimating prediction error. However, the strength of the relationships between observed and diatom–inferred variables remain high after cross-validation, suggesting that conductivity and alkalinity (%$HCO_3 + CO_3$) can both be accurately inferred using these transfer functions.

Figure 7 shows the weighted-average optima and tolerances for conductivity and alkalinity (% $HCO_3+ CO_3$) of the most abundant taxa in the dataset. Species with the lowest conductivity optima, such as *Achnanthes levanderi* and *Cymbella perpusilla* also display the narrowest tolerance ranges, whilst a number of species are found in waters of varying E.C. These include *Nitzschia paleacea*, *Nitzschia palea* and *Fragilaria* (=*Staurosira*) *construens*. Species found in waters with the highest bicarbonate–carbonate content have narrow tolerances, whilst most of the taxa in the dataset display broad ranges, such as *Cyclotella meneghiniana*, *Aulacoseira ambigua* and *Fragilaria* (= *Staurosirella*) *pinnata* var. *lancettula*.

Discussion

Although this calibration dataset is relatively small (53 samples), the apparent predictive ability of both the conductivity and alkalinity transfer functions (r^2=0.91 and r^2=0.90, respectively) is high. These values compare well with transfer functions developed for lakes in Spain (conductivity: r^2=0.91) (Reed, 1998) and Africa (conductivity: r^2=0.87, carbonate–bicarbonate and sulphate + chloride ions: r^2=0.82) (Gasse et al.,

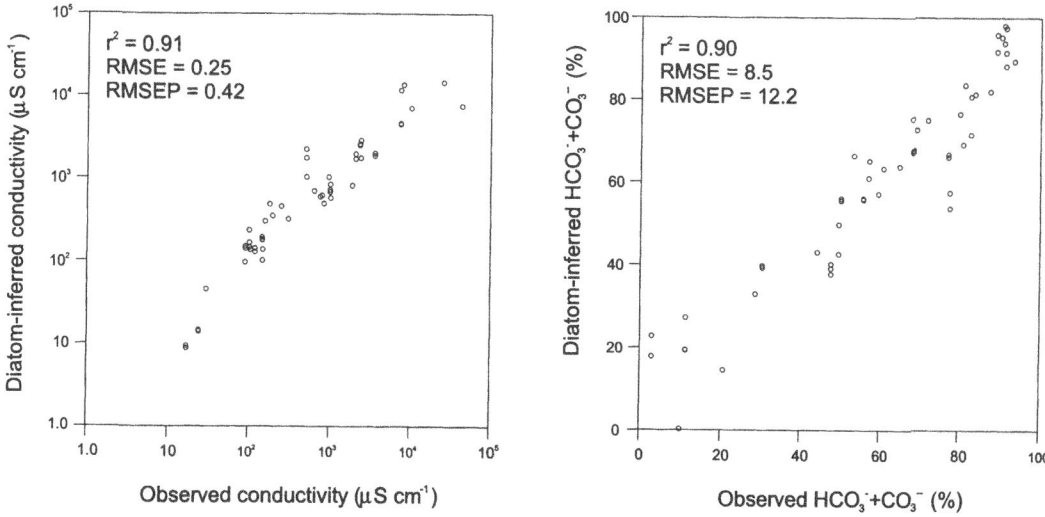

Figure 6. Relationship between observed and diatom-inferred conductivity and % (HCO$_3$+ CO$_3$).

1995). The trend in chemical composition of sites in the Central Mexican Diatom Dataset is similar to that reported for lakes in East Africa (Gasse et al., 1983), which have since been incorporated into the larger African dataset (Gasse et al., 1995). The British Columbia dataset (Cumming and Smol, 1993) also contains a significant number of bicarbonate and carbonate lakes. Carbonate lakes are poorly represented in the Northern Great Plains (Fritz et al., 1993) and the Spanish (Reed, 1998) datasets, which are largely dominated by sites of sulphate or chloride composition. By comparing this datset with those from other regions it is possible to explore the degree of consistency between datasets in the ecological optima derived for different species. Direct comparison is not possible with the Northern Great Plains and British Colombia datasets as transfer functions are based on salinity rather than conductivity. It is possible, however, to examine the broad trends in the data.

In general, the conductivity and salinity optima of species common to the datasets lie in similar positions along the ecological gradient identified in the different regions. Derived conductivity optima compare well between the Central Mexican and African datasets, particularly at the lower end of the conductivity spectrum. Estimated optima for *Fragilaria* (=*Staurosira*) *pinnata* are 205 μS cm^{-1} (Mexico) and 145 μS cm^{-1} (Africa), whilst *Aulacoseira granulata* var. *angustissima* has optima of 375 μS cm^{-1} (Mexico) and 364 μS cm^{-1} (Africa). There is also good agreement with those species found at the lower end of the salinity spectrum in the British Columbia datset. At the more concentrated end of the conductivity gradient, optima are similar for some species. For example, *Anomoeoneis costata* has estimated optima of 12 920 μS cm^{-1} (Mexico) and 12 280 μS cm^{-1} (Africa), although it is slightly lower in the Spanish dataset (9760 μS cm^{-1}). This species has high salinity optima in the Northern Great Plains and British Columbia datasets.

There are, however, a number of significant discrepancies between the datasets. Reed (1998) has highlighted the large difference in the conductivity optima of *Cocconeis placentula* in different datasets. In Central Mexico, *C. placentula* tends to be found in fresh, slightly alkaline water (E.C. opt. = 433 μS cm^{-1}). This is consistent with an optimum of 469 μS cm^{-1} in the African dataset and a salinity optimum of 0.2 g l^{-1} in British Columbia. In the Spanish and Northern Great Plains datasets, however, this species has optima of 20 700 μS cm^{-1} and 16 800 μS cm^{-1} respectively (Reed, 1998). Derived conductivity optima for *Navicula halophila* are 19 050 μS cm^{-1} (Mexico), 13 760 μS cm^{-1} (Spain) and 2980 μS cm^{-1} (Africa). In British Columbia, this species has a salinity optimum of 0.9 g l^{-1}, whilst in the Northern Great Plains, its optimum salinity is 5.3 g l^{-1}.

Discrepancies between the regions could be due to the relatively small size of most of the datasets. The African dataset is the largest, with 282 samples, whilst the Central Mexican, Spanish, British Columbia and Northern Great Plains datasets have 53, 74, 59 and 63 samples respectively. Therefore, it is possible that the full ecological range of certain species has not been encountered in the individual regions. Differ-

210

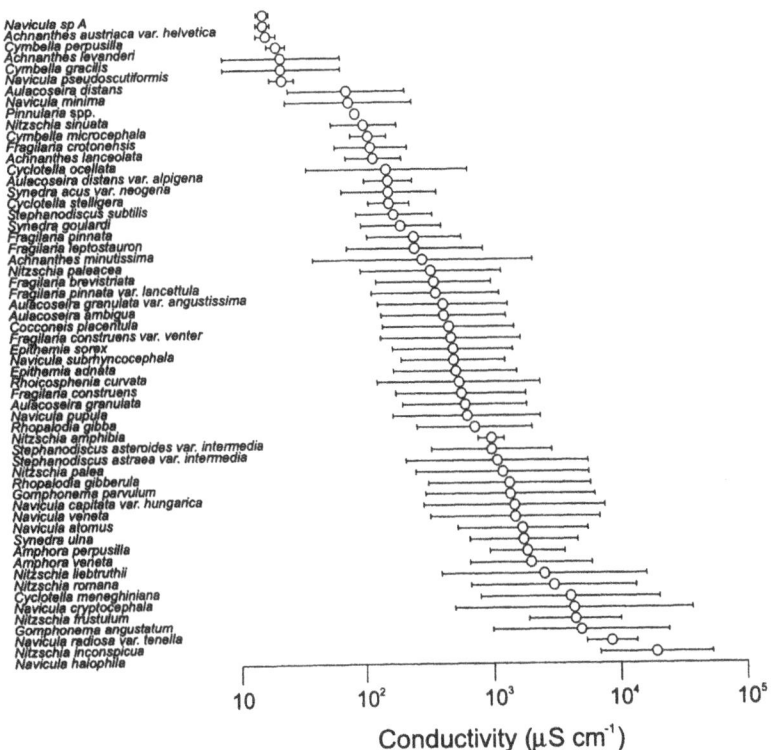

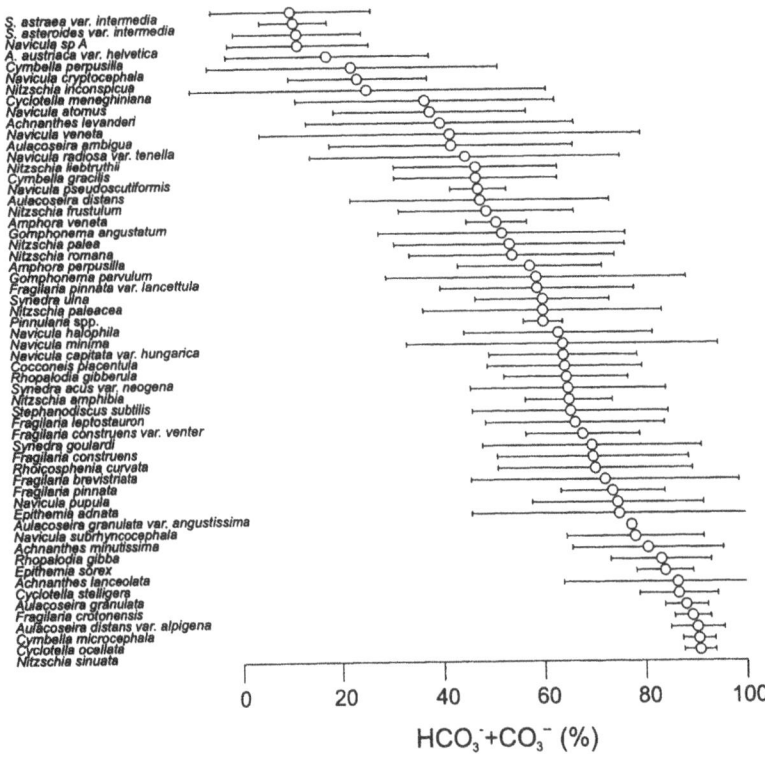

Figure 7. Conductivity and% (HCO₃+ CO₃) optima and tolerances for selected taxa.

ences may also be due to the sensitivity of taxa to different brine types, rather than to electrical conductivity/salinity alone. It is interesting to note that the Central Mexican dataset is most similar to the African dataset, which contains lakes of similar chemical composition. A further issue for consideration is that derived optima may be biased by the ecological gradient covered by the individual datsets. The Central Mexican dataset has a clear bias towards sites with low to medium conductivity. Twenty-two of the 31 different sites in the dataset have a conductivity lower than 1000 μS cm^{-1}. The Northern Great Plains and Spanish datasets have few freshwater samples and are biased to more saline lakes (Fritz et al., 1993; Reed, 1998). It has been suggested that bias may also be introduced due to the fact that water chemistry measurements have been made at one time and therefore may not be truly representative of the range of conditions in which the diatom assemblage has been living (Fritz et al., 1993). However, Mexican lakes sampled fifteen years apart show remarkable consistency in the environmental variables measured, suggesting that this factor is not significant with regard to the Central Mexican dataset. This preliminary comparison of species optima from different calibration datasets appears to confirm that merging the Mexican and African datasets would be a sensible step.

There are a number of limitations to quantitative palaeoenvironmental reconstructions in Mexico using the current dataset of 53 samples. Firstly, a number of important fossil species in Mexican lake sediments are either absent or poorly represented in the dataset. For example, the fresh water planktonic species *Stephanodiscus niagarae* is abundant in Late Pleistocene material (Metcalfe, 1992; Metcalfe et al., 1997; Caballero & Ortega, 1998; Caballero et al., 1999) but has not been reported in the present day environment. Similarly, *Cyclotella stelligera* is common in lake sediments (Metcalfe, 1992), but is rare in the dataset. This will lead to problems in applying the transfer functions to palaeoecological data. It is thought that this may be due to important climatic changes in the region as well as to prolonged and significant human impact on many lakes in central Mexico (Metcalfe, 1988). Many lake basins in the TMVB have experienced significant human impact during the last 3000 years (Metcalfe et al., 1989). More recently, deliberate drainage, excessive groundwater abstraction and cultural eutrophication have all contributed to the significant alteration of many lakes from their natural state (Alcocer & Escobar, 1996). Given the extent of

human impacts on Mexican lakes, further sampling of Mexican lakes may not actually solve the problem of no-analogue situations.

Some potentially important environmental variables, such as nitrate and phosphorus content, have not been considered in this study. It is possible that their inclusion may lead to the masking of the importance of ionic composition in the same way that habitat preferences interfered in Metcalfe's (1988) dataset. However, these variables are likely to be significant and future research may focus on developing a dataset for nutrient transfer functions. Such a dataset would provide valuable information on the response of lake ecosystems to anthropogenic nutrient inputs in the absence of an extensive monitoring network.

Summary and conclusions

This survey of central Mexican lakes has revealed a clear evolutionary trend in ionic composition following pathway IIIA of Eugster & Hardie's (1978) model. Fresh waters have high proportions of calcium, magnesium and bicarbonate ions, whilst more saline systems are richer in sodium and chloride ions. Our results demonstrate a strong correlation between diatom species distributions and ionic strength and composition, illustrating the potential for numerical palaeoenvironmental reconstructions. Transfer functions have been successfully developed for conductivity and alkalinity ($\%HCO_3 + CO_3$). However, significant environmental changes associated with climatic fluctuations and anthropogenic disturbance in many lakes in central Mexico has meant that a number of important species in the palaeolimnological record do not have modern analogues. Further sampling of sites is ongoing and it is hoped to merge this dataset with the much larger, but chemically similar, African dataset (Gasse et al., 1995), which as Reed (1998) suggests, will improve the quality of transfer functions and provide more analogues.

Acknowledgements

SJD was in receipt of a departmental PhD studentship from the Department of Geography, University of Edinburgh. MCM received a PhD scholarship from the DGAPA at the Universidad Nacional Autónoma de México. Fieldwork has been funded by the Carnegie Trust for the Universities of Scotland and the Dudley

Stamp Memorial Trust (SJD), NERC the Royal Society, the French Government, the Gilchrist Educational Trust and the Godman Fund (SEM). We are grateful to Dr Ma. Aurora Armienta and her staff at the analytical chemistry laboratory in the Instituto de Geofísica, UNAM, who carried out many water analyses for this project. MCM thanks Teodoro Hernández for his help during field work. Our thanks go to Dr J. Platt Bradbury, Dr Françoise Gasse, Dr Hannelore Hakansson and Prof. David Mann for their help with diatom identification. We thank two anonymous reviewers for their helpful comments on an earlier version of this paper.

References

Alcocer, J. & E. Escobar, 1996. Limnological regionalization of México. Lakes and Reservoirs: Research and Management 2: 55–69.

Armienta, M. A., V. Zamora & F. Juaréz, 1987. Manual para análisis químicos de aguas naturales en el campo y en el laboratorio. Comunicaciones Técnicas: Docencia y Divulgación No. 4. Instituto de Geofísica, Universidad Nacional Autonoma de México. México.

Battarbee, R. W., 1986. Diatom Analysis. In Berglund (ed.), Handbook of Holocene Palaeoecology and Palaeohydrology. John Wiley, Chichester: 527–570.

Bennion, H., 1994. A diatom-phosphorous transfer function for shallow eutrophic ponds in southeast England. Hydrobiologia 275/276: 391–410.

Birks, H. J. B., 1995. Quantitative palaeoenvironmental reconstructions. In Maddy, D. & J. S. Brew (eds), Statistical Modelling of Quaternary Science Data. Technical Guide 5, Quaternary Research Association, Cambridge: 161–254.

Birks, H. J. B., 1998. Numerical tools in palaeolimnology – Progress, potentialities and problems. J. Paleolimnol. 20: 307–332.

Birks, H. J. B., J. M. Line, S. Juggins, A. C. Stevenson & C. J. F. ter Braak, 1990. Diatoms and pH reconstructions. Phil. Trans. r. Soc. Lond. B 327: 263–278.

Bradbury, J. P., 1971. Paleolimnology of Lake Texcoco, Mexico. Evidence from diatoms. Limnol. Oceanogr. 16: 180–200.

Bradbury, J. P., 1989. Late Quaternary paleoenvironments in the Cuenca de México. Quat. Sci. Rev. 8: 75–100.

Caballero, M. E., 1995. Late Quaternary palaeolimnology of Lake Chalco, the Basin of México: new evidence for palaeoenvironmental and palaeoclimatic change in central México during the last 40 000 years. PhD thesis, University of Hull.

Caballero, M. E., 1996. The diatom flora of two acid lakes in central Mexico. Diatom Res. 11: 227–240.

Caballero, M. & B. Ortega-Guerrero, 1998. Lake levels since about 40 000 years ago at Lake Chalco, near Mexico City. Quat. Res. 50: 69–79.

Caballero, M. E., S. Lozano, B. Ortega, J. Urrutia & J. L. Macias, 1999. Environmental characteristics of Lake Tecocomulco, northern Basin of Mexico, for the last 50 000 years. J. Paleolimnol. 22: 399–411.

Cavazos, T., & S. Hastenrath, 1990. Convection and rainfall over Mexico and their modulation by the Southern Oscillation. Int. J. Climatol. 10: 377–386.

Cumming, B. F. & J. P. Smol, 1993. Development of diatom-based salinity models for paleoclimatic research from lakes in British Columbia (Canada). Hydrobiologia 269/270: 179–196.

De Deckker, P. & R. M. Forester, 1988. The use of ostracods to reconstruct continental palaeoenvironments. In De Deckker, P., J. P. Coln & J. P. Peypouquet (eds), Ostracoda in the Earth Sciences. Elsevier: 175–199.

Dixon, P. M., 1993. The bootstrap and the jackknife: describing the precision of ecological indices. In Scheiner, S. M. & J. Gurevitch (eds), Design and Analysis of Ecological Experiments. Chapman and Hall, New York: 290–318.

Eugster, H. P. & L. A. Hardie, 1978. Saline Lakes. In Lerman, A. (ed.), Lakes: Chemistry, Geology, Physics. Springer-Verlag, New York: 237–294.

Fritz, S. C., 1990. Twentieth-century salinity and water-level fluctuations in Devil's Lake, N. Dakota: a test of a diatom-based transfer function. Limnol. Oceanogr. 35: 1771–1781.

Fritz, S. C., B. F. Cumming, F. Gasse & K. Laird, 1999. Diatoms as indicators of hydrological and climatic change in saline lakes. In Stoermer, E. F. & J. P. Smol (eds), Diatoms: Applications for the Environmental and Earth Sciences. Cambridge University Press, Cambridge: 41–72.

Fritz, S. C., S. Juggins & R. W. Battarbee, 1993. Diatom assemblages and ionic characterization of lakes of the northern Great Plains, North America: a tool for reconstructing past salinity and climate fluctuations. Can. J. Fish. aquat. Sci. 50: 1844–1856.

Fritz, S. C., S. Juggins, R. W. Battarbee & D. R. Engstrom, 1991. Reconstruction of past changes in salinity and climate using a diatom–based transfer function. Nature 352: 706–708.

Gasse, F., 1980. Les diatomées lacustres Plio-Pléistocenes de Gadeb (Éthiope): systématique, paleoécologie, biostratigraphie. Revue Algologique, mémoire hors-series 3: 249 pp.

Gasse, F., 1986. East African Diatoms. Taxonomy, ecological distribution. Bibliotheca Diatomologica. Vol. 11. J. Cramer, Stuttgart: 202 pp.

Gasse, F. & F. Tekaia, 1983. Transfer functions for estimating paleoecological conditions (pH) from East African diatoms. Hydrobiologia 103: 85–90.

Gasse, F., P. Barker, P. A. Gell, S. C. Fritz & F. Chalié, 1997. Diatom-inferred salinity in palaeolakes: an indirect tracer of climate change. Quat. Sci. Rev. 16: 547–563.

Gasse, F., S. Juggins & L. Ben Khelifa, 1995. Diatom-based transfer functions for inferring hydrochemical characteristics of African palaeolakes. Palaeo. Palaeo. Palaeo. 117: 31–54.

Gasse, F., J. F. Talling & P. Kilham, 1983. Diatom assemblages in East Africa: classification, distribution and ecology. Rev. Hydrobiol. Trop. 116: 3–34.

Germain, H., 1981. Flore des Diatomées. Diatomophycées d'eaux douces et saumâtres du Massif Armoricain et des contrées voisines d'Europe occidentale. Soc. Nouv. des Éditions Boubée, Paris: 444 pp.

Haberyan, K., S. P. Horn, & B. F. Cumming, 1997. Diatom assemblages from Costa Rican lakes: an initial ecological assessment. J. Paleolimnol. 17: 263–274.

Hill, M. O., 1979. TWINSPAN: A FORTRAN program for arranging multivariate data in an ordered two-way table by classification of the individuals and attributes. Ecology and Systematics, Cornell University, New York: 48 pp.

Hustedt, F., 1930, Die Kieselalgen Deutschlands, Österreichs und der Schweiz, Vol. 1. Reprinted 1977, Otto Koeltz, Koenigstein: 920 pp.

Hustedt, F., 1959, Die Kieselalgen Deutschlands, Österreichs und der Schweiz, Vol. 2. Reprinted 1977, Otto Koeltz, Koenigstein: 845 pp.

Hustedt, F., 1961–1966. Die Kieselalgen Deutschlands, Österreichs und der Schweiz, Vol. 3. Reprinted 1977, Otto Koeltz, Koenigstein: 816 pp.

Jones, V. J., S. Juggins & J. C. Ellis-Evans, 1993. The relationship between water chemistry and surface sediment diatom assemblages in maritime Antarctic lakes. Antarct. Sci. 5: 339–348.

Juggins, S. & C. J. F. ter Braak, 1999. CALIBRATE Version 1.0 – a program for species–environment calibration by [weighted averaging] partial least squares regression. Unpublished computer program, Department of Geography, University of Newcastle: 25 pp.

Krammer, K. & H. Lange-Bertalot, 1986. Süsswasserflora von Mitteleuropa. Bacillariophyceae. 1. Teil: Naviculaceae. Vol. 2/1. Gustav Fischer Verlag, Stuttgart: 876 pp.

Krammer, K. & H. Lange-Bertalot, 1988. Süsswasserflora von Mitteleuropa. Bacillariophyceae. 2. Teil: Epithemiaceae, Bacillariaceae, Surirellaceae. Vol. 2/2. Gustav Fischer Verlag, Stuttgart: 596 pp.

Krammer, K. & H. Lange-Bertalot, 1991a. Süsswasserflora von Mitteleuropa. Bacillariophyceae. 3. Teil: Centrales; Fragilariaceae, Eunotiaceae. Vol. 2/3. Gustav Fischer Verlag, Stuttgart: 576 pp.

Krammer, K. & H. Lange-Bertalot, 1991b. Süsswasserflora von Mitteleuropa. Bacillariophyceae. 4. Teil: Achnanthaceae. Vol. 2/4. Gustav Fischer Verlag, Stuttgart: 437 pp.

Manly, B. F. J., 1992. The Design and Analysis of Research Studies. Cambridge, Cambridge University Press: 353 pp.

Metcalfe, S. E., 1985. Late Quaternary environments of central Mexico: a diatom record. D.Phil. thesis, University of Oxford.

Metcalfe, S. E., 1988. Modern diatom assemblages in Central Mexico: the role of water chemistry and other factors as indicated by TWINSPAN and DECORANA. Freshwat. Biol. 217–233.

Metcalfe, S. E., 1992. Changing environments of the Zacapu Basin, Mexico: a diatom-based record. Research Paper No. 48, School of Geography, Oxford: 38 pp.

Metcalfe, S. E., 1995. Holocene environmental change in the Zacapu Basin, Mexico: a diatom-based record. The Holocene 5: 196–208.

Metcalfe, S. E. & P. E. Hales, 1994. Holocene diatoms from a Mexican crater lake – La Piscina de Yuriria. In Kociolek, P. (ed.), Proceedings of the Eleventh International Diatom Symposium, San Francisco, 1990. California Academy of Sciences No. 17: 501–515.

Metcalfe, S. E., A. Bimpson, A. J. Courtice, S. L. O'Hara, & D. M. Taylor, 1997. Climate change at the monsoon/westerly boundary in northern Mexico. J. Paleolimnol. 17: 155–171.

Metcalfe, S. E., F. A. Street-Perrott, R. B. Brown, P. E. Hales, R. A. Perrott & F. M. Steininger, 1989. Late Holocene human impact on lake basins in Central México. Geoarchaeology 4: 119–141.

Metcalfe, S. E., F. A. Street-Perrott, R. A. Perrott & D. D. Harkness, 1991. Palaeolimnology of the Upper Lerma Basin, Central Mexico: a record of climatic change and anthropogenic disturbance since 11 600 yr BP. J. Paleolimnol. 5: 197–218.

Mosiño-Aleman, P. & E. García, 1974. The Climate of Mexico. In Bryson, R. A. & F. K. Hare (eds), World Survey of Climatology Vol. 11: Climates of North America, Elsevier, Amsterdam: 345–404.

Patrick, R. & C. W. Reimer, 1966. The diatoms of the United States exclusive of Alaska and Hawaii. Vol. 1. The Academy of Natural Sciences of Philadelphia, Philadelphia, Monograph 13: 668 pp.

Patrick, R. & C. W. Reimer, 1975. The diatoms of the United States exclusive of Alaska and Hawaii. Vol. 2. Part 1. The Academy of Natural Sciences of Philadelphia, Philadelphia, Monograph 13: 213 pp.

Reed, J. M., 1998. A diatom–conductivity transfer function for Spanish salt lakes. J. Paleolimnol. 19: 399–416.

Round, F. E., R. M. Crawford & D. G. Mann, 1990. The Diatoms: Biology and Morphology of the Genera. Cambridge University Press. Cambridge: 747 pp.

Servant-Vildary, S. & M. Roux, 1990. Multivariate analysis of diatoms and water chemistry in Bolivian saline lakes. Hydrobiologia 197: 267–290.

Street-Perrott, F. A. & S. P. Harrison, 1985. Lake levels and climate. In Hecht, A. D. (ed.), Paleoclimate Analysis and Modeling. Wiley, New York: 291–340.

ter Braak, C. J. F., 1986. Canonical correspondence analysis: a new eigenvector technique for multivariate direct gradient analysis. Ecology 67: 1167–1179.

ter Braak, C. J. F. & S. Juggins, 1993. Weighted averaging partial least squares regression (WA-PLS): an improved method for reconstructing environmental variables from species assemblages. Hydrobiologia 269/270: 485–502.

ter Braak, C. J. F. & P. Šmilauer, 1998. CANOCO Reference Manual and User's Guide to CANOCO for Windows: Software for Canonical Community Ordination (version 4). Ithaca, NY, U.S.A., Microcomputer Power: 352 pp.

Hydrobiologia **467**: 215–228, 2002.
J. Alcocer & S.S.S. Sarma (eds), Advances in Mexican Limnology: Basic and Applied Aspects.
© 2002 *Kluwer Academic Publishers.*

215

Hydrogeochemical and biological characteristics of cenotes in the Yucatan Peninsula (SE Mexico)

J.J. Schmitter-Soto[1], F.A. Comín[2,3], E. Escobar-Briones[4], J. Herrera-Silveira[2], J. Alcocer[5],
E. Suárez-Morales[1], M. Elías-Gutiérrez[1], V. Díaz-Arce[2], L.E. Marín[6] & B. Steinich[7]
[1]*ECOSUR. Apdo. Postal 424. Chetumal. Quintana Roo. Mexico*
E-mail: jschmit@ecosur-qroo.mx
[2]*Centro de Investigación y de Estudios Avanzados del IPN. Unidad Mérida. Apdo. Postal 73 Cordemex.*
Mérida. Yucatán. Mexico
E-mail: jherrera@mda.cinvestav.mx
[3]*Departamento de Ecología. Universidad de Barcelona. Spain*
E-mail: comin@porthos.bio.ub.es
[4]*Instituto de Ciencias del Mar y Limnología. Universidad Nacional Autónoma de México.*
Apdo. Postal 70-305. MX-04510. México D.F.
E-mail: escobri@mar.icmyl.unam.mx
[5]*Escuela Nacional de Estudios Profesionales Iztacala, Laboratorio de Limnología. Proyecto CyMA,*
UIICSE. MX-54090 Tlalnepantla. Mexico
E-mail: jalcocer@servidor.unam.mx
[6]*Instituto de Geofísica, Universidad Nacional Autónoma de México. Ciudad Universitaria. MX-04510.*
México D.F.
E-mail: Imarin@tonatiuh.igeofcu.unam.mx
[7]*Instituto de Geofísica. Unidad de Ciencias de la Tierra-UNAM. Campus Juriquilla. Querétaro. MX 76001. Mexico*
E-mail: birgit@mail.unicit.unam.mx

Key words: karst, tropical limnology, sinkholes, nutrients, chlorophyll, biodiversity

Abstract

Cenotes (sinkholes) are the most peculiar aquatic ecosystem of the Yucatan Peninsula (SE Mexico). They are formed by dissolution of the carbonate rock in the karstic platform of the Yucatan Peninsula. A wide morphological variety is observed from caves filled with ground water to open cenotes. In some cenotes, particularly those close to the sea, underneath the fresh water one finds saltwater, where meromixis can take place. This occurs because in the Yucatan Peninsula there is a thin lens (10s of meters thick) that floats above denser saline water. In these cenotes, a relative enrichment of sodium related to calcium is observed while conductivity increases. In contrast, a higher increase of calcium associated to sulfate is observed in cenotes located in SE Yucatan Peninsula. A marked vertical stratification of the water is established during the warm and rainy season of the year (May–October). In cenotes with good hydraulic connection with the rest of the aquifer, the water remains clear during most of the year. However, cenotes with poor hydraulic connection with the aquifer are characterized by turbid waters and very low light transparency. In this group of cenotes, the water column contains a high concentration of chlorophyll (mostly due to chlorophyceans, cyanobacteria, diatoms and dinoflagellates); the hypolimnion and the sediment are rich in organic matter and anaerobic bacteria mediated biogeochemical processes are dominant. The upper part of the cenotes walls is well illuminated and covered by a rich microbial mat. Floating macrophytes may also occupy part of the water surface in oligotrophic cenotes. A great variety of food web paths are represented in the habitats occurring in the cenotes, in which few trophic levels are involved. A few endemic species (crustaceans and fishes) have been reported from cenotes found in the Yucatan Peninsula. Because of the high organic matter input (alochthonous) and production (autochthonous) and the low water flow, cenotes can be considered heterotrophic systems.

Introduction

Regional studies on the ecology of aquatic ecosystems have been constant during the development of limnology in many countries (Margalef, 1983). There is still a need to continue and to increase the scientific knowledge about peculiar ecosystems and unexplored limnological regions (Gopal & Wetzel, 1995; Wetzel & Gopal, 1999). Classifying the observed characteristics and typifying the water masses in relation with other already known aquatic ecosystems are major objectives of this type of studies (Hutchinson, 1957).

The Yucatan Peninsula (SE Mexico) is subject to rapid urban development, explosive in the coastal zone, particularly along the Caribbean littoral; proper management of this large ecosystem is imperative and freshwater sustainability as well as coastal water quality are subjects of major concern. The development of limnology in the Yucatan Peninsula is increasing. Up to date, at least 100 papers have been published on fresh and saline waters, but our knowledge is nonetheless sparse and fragmentary (Alcocer & Escobar, 1996; Comín et al. 1996; Herrera-Silveira et al., 1998). Part of this limnological delay (in comparison with other regions) is due to the isolation, size and difficult access to most Yucatan Peninsula inland aquatic ecosystems.

Because of its karstic nature, there are no rivers in the Yucatan Peninsula, and only 12 lakes with water volume higher than $5 \cdot 10^5$ m^3 exist, none of which occurs in the northern half (Doehring & Butler, 1974; Alcocer & Escobar, 1996). Many solution features form small ponds (sinkholes) (in Spanish *cenotes*, from the Maya word *ts'onot*), caves, and minor cavities (named locally *sartenejas*), all of them caused by the percolation of CO_2-laden water through limestone (Alcocer & Escobar, 1996; Steinich, 1996). They all are solution lakes in the terms described by Hutchinson (1957). More than 7000 solution features have been mapped just in northwest Yucatan Peninsula (Steinich, 1996).

This paper is an overview of the general limnological characteristics of cenotes in the Yucatan Peninsula. The aim is to present a classification, their characteristics and an interpretation of their functional ecology.

General characteristics of the study area

The Yucatan Peninsula is located between 19° 40′ and 21° 37′ N and 87° 30′ and 90° 26′ W, surrounded by the Gulf of Mexico and the Caribbean Sea (Fig. 1). It extends over an area of 39 340 km^2, representing 2% of the surface of the Mexican Republic. The climate has three characteristic seasons: (1) warm and dry season (March–May), (2) winter storm season with occasional short showers (November–February), and (3) rainy season from June to October.

Winds are highly seasonal, being strongest from November to February while calm condition lasts from January to October. Mean annual air temperature is 26.1 °C with a minimum of 5 °C and a maximum of 42.5 °C. Annual rainfall varies from 760 mm.yr^{-1} to 1198 mm.yr^{-1} in the north portion and from 1138 mm.yr^{-1} to 1440 mm.yr^{-1} in the southern portion. The highest precipitation occurs in September, with an average of 232 mm.

The Yucatan Peninsula is a calcareous platform which originated in the Cenozoic that averages ten meters above sea level with just one small prominent Sierra in the center of the Peninsula, where a maximum altitude of 150 m is reached (Stringfield & LeGrand, 1974). The Peninsula attained its present shape in the late Pliocene (López Ramos, 1975); however, large eolianites were deposited on the coast during the Holocene, and reefs are still developing in the north and east (Ward et al., 1985). There are Pleistocene marine deposits in the east, north and northeast coasts of the Peninsula, in Laguna de Términos (SW) and some interior paleolakes. In Campeche, the Eocenic terrain reaches the coast (López Ramos, 1975). The maximum interglacial sea level was at the time 30 m higher than today (Back, 1985). Present sea level was attained only 5500 years ago (Ward et al., 1985); in the early Holocene sea level was some 100 m lower than today (Buskirk, 1985).

Wilson (1980) classified the Peninsula in 14 physiographic districts. Eight of them are represented in the northern half (Fig. 1), where cenotes are more abundant: (1) the coastal zone, geologically the youngest, where most of the anchialine cenotes appear; (2) the district of Mérida, within the Ring of Cenotes (Marín, 1990); (3) the district of Chichén Itzá, with more cenotes and a coarser relief than Mérida; (4) the Puuc district, on the Sierrita de Ticul; (5) the Bolonchén district, with a less developed karst landscape; (6) the district of Cobá, with geologic faults, some of them filled with water; (7) the district of Río

217

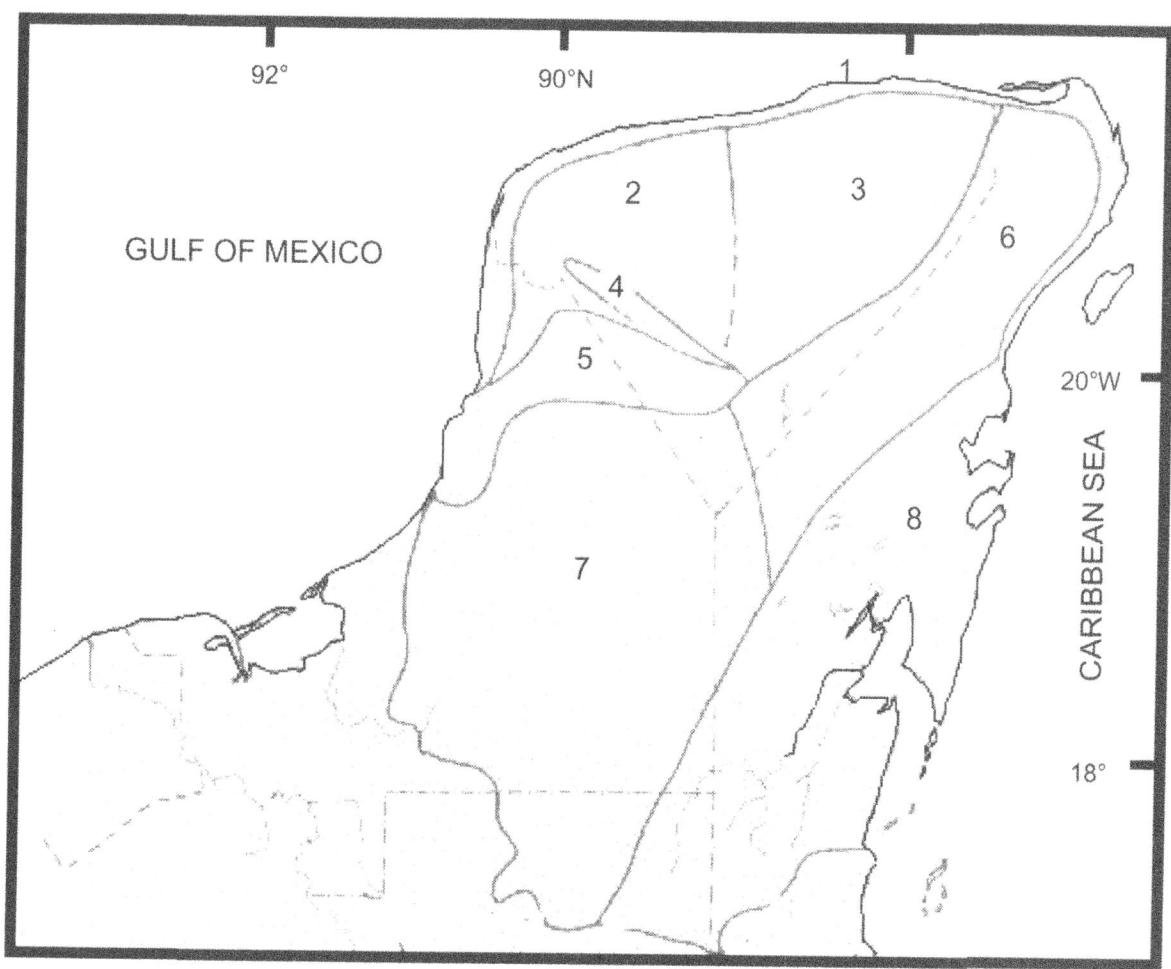

Figure 1. Map of the Yucatan Peninsula, with the physiographic districts within the study area (modified from Wilson, 1980). **1**, coastal zone. **2**, Mérida. **3**, Chichén Itzá. **4**, Puuc. **5**, Bolonchén. **6**, Cobá. **7**, Río Bec. **8**, Río Hondo.

Bec, partly outside the study area, and differing from it by a higher altitude and the presence of rivers; and (8) the Hondo river district (basin), partly outside the study area.

The Yucatan Peninsula receives on the average $172\,158 \times 10^6$ m^3 of rainwater per year. Average annual rainfall increases from the northwest (500 mm) towards the southeast (2000 mm). About 85% of the precipitation is evapotranspired. The aquifer of the Yucatan Peninsula is a karstic aquifer characterized by its high permeability. Karst features such as underground channels and caverns (cenotes) are widely present throughout the Peninsula (Steinich et al., 1996). The aquifer is unconfined, except for a narrow band parallel to the coast, where it is confined (Perry et al., 1989). Ground water flows through a dual-porosity medium: flow occurs through the rock matrix and through frac-

tures, joints and dissolution features. Discharge of the aquifer occurs both as diffuse flow throughout the coast and as springs (found both inshore and offshore). Many cenotes in the North of the Peninsula are located along a semicircle, known as the Ring of Cenotes, which is centered in Chicxulub (a village in the north coast, on the east of Progreso) (Marín et al., 1990). It has been proposed that this distribution of cenotes is associated with the Chicxulub Impact Crater (Sharpton et al., 1992, 1993). The density of cenotes along the ring varies between one and a few cenotes per kilometer (Marín et al., 1990). This zone has been shown to act as an underground river or groundwater trough (Marín, 1990; Velázquez, 1995; Perry et al., 1995; Steinich & Marín, 1996; Steinich et al., 1996).

The two intersections of the ring with the coast give rise to a high density of submarine springs (Marín et al., 1990). The water table throughout the Yucatan Peninsula is less than two meters above mean sea level (Marín, 1990; Steinich & Marín, 1996). The hydraulic gradient is very low, on the order of 7–10 mm per kilometer (Marín, 1990). As a result, the aquifer is a thin freshwater lens that floats above denser, saline water. Steinich & Marín (1996) have shown that salt water is found more than 110 km from the coast. This salt water has a dual origin: salt water intrusion and dissolution of evaporites (Perry et al., 1995; Velázquez, 1995).

Parallel to the coast, in the 'ciénagas' or wetlands, another type of cenote – locally known as 'petén' – is found where ground water dissolves the limestone as it travels along the flow lines. When it approaches the coast, CO_2 escapes, and as a result the groundwater precipitates calcium carbonate sealing the intergranular space in the rocks. This is the process that is confining the aquifer parallel to the coast (Perry et al., 1989). However, within this area, there are large cenotes that formed when sea level was lower than today. Since these cenotes range between 5 and 15 m in diameter, the precipitation of calcium carbonate can not seal these large cavities. Thus, they act as artesian springs (Marín et al., 1988). As brackish water is discharged into the ciénaga, it mixes with the waters from the swamp, creating radial patterns in salinity which in turn are colonized by different mangrove species that can tolerate different salinity concentrations. The 'petenes', when observed from the distance, show up as islands in the middle of the 'ciénaga'.

There are 16 kinds of vegetation characterizing the Peninsula, the most widely distributed are: the low caducifolium forest, medium subcaducifolium forest and medium subperennifolium forest, the two first covering the north and central regions of the Yucatan State and the other covering mostly Campeche and Quintana Roo. Redzine and litosol are among the dominant soils than cover the Peninsula (Flores & Espejel, 1994).

Origin and types of cenotes

The main process in the formation of cenotes is the dissolution of the limestone by carbonic acid. In areas where there is a significant soil cover, the CO_2 concentration may increase by orders of magnitude (in the Yucatan Peninsula it takes place towards the southeast), resulting in waters that are more aggressive (i.e. with a higher capacity to dissolve the rocks). The CO_2 involved in the process may not be alochthonous, but organically generated *in situ* (Gaona-Vizcayno et al.,

1980). A second process that can contribute to the dissolution of the carbonate rocks is the mixture of fresh and salt water, which enhances the reactivity on aragonite and calcite (Stoessel et al., 1989). The third process is local and of larger importance. A high concentration of H_2S has been observed in the water of several cenotes as a consequence of the reduction of the accumulated organic matter. The H_2S may, then, dissolve the rock within these horizons (Stoessel et al., 1993). Microbial activity associated to all these processes contributes to the formation of cenotes (Martin & Brigmon, 1994).

We propose here that in younger (lotic) cenotes, the water is well interconnected with the ground water through fractures, and dissolution features, and its residence time is short. Older cenotes have a lentic condition with slow flow and turnover through sedimentation and blocking of the water source and the siphon. Although we may recognize that groundwater continues to flow through the cenote, many of the pathways are blocked, and the exchange of ground and free-overlying water in the cenote is restricted. Two processes may restrict groundwater flow to and from the cenote: roof or wall collapse and sedimentation.

This idea is proposed on the marked geochemical changes observed. Thus, it is suggested that the lentic cenotes are fed primarily by diffuse flow (low groundwater velocities, thermal stratification and other processes) and that the lotic cenotes are primarily fed by ground water flowing through fracture and dissolution cavities. Lotic cenotes have clear waters, clean, sandy or rocky bottoms, and a homogeneous, well-oxygenated water column. Lentic or near-lentic cenotes are turbid, thermally stratified; the surface water layer is alkaline, oversaturated with oxygen, while water near the bottom is acid, devoid of O_2, and with H_2S.

The cenotes have been classified according to the stages in the process described above (Hall, 1936) as: caves, jug-shaped, cylindrical, and plate-shaped cenotes. Navarro-Mendoza (1988) and Marín et al. (1990) have suggested differences between coastal and inland cenotes. The former are shallower, 3–35 m deep; their walls are rocky, often with compacted organic matter among mangrove roots. The latter with depths greater than 100 m, and walls with up to 20 m high. In both cases the diameter may go from a few meters to more than 100 m.

The 'petenes', and deep inland cenotes, whose walls penetrate below the salt/freshwater interface, have water that is stratified based on density. This saline stratification produces a meromixis, that is, the partial mixing of the water column, in contrast to the holomixis, a thorough mixture of the water mass, usually the case in lotic cenotes. Between the freshwater layer in the surface, the mixolimnion, and the saline, denser, bottom layer, the monimolimnion, an abrupt transition zone occurs, the halocline. There could be also thermal stratification (i.e. thermocline). Herrera-Silveira & Comín (2000) showed that in both, lotic and lentic, types of cenotes a thermal stratification can be established during the dry and rainy seasons (March–October), while the water column remains mixed during the winter storms season (November–February). The length of the thermal stratification in lotic cenotes could vary from hours up to several days according to the sheer velocity, the water column depth, and other factors. Differences between cenotes related to the thermocline depth may be related to the transmission of convective heat between atmosphere and water.

Under chemical stratification, the monimolimnion may be stagnant and in anoxic conditions, or it may slowly flow according to groundwater input, tides and storms through tunnels and crevices. Most of the cenotes found throughout the Peninsula will usually occur in an intermediate position between these two extreme types described (Van der Kamp, 1995).

Physical and chemical characteristics of the water in the cenotes

The temperature is stable in lotic cenotes and it is controlled by the geometry of the flow system. Thermal stability reflects the constancy of water temperature below depths of 10 m (Van der Kamp, 1995). In lentic cenotes there are horizontal and vertical variations along the year. Mean water temperature in cenotes, 24–29 °C, is similar to the mean air temperature (Alcocer et al., 1998).

The pH is also homogeneous and stable in lotic cenotes, generally with acid values (<7). In lentic cenotes, there is a pH-gradient along the water column. The epilimnion is usually basic. The hypolimnion is acid, because respiration predominates, as well as the formation of H_2S under anoxic conditions. Thus, the range of pH goes from 6.7 to 8.0 in coastal cenotes, and up to 8.6 in inland cenotes (Hall, 1936).

Alkalinity fluctuates widely in cenotes, because of the variable input of meteoric water, rich in carbonates and bicarbonates (up to 696 mg l^{-1} $CaCO_3$), and of rainwater, which lowers the concentration of these

ions by dilution and by neutralization with humic acids and tannins from the mangrove (Navarro-Mendoza, 1988; Alcocer et al., 1998).

When water enters the water table, it acquires CO_2 from the soil and from the oxidation of dissolved and particulate organic matter. These processes decrease the concentration of O_2 and increase acidity, which in turn is neutralized by solution of the limestone. The total dissolved solids in rainwater are concentrated by evapotranspiration and, combined with the dissolution of minerals found in the soils, they contribute to raise the amount of total dissolved solids (TDS) in the cenote water (Van der Kamp, 1995). TDS in cenotes have a uniform concentration, typical of freshwaters (<3 g l^{-1}), except in those cenotes with marine influence. Conductivity measured in a number of cenotes (Fig. 2) ranged between 42.5 and 7390 μS cm^{-1}. Salinity of ground water in the Yucatan Peninsula lies within 0.4 and 2.9 g l^{-1}(Velázquez, 1995). In meromictic cenotes, water may go from fresh in the mixolimnion to marine in the monimolimnion. The thickness of the halocline increases in cenotes closer to the coast, because of the mixture produced by the friction between the freshwater mass going to the sea and the marine water advancing inland.

A study of 71 inland water-bodies distributed throughout the Yucatan Peninsula showed the strong influence of the process of rock dissolution on the major ionic composition of the water in the Yucatan Peninsula (Comín et al., 1996). However, the two processes of salt enrichment can be important (Fig. 3). One process, observed in localities with a higher enrichment of sodium compared to calcium, is associated with the direct influence of seawater – via groundwater – in localities close to the coast. The second process occurs in places located in the southeastern zone of the Peninsula, which includes increasing values of TDS associated with the dissolution of sulfate-rich deposits. In this case, the enrichment in calcium relative to sodium takes place as the total amount of dissolved salts increases (Herrera-Silveira et al., 1998). In some cenotes, the sulfate concentration is high (up to 2400 mg l^{-1}) due to gypsum beds, but in others it lies close to 170 mg l^{-1}; precipitation may lower it to 30 mg l^{-1}. Chloride increases from 70 mg l^{-1} in distance from the sea (Hall, 1936) to 16 200 mg l^{-1} in cenotes with marine influence.

The presence of organic matter defines the geochemical equilibrium, as shown by the recristalization observed in cenotes with low organic activity, but absent from those with high organic activity (Gaona-

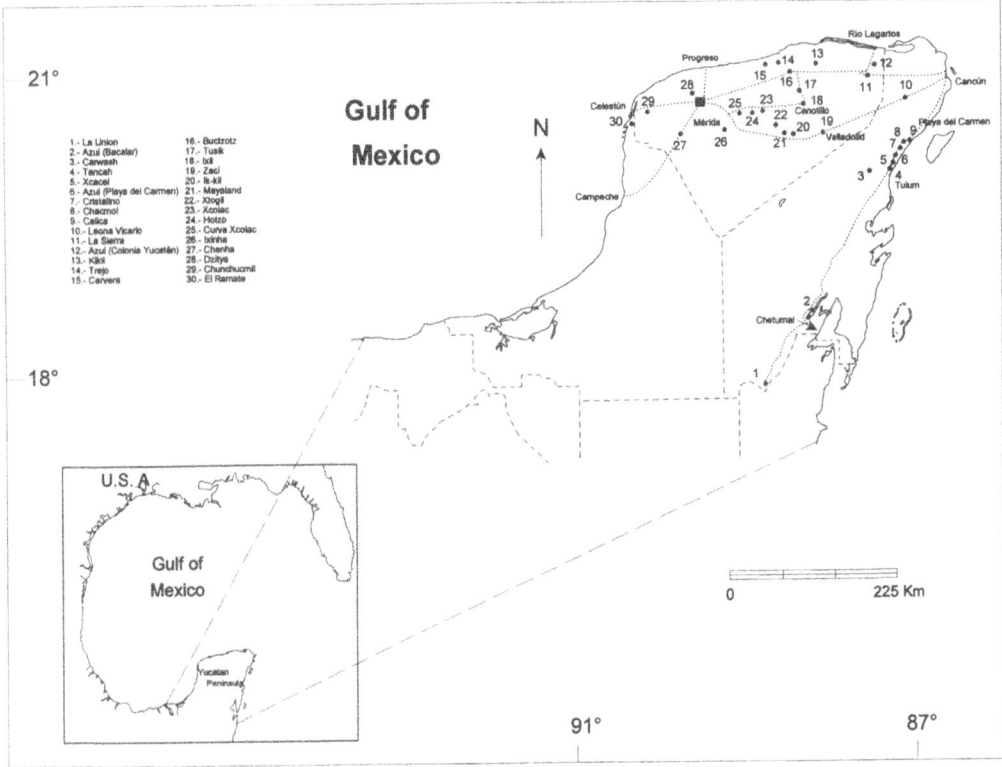

Figure 2. Localization of the Yucatan Peninsula and some of the cenotes sampled.

Vizcayno et al., 1980). Its concentration depends on the lentic or lotic character of the cenote as well as on the size of the opening, because it determines how much alochthonous matter can enter the cenote transported by rainwater. *In situ* photosynthetic production depends on exposure to light, and thus varies according to the type of cenote. Lentic cenotes easily increase their trophic state, favoring the production of large amounts of organic matter (e.g. as phytoplankton). The process brings along an increase in pH, turbidity, dissolved oxygen concentration at the surface and anoxic, acid conditions at the bottom of the cenote.

Nutrient concentration in lotic cenotes is expected to be lower than in lentic ones, because of the difference in turnover rate. On the other hand, vertical stratification concentrates nutrients in the hypolimnion, where remineralization occurs, and impoverishes the epilimnion, where primary producers consume the nutrients; in lotic cenotes, the nutrient-rich waters at the bottom are carried to the surface, where they are once again available to primary producers.

Phosphorus is scarce in cenotes, because the calcareous rocks favor its co-precipitation with calcium, abundant in the karstic environment. High concentra-

Table 1. Averages (AVG), standard error (SE), minims (MIN) and maxims (MAX) of limnological variables analyzed in 30 cenotes (Fig. 2) once during each of the seasons of the year ('nortes', dry and rainy) (from Herrera-Silveira et al., 1998)

	Avg	SE	Min	Max
Temperature (°C)	26.4	0.3	22	33.5
Suspended. Solids (mg l^{-1})	59.6	9	0.3	590.7
DO (mg l^{-1})	4.46	0.3	0.82	10.6
Conductivity (μS cm^{-1})	1645	150	42.5	7390
pH	7.5	0.08	6.31	10.36
Alkalinity (meq l^{-1})	4.33	0.2	0.8	8.51
Cl$^-$ (meq l^{-1})	2.47	0.5	0.11	33.33
SO$_4^-$ (meq l^{-1})	2.12	0.8	0.06	42.2
Ca^{++} (meq l^{-1})	6.93	0.8	0.99	36.5
Mg++ (meq l^{-1})	3.97	0.5	0.29	23.05
Na$^+$ (meq l^{-1})	5.96	1.2	0.20	75.52
K$^+$ (meq l^{-1})	0.27	0.06	0.03	3.48
N–NO$_3^-$ (μM)	63.3	8.3	0.52	500
N–NO$_2^-$ (μM)	0.97	0.2	0.02	15
N–NH$_4^+$ (μM)	6.57	1.4	0.09	84.9
SRP (μM)	1.59	0.4	0.02	20
SRSi (μM)	227.3	19.7	1.48	550
Chlorophyll-a (mg m^{-3})	11.47	2.6	0.11	97.4

tion of nutrients is frequent near urban areas. Concentrations of soluble reactive P range from 0.02 to 20 μg l^{-1}, with a mean of 1.59 μg l^{-1} (Herrera-Silveira et al., 1998). Localities can be divided in two types based on nitrate concentration (Pacheco & Cabrera, 1997). Close to urban developments, farms, and agriculture, high nitrate concentrations have been observed. In areas with little human activity, nitrates have been observed to be more abundant affecting coastal cenotes than in inland ones. The nitrates of the latter environments come mostly from the surrounding vegetation. In contrast, nitrites and ammonia are scarce (Sánchez et al., 1998). Minimum and maximum concentrations of micronutrients observed in a number of cenotes are summarized in Table 1 (Fig. 2).

Biota of the cenotes

Bacterioplankton, phytoplankton and primary production

Our knowledge of the bacterioplankton in cenotes is sparse (Edler & Dodds, 1992; Brigmon et al., 1994), however available information from cenotes and anchialine caves of Quintana Roo shows extremely low bacterioplankton densities even for oligotrophic environments (Alcocer et al., 1999). Chemoautotrophic bacteria are associated to the bottom, walls, and the halocline. Their appearance is a white-grayish mat or floating filament (Brigmon et al., 1994; Martin et al., 1995).

Cenotes of the Yucatan Peninsula can be considered as islands of aquatic life. A dense and tall vegetation made of big trees (e.g. *Ficus cotinifolia*) is frequently found in inland cenotes (Reddell, 1981), while 'petenes', near the coast, may be surrounded by mangrove, especially *Rhizophora mangle*. The small mangroves tree *Conocarpus erecta* and the emergent macrophytes *Cladium jamaicense* and *Phragmites australis* may also border 'petenes'. Other aquatic macrophytes observed in cenotes are *Typha domingensis*, *Acrostichum danaefolium*, *Nymphaea ampla*, *Sagittaria lancifolia*, *Cabomba palaeformis*, *Sesbania emerus*, *Rhabdadenia biflora*, *Thrinax radiata* and *Bravaisia tubiflora*. In shallow cenotes, the algae *Chara* is common (Esquivel, 1991; Sánchez et al., 1991; Cabrera-Cano & Sánchez-Vázquez, 1994).

The phytoplanktic flora of cenotes is largely unknown (Hernández & Pérez, 1991) compared to other aquatic ecosystems of the world. However, a relatively long list of species has been compiled during the last decade (Table 2). Almost 150 species have been

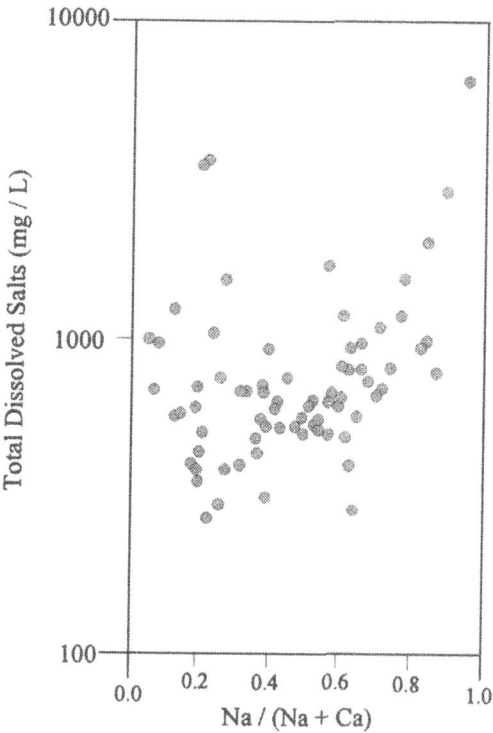

Figure 3. Total dissolved solids vs. the Na(Na + Ca) ratio (from weight data) for the localities studied in the Yucatan Peninsula.

recorded in sinkholes, where chlorophyceans, cyanophyceans and diatoms were dominant (López-Adrián & Herrera-Silveira, 1994; Díaz-Arce, 1999; Sánchez et al., 2002). The species composition is quite similar to the one found in other tropical and temperate lakes (Sánchez-Molina, 1985; Esquivel, 1991; López-Adrián et al., 1993; López-Adrián & Herrera-Silveira, 1994; Herrera-Silveira et al., 1998). Among the green algae, the genus *Monoraphidium* (*M. caribeum* and *M. tortile*) is the most common, while *Aphanocapsa* (*A. pulchra*), *Chroococcus* (*C. dispersus*) and *Microcystis* (*M. aeruginosa*) were common cyanophyceans. The genus *Microcystis*, found in eutrophic temperate lakes during summer, is also common in freshwater inland waters from Yucatan (Díaz-Arce, 1999). The most frequently found diatom species are *Achnanthes gibberula*, *Amphora ventricosa*, *Cocconeis placentula* and *Gomphonema lanceolatum*.

The highest number of cyanobacteria and chlorophycean species are found in the winter storm and rainy seasons in the lentic cenotes. Euglenophyceans are poorly represented; a few species of dinophyceans can be found in those sinkholes enriched with allocthonous organic matter.

Table 2. List of representative species recorded in cenotes of the Yucatan Peninsula

Cyanobacteria	*Staurastrum pentasterias*	**Bacillariophyta**	*Stephanodiscus niagarae*
Chroococcidiopsis indica	*Chlamydomonas paraserbinowi*	*Achnanthes gibberula*	*Fragilaria capucina*
Aphanocapsa montana	*Sphaerella lacustris*	*Cocconeis disculus*	*Synedra tenera*
Aphanocapsa pulchra	*Pandorina morum*	*Cocconeis placentula*	*Synedra ulna*
Chroococcus dispersus	*Micromonas pusilla*	*Cylindrotheca closterium*	*Terpsinoe musica*
Gloeocapsa polydermatica		*Denticula kuetzingii*	
Gloeocapsa rupestris	**Chrysophyta**	*Hantzschia amphioxys*	**Higher plants**
Gomphosphaeria aponina	*Chrysococcus minutus*	*Nitzschia closterium*	*Acrostichum danaefolium*
Merismopedia tenuissima	*Chrysococcus vulneratus*	*Nitzschia longissima*	*Cladium jamaicense*
Microcystis aeruginosa	*Dinobyron sertularia*	*Nitzschia scalaris*	*Conocarpus erecta*
Microcystis inserta	*Salpingoeca ringens*	*Cymbella amphicephala*	*Acoelorrhaphe wrightii*
Synechocystis pevalekii	*Mallomonas pulchella*	*Cymbella turgida*	*Bravaisia tubiflora*
Anabaena fertilissima		*Eunotia maior*	*Cabomba palaeformis*
Nostoc commune	**Euglenophyta**	*Eunotia monodon*	*Ficus cotinifolia*
Oscillatoria nigra	*Euglena agilis*	*Eunotia praerupta*	*Nymphaea ampla*
Trichodesmium thiebautii	*Euglena sanguinea*	*Mastogloia smithii*	*Phragmites australis*
	Phacus onyx	*Melosira granulata*	*Rhabdadenia biflora*
	Trachelomonas volvocinopsis	*Diploneis elliptica*	*Rhizophora mangle*
Cryptophyta		*Diploneis puella*	*Sagittaria lancifolia*
Cryptomonas acuta		*Gomphonema acuminatum*	*Sesbania emerus*
Cryptomonas erosa	**Pyrrophyta**	*Gomphonema angustatum*	*Thrinax radiata*
Rhodomonas pusilla	*Gonyaulax scrippsae*	*Gomphonema lanceolatum*	*Typha domingensis*
	Amphidinium crassum	*Amphiprora paludosa*	
	Gymnodinium grammaticum	*Anomoeneis vitrea*	
Chlorophyta	*Sphaerodinium polonicum*	*Frustalia vulgaris*	
Dictosphaerium botrytella	*Peridinium simplex*	*Gyrosigma exilis*	
Coelastrum microporum	*Peridinium umbonatum*	*Navicula cryptocephala*	
Ankyra ancora	*Scrippsiella tochoidea*	*Navicula recens*	
Chlorella vulgaris	*Procentrum lima*	*Pinnularia intermedia*	
Monoraphidium caribeum		*Stauroneis undulata*	
Monoraphidium circinale		*Amphora copulata*	
Selenastrum capricornutum	**Xanthophyta**	*Amphora ovalis*	
Ceraterias staurastroides	*Characiopsis callosa*	*Amphora ventricosa*	
Scenedesmus circumfusus	*Chloropedia plana*	*Chaetoceros gracilis*	
Scenedesmus opoliensis	*Merismogloea polychloris*	*Chaetoceros muelleri*	
Tetrachlorella alternans	*Chlorogibba trochisciaeformis*	*Rhizosolenia setigera*	
Cosmarium portianum	*Rhizochloris stigmatica*	*Cyclotella meghiniana*	
Cosmarium punctulatum	*Neonema quadratum*		
Staurastrum muticum	*Tribonema ambiguum*		

Continued on p. 223

Using the guidelines of Carlson (1977) and based on the phytoplankton chlorophyll *a* concentration measured in 30 open cenotes, the trophic index of these cenotes can be classified into three major groups: oligotrophic (<3 mg Ch. *a* m^{-3}), mesotrophic (3–20 mg Ch. *a* m^{-3}) and eutrophic (20–150 mg Ch. *a* m^{-3}) (Fig. 4). Most of these cenotes remained in the oligo-mesotrophic range, and 15% of them were eutrophic.

Primary production studied in cenote Noc Ac in 1995 was low due to poor biomass and chlorophyll *a*

concentrations (Ávila et al., 1995). The limited data existing on primary production prevent generalizations and comparisons with other tropical or temperate waters.

Invertebrates

The knowledge of protozoan, hydrozoans, gastrotrichs, tardigrads, free-living nematodes, and annelids is scarce (Suárez-Morales & Rivera-Arriaga, 1998). Reports of other invertebrates (e.g. sponge: *Spongilla*

Table 2. contd.

Rotifera	Decapoda	*Strongylura notata*
Brachionus spp.	*Agostocaris bozanci*	*Thorichthys meeki*
Keratella americana	*Calliasmata nohochi*	
Lecane aculeata	*Creaseria morleyi*	**Tetrapods**
Lecane furcata	*Janicea antiguensis*	*Bufo marinus*
Lecane luna	*Parahippolyte sterreri*	*Crocodylus moreleti*
Lepadella spp.	*Procaris* nov. sp.	*Ctenosaura similis*
Polyarthra vulgaris	*Somersiella sterreri*	*Chrysemys scripta*
	Typhlatya campechae	*Dermatemys mawii*
Branchiopoda	*Typhlatya mitchelli*	*Kinosternon creaseri*
Alona spp.	*Typhlatya pearsei*	*Kinosternon leucostomum*
Dunhevedia spp.	*Yagerocaris cozumel*	*Kinosternon scorpoides*
Euryalona spp.	Remipedia	*Leptodactylus labialis*
Macrothrix spp.	*Speleonectes tulumensis*	*Rhinoclemys areolata*
Moina spp.	Thermosbaenacea	*Lophogobius cyprinoides*
Moinadaphnia spp.	*Tulumella unidens*	*Ophisternon aenigmaticum*
Scapholeberis spp.		
Simocephalus spp.	**Fishes**	
Anguilla rostrata	*Archocentrus octofasciatus*	
	Astyanax aeneus	
Copepoda	*Astyanax altior*	
Arctodiaptomus dorsalis	*Belonesox belizanus*	
Leptodiaptomus novamexicanus	*'Cichlasoma' synspilum*	
Mastigodiaptomus spp.	*'Cichlasoma' urophthalmus*	
Mesocyclops spp.	*Eleotris pisonis*	
Pseudodiaptomus marshi	*Floridichthys polyommus*	
	Gambusia yucatana	
Amphipoda	*Gerres cinereus*	
Hyalella azteca	*Gobiomorus dormitor*	
Mayaweckelia cenoticola	*Lutjanus griseus*	
Quadriviso lutzi	*Megalops atlanticus*	
	Ogilbia pearsei	
Isopoda	*Ophisternon infernale*	
Bahalana mayana	*Petenia splendida*	
Creaseriella anops	*Poecilia mexicana*	
	Poecilia orri	
Mysidacea	*Poecilia velifera*	
Antromysis cenotensis	*Rhamdia guatemalensis*	

cenota by Poirrier, 1976) were published in the seventies. Most studies have dealt with macrocrustaceans and zooplankton.

Rotifers are one of the most diversified group, with 102 species in only 12 sampled localities of the Peninsula (Sarma & Elías-Gutiérrez, 1999). Eutrophic systems are characterized by the dominance of brachionids, such as *Brachionus* and *Keratella*. In oligotrophic cenotes, the number of species is higher, including species of the groups *Lecane*, *Lepadella* and bdelloids.

Up to 30 species of branchiopods have been recorded (Elías-Gutiérrez et al., 1999) in the Yucatan Peninsula; the species are smaller in size, a fact attributed to a more intense predation.

Ostracods were initially studied by Furtos (1936), who described seven species, three of them within the cave connected to the cenotes: *Cypridopsis inaudita*, *C. mexicana* and *C. yucatanensis*. *Danielopolina mexicana* was recently described by Kornicker & Iliffe (1989); it is the most primitive species in the genus (Danielopol, 1990).

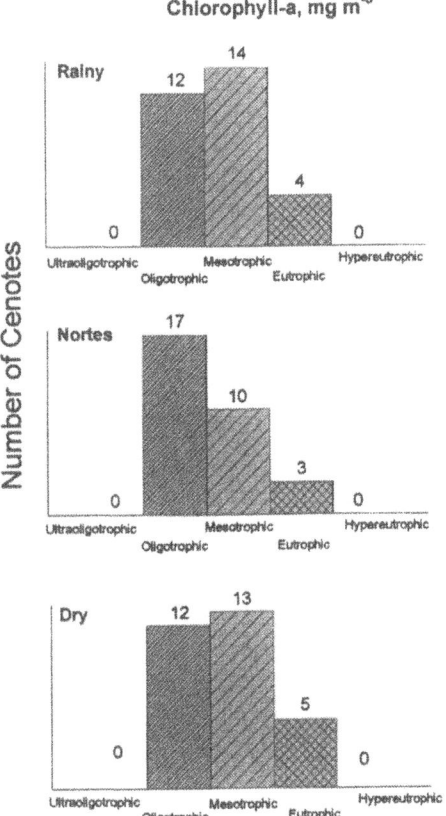

Chlorophyll-a, mg m⁻³

Figure 4. Trophic classification of 30 cenotes based on the phytoplankton chlorophyll *a* concentration.

Thirty-five copepod species have been recorded in the Yucatan Peninsula (Suárez-Morales & Reid, 1998). The biogeography of *Arctodiaptomus* and *Mastigodiaptomus* suggests an affinity of the calanoids of the Yucatan Peninsula with those from the insular Caribbean (Suárez-Morales et al., 1996). Endemism is characteristic to the region, not one copepod species is distributed all over the Peninsula (Reid, 1990; Suárez-Morales et al., 1996; Suárez-Morales & Reid, 1998). Most cyclopoids are benthic, two *Mesocyclops* species from cenotes in the Yucatan Peninsula are adapted to planktic life (Fiers et al., 1996). Marine forms occur in coastal cenotes.

Most amphipods occurring in cenotes derive from marine ancestors. Some of them are cosmopolitan species, such as *Hyalella azteca* and *Quadriviso lutzi*; others have restricted distributions, indicating a strong isolation among cenotes (Fiers et al., 1996). The same is true for cirolanid isopods, such as *Bahalana mayana* and *Creaseriella anops* (Wilkens, 1982), and mysids, among them *Antromysis cenotensis* (Iliffe, 1992).

In the case of decapods, diverse species, among which *Creaseria morleyi, Typhlatya mitchelli, T. pearsei* and *T. campechae*, seem to derive from Caribbean marine ancestors (Hobbs & Hobbs, 1976; Reddell, 1977; Wilkens, 1982). *Somersiella sterreri* is considered a Tethyan relict (Iliffe et al., 1983), as is the thermosbaenacean *Tulumella unidens* (Cals & Monod, 1988).

Other macrocrustaceans from the Peninsula include *Agostocaris bozanci, Yagerocaris cozumel, Janicea antiguensis, Calliasmata nohochi* (Escobar-Briones et al., 1997) and an undescribed *Procaris* (Iliffe, 1992). The remipedian *Speleonectes tulumensis*, one of the most primitive crustaceans, is endemic to some localities in the region (Yager, 1987); a new remipedian is to be described this year in caves near Mérida (Álvarez, pers. com).

Hubbs (1936) was the first to explore systematically the ichthyofauna of the Yucatan Peninsula; he described most of the endemic species. Fish diversity increases southwards and in more coastal cenotes (Wilkens, 1982; Schmitter-Soto, 1998a). The most isolated sites are located in the geologically oldest areas, that remained dry during the Quaternary transgressions, which have been colonized only by two species: *Rhamdia guatemalensis* and *Gambusia yucatana*. *R. guatemalensis*, a fish with nocturnal habits, could have reached this cenote through underground tunnels (Wilkens, 1982), while *G. yucatana*, a small livebearer, tolerant to environmental extreme conditions, could have arrived as hurricane-transported gravid females.

In coastal cenotes cichlids are dominant. Other frequent and abundant species are *Astyanax aeneus* and *R. guatemalensis*, together with poeciliids. In systems close to the sea associated with sea inlets diverse marine invaders occur, mostly as juveniles of gerreids, lutjanids, gobiids, eleotrids, belonids and even the tarpon, *Megalops atlanticus*, and the eel, *Anguilla rostrata* (Navarro-Mendoza, 1988; Schmitter-Soto, 1998a). In between these coastal cenotes and those in more ancient zones, the fish fauna is dominated by *Astyanax altior, 'Cichlasoma' urophthalmus, Poecilia mexicana* and *P. velifera* (Wilkens, 1982), as well as the ubiquitous *Rhamdia* and *Gambusia*.

Many fish populations in cenotes have peculiar morphological features, and they have been described as subspecies. The status of some of these taxa has been questioned, but some have been included among the vulnerable, given their reduced, isolated habitats (Williams et al., 1989). Some subspecies include *R.*

guatemalensis decolor, R. g. depressa, R. g. sacrificii, R. g. stygaea, 'C.' urophthalmus conchitae (extinct by disappearance of its only known locality, a cenote within the city of Mérida), *'C.' u. ericymba, 'C.' u. mayorum* and *'C.' u. zebra* (Hubbs, 1936, 1938). The anchialine cenotes of Tulum (northeast Yucatan Peninsula) share four endemic species with localities in northwestern Yucatan: *A. altior, P. velifera* (Schmitter-Soto, 1998a, b), *Ophisternon infernale* and *Ogilbia pearsei* (Navarro-Mendoza & Valdés-Casillas, 1990).

In addition to fishes, vertebrates observed in the waters of cenotes include crocodiles, iguanas, turtles and anurans (Navarro-Mendoza, 1988; Pozo et al., 1991). Many birds and bat species live temporarily in the walls and trees of the cenotes.

Energy flow

The organic matter (OM) in cenotes has both autochthonous and alochthonous origins. The former enters the cenotes through solar radiation and is incorporated into organic matter via aquatic primary producers (mostly phytoplankton, rooted and floating macrophytes) and vegetation in the shores of the cenotes.

The latter enters the cenotes during the rainy season by soil lixiviation, the weathering of logs, leaves and transport of animal carcasses, and anthropogenic sewage. This OM is not incorporated immediately into the trophic web by the cenote fauna. Its main components, chitin and cellulose, are slowly degraded by fungi and bacteria. Eventually this OM dissolves and is utilized by bacteria found on the walls of the cenote, in the halocline and throughout the water column. Larger particulate matter is fragmented by the biological activity and enters the detritus pathway. The larger input of external organic (and inorganic) matter and the higher sedimentation rate in the cenotes autochthonous OM is photoautotrophic and chemoautotrophic in origin. Sulfate-reducer bacteria in the bottom of the cenote supports chemoautotrophy in the water-sediment interface.

Most lotic cenotes have a continuous water flow (1–3 cm s^{-1}) and their photosynthetic production is low, due to the limited availability of nitrogen and phosphorus; it is based on the phytobenthos and the epiphyton in the border of the cenote. Their waters are transparent. In contrast lentic cenotes contain nutrient and phytoplankton rich waters which give brilliant colors to the waters when they are observed from the shores.

In lentic cenotes, cyanobacteria dwell in the photic layer. Some insect larvae, filtering bivalves and occasionally some fishes eat these bacteria, which give a greenish tint to the cenote waters. This production is not fully utilized. Cyanobacteria are replaced near the bottom or at the halocline (chemocline) by purple bacteria, which may form a turbid layer, often mistaken as a false bottom. At the true bottom, a marine, transparent, anoxic layer is found, where non-described associations of bacteria and fungi occur, forming large mats. Stable isotope data confirm the origins of the two organic carbon sources in cenotes (Pohlman, 1995).

Suspended particulate OM remaining in the cenotes is transported by the slow flow into the tunnels, where the concentrations of organic carbon nitrogen decrease because of its use by bacteria. Some crustaceans and fish, which specialize in particulate OM (ostracods, mysids and carideans), bacterial film (carideans), and corpses and detritus (amphipods, isopods, thermosbaenaceans, carideans, benthic fishes) are the final recipients of these pathways. This fauna supports predators (carideans, remipedians, and fishes). There is a clear-cut niche separation in the cenote (e.g. between walls and bottom), which continues into the submerged tunnels. Potential prey and flow velocity are the two main factors in faunal distribution in cenotes (Culver, 1985).

Food webs in cenotes are relatively simple; few trophic levels and an efficient energy transfer characterize them. Bacteria, fungi, algae, and protozoa are the first levels, and non-specialized micro and macroinvertebrates consume them. Most species are polyphagous, and some are top consumers; as a response to oligotrophy, all of them are starvation-resistant and efficient in food processing (Culver, 1985). These food webs are fragile and easily altered in higher trophic levels.

Epiphyte algae and macrophytes support the herbivorous food web in lotic cenotes, where crustaceans and insect larvae are primary consumers. Isotopic records confirm that this web is complemented by particulate matter from outside the cenote, enriched by bacteria and assimilated by copepods, which in turn are predated by *Astyanax* and other fishes. *Astyanax* is a prey of top predators, notably the eel and *Rhamdia*, which in other habitats tends to be an omnivore rather than a predator (Navarro-Mendoza, 1988).

The accumulation of OM in the sediment, as well as the anoxic or near-anoxic conditions at the bottom, leads to the generation of H_2S or HS^- (Stoessel et al., 1993). This is transformed to S^0 and subsequently

226

to sulfide by bacteria (*Beggiatoa*, *Thiobacillus*, *Thiothrix*) (Jorgensen, 1983). In cenotes of Quintana Roo there are mollusks associated to bacterial patches, probably consuming them.

Acknowledgements

The works referred to in this paper were partly supported by UNAM-DGAPA-PAPIIT (Project IN203894), CONABIO-Mexico (Project M011) and AECI-Spain (Programa CCIB). L. Capurro, CINVESTAV and P. Beddows, Bristol are recognized for their helpful comments to the manuscript. The authors thank E. C. Perry and J. Pacheco for reviewing this manuscript. We thank H. V. Grey for her valuable linguistic revision.

References

Alcocer, J. & E. Escobar, 1996. Limnological regionalization of México. Lakes Reserv. Res. Mgmt 2: 55–69.

Alcocer, J., A. Lugo, L. E. Marín & E. Escobar, 1998. Hydrochemistry of waters from five cenotes and evaluation of their suitability for drinking-water supplies, northeastern Yucatan, Mexico. Hydrogeol. J. 6: 293–301.

Alcocer, J., A. Lugo, M. R. Sánchez, E. Escobar & M. Sánchez, 1999. Bacterioplankton from cenotes and anchialine caves of Quintana Roo, Yucatan Peninsula, Mexico. Rev. Biol. trop. 47: 73–80.

Ávila, J., L. Canto, J. Canche, M. Chavez, G. Ferrer, F. Escamirosa, J. Ku & D. Vázquez, 1995. Productividad primaria y respiracion en el cenote de Noc-Ac. Boletín FIUADY. 27: 25–40.

Back, W., 1985. Hydrogeology of the Yucatan. In Swinehart, J. & D. Loope (eds), Yucatán 1990. A Source Book on Coastal Quintana Roo. Mexico. Univ. Nebraska, Lincoln: 25–40.

Brigmon, H., W. Martin, T. L. Morris, G. Britton & S. G. Zam, 1994. Biogeochemical ecology of *Thiothrix* spp. in underwater limestone caves. Geomicrobiol. J. 12: 141–159.

Buskirk, R. E., 1985. Zoogeographic patterns and tectonic history of Jamaica and the northern Caribbean. J. Biogeogr. 12: 445–461.

Cabrera-Cano, E. & A. Sánchez-Vázquez, 1994. Comunidades vegetales en la frontera México-Belice. In Suárez-Morales, E. (ed.), Estudio Integral de la Frontera México-Belice. Recursos Naturales. CIQRO, Chetumal: 17–35.

Cals, P. & T. Monod, 1988. Évolution et biogeographie des Crustacés Thermosbaénacés. C. r. Acad. Sci. Paris 307: 341–348.

Carlson, R., 1977. A trophic state index for lakes. Limnol. Oceanogr. 22: 361–368.

Comín, F. A., J. A. Herrera, C. García & M. Martín, 1996. Caracterízación física y química de los cenotes de Yucatán (SE México). In Antigüedad, I. & A. Eraso (eds), Recursos Hídricos en Regiones Kársticas. Vitoria: 357–366.

Culver, D. C., 1985. Trophic relationships in aquatic cave environments. Stygologia 1: 43–53.

Danielopol, D. L., 1990. The origin of anchialine cave fauna, the 'deep sea *versus* the 'shallow water' hypothesis tested against the empirical evidence of the Thaumatocyprididae. Bijdr. Dierk. 60: 137–143.

Díaz-Arce, V., 1999. Análisis de algunas características físicas y químicas y su relación con la biomasa fitoplánctica del cenote Ixin-ha, Yucatán durante un ciclo anual. UADY. Bachelor Degree Thesis. Mexico: 99 pp.

Doehring, D. O. & J. H. Butler, 1974. Hydrogeologic constraints on Yucatan's development. Science 186: 591–595.

Edler, C. &. W. K. Dodds, 1992. Characterization of a groundwater community dominated by *Caecidotea tridentata* (Isopoda). First International Conference on Groundwater Ecology. USEPA & American Water Resources Association. Kansas: 91–99.

Elías-Gutiérrez, M., J. Ciros Pérez, E. Suárez-Morales & M. Silva-Briano, 1999. The freshwater cladocera (Crustacea, Ctenopoda and Anomopoda) of Mexico, with comments on selected taxa. Crustaceana 72: 171–186.

Escobar-Briones, E., M. E. Camacho & J. Alcocer, 1997. *Calliasmata nohochi*, new species (Decapoda: Caridea: Hyppolitidae), from anchialine cave systems in continental Quintana Roo, Mexico. J. Crust. Biol. 17: 733–744.

Esquivel, M. D., 1991. Flora béntica. In Camarena-Luhrs, T. & S. I. Salazar-Vallejo (eds), Estudios Ecológicos Preliminares de la Zona Sur de Quintana Roo. CIQRO, Chetumal: 85–91.

Fiers, F., J. W. Reid, T. M. Iliffe & E. Suárez-Morales, 1996. New hypogean cyclopoid copepods (Crustacea) from the Yucatan Peninsula, Mexico. Contr. Zool. 66: 65–102.

Flores, S. & I. Espejel, 1994. Tipos de vegetación de la Península de Yucatán. Etnoflora Yucatanense. Fasc. 3: 135 pp.

Furtos, N. C., 1936. On the ostracods from the cenotes of Yucatan and vicinity. Carnegie Inst. Wash. Publ. 457: 89–115.

Gaona-Vizcayno, S. T. Gordillo de Anda & M. Villasuso-Pino, 1980. Cenotes, karst característico: mecanismo de formación. Inst. Geol. Rev. 4: 32–36.

Gopal, B. & R. G. Wetzel, 1995. Limnology in developing countries. SIL- International Association for Limnology, Vol. 1: 230 pp.

Hall, F. G., 1936. Physical and chemical survey of cenotes of Yucatan. Carnegie Inst. Wash. Publ. 457: 5–16.

Hernández B., D. U. & F. Pérez C., 1991. Flora planctónica y producción primaria. In Camarena-Luhrs, T. & S. I. Salazar-Vallejo (eds), Estudios Ecológicos Preliminares de la Zona Sur de Quintana Roo. CIQRO, Chetumal: 79–84.

Herrera, J., 1993. Ecología de los productores primarios de la laguna de Celestún, México. Patrones de variación espacial y temporal. Ph.D. Thesis. Barcelona: 233 pp.

Herrera-Silveira, J. A., F. A. Comín, S. López & I. Sánchez, 1998. Limnological characterization of aquatic ecosystems in Yucatan Peninsula (SE Mexico). Verh. int. Ver. Limnol. 26: 1348–1351.

Herrera-Silveira J. & F. Comín, 2000. An introductory account of the types of aquatic ecosystems of Yucatan Peninsula (SE México). In Munawar, M., S. G. Lawrence, I. F. Munawar & D. F. Malley (eds), Aquatic Ecosystems of Mexico: Status and Scope. Backhuys. Leiden: 213–227.

Hobbs, H. H., III & H. H. Hobbs Jr, 1976. On the troglobitic shrimps of the Yucatán peninsula, México. Smiths. Contr. Zool. 240: 1–23.

Hubbs, C. L., 1936. Fishes of the Yucatan Peninsula. Carnegie Inst. Wash. Publ. 457: 157–287.

Hubbs, C. L., 1938. Fishes from the caves of Yucatan. Carnegie Inst. Wash. Publ. 491: 261–296.

Hutchinson, G. E., 1957. A Treatise on Limnology, Vol I. Wiley, New York: 1015 pp.

Iliffe, T. M., 1992. An annotated list of the troglobitic anchialine and freshwater fauna of Quintana Roo. In Navarro, D. & E. Suárez-Morales (eds), Diversidad Biológica en la Reserva de la Biosfera de Sian Ka'an, Quintana Roo, México. Vol. II. CIQRO/SEDESOL, Chetumal: 197–217.

Iliffe, T. M., C. W. Hart Jr & R. B. Manning, 1983. Biogeography and the caves of Bermuda. Nature 302: 141–142.

Jorgensen, B. B., 1983. The microbial sulphur cycle. In Krumberlein, W. E. (ed.), Microbial Geochemistry. Blackwell, Boston: 91–124.

Kornicker, L. S. & T. M. Iliffe, 1989. New Ostracoda (Halocyprida: Thaumatocyprididae and Halocyprididae) from anchialine caves in the Bahamas, Palau and Mexico. Smiths. Contr. Zool. 470: 1–47.

López-Adrián, S., I. Sánchez, R. Tavera, J. Komárek, J. Komarkova & M. Villasuso, 1993. Estudio ecológico de los cuerpos de agua continentales de la Península de Yucatán. Aspectos ficológicos. Programa de Ecología Terrestre. SEP, UAY, FMVZ, Mérida.

López-Adrián, S. & J. A. Herrera-Silveira, 1994. Plankton composition in a cenote, Yucatán, México. Verh. int. Ver. Limnol. 25: 1402–1405.

López Ramos, E., 1975. Geological summary of the Yucatan peninsula. In Nairn, A. E. M. & F. G. Stehli (eds), The Ocean Basins and Margins. III. The Gulf of Mexico and the Caribbean. Plenum, New York: 257–282.

Margalef, R., 1983. Limnología. Omega, Barcelona: 1010 pp.

Marín, L. E., 1990. Field investigations and numerical simulation of groundwater flow in the karstic aquifer of northwestern Yucatan, Mexico, Ph.D. thesis, Northern Illinois University, DeKalb, IL: 176 pp.

Marín, L. E., R. Sanborn, A. Reeve, T. Felger, J. Gamboa, E. C. Perry & M. Villasuso, 1988. Petenes: a key to understanding the hydrogeology of Yucatan, Mexico. International Symposium of Hydrology of Wetlands in Semiarid and Arid Regions, Sevilla, Spain, May 9–12.

Martin, H. W. & R. L. Brigmon, 1994. Biogeochemistry and sulfide oxidizing bacteria in phreatic karst. In Sasowsky, I. D. & M. V. Palmer (eds), Breakthroughs in Karst Geomicrobiology and Redox Geochemistry Symposium. Karst Waters Inst., Charles Town: 49–51.

Martin, H. W., R. L. Brigmon & T. L. Morris, 1995. Diving protocol for sterile sampling of aquifer bacteria in underwater caves. Nat. Speleol. Soc. Bull. 57: 24–30.

Navarro-Mendoza, M., 1988. Inventario íctico y estudios ecológicos preliminares en los cuerpos de agua continentales en la reserva de la biósfera de Sian Ka'an y áreas circunvecinas en Quintana Roo, México. Tech. Rept., CIQRO/CONACYT/USFWS, Chetumal.

Navarro-Mendoza, M. & C. Valdés-Casillas, 1990. Peces cavernícolas de la península de Yucatán en peligro de extinción, con nuevos registros para Quintana Roo. In Camarillo, J. L. & F. Rivera A. (eds), Áreas Naturales Protegidas en México y Especies en Extinción. ENEP-Iztacala, UNAM, México: 218–241.

Pacheco, J. & A. Cabrera, 1997, Ground water contamination by nitrates in the Yucatan Peninsula, Mexico. J. Hydrogeol. 5: 47–53

Perry, E., J. Swift, J. Gamboa, A. Reeve, R. Sanborn, L. Marín & M. Villasuso, 1989. Geologic and environmental aspects of surface cementation, north coast, Yucatan, Mexico. Geology 17: 818–821.

Perry, E., L. Marín, J. McClain & G. Velázquez, 1995. Ring of cenotes (sinkholes), northwest Yucatan, Mexico: its hydrogeologic characteristics and possible association with the Chicxulub impact crater. Geology 23: 17–20.

Pohlman, J. W., 1995. Analysis of the ecology of anchialine caves using carbon and nitrogen isotopes. M.Sc. Thesis, Texas A&M Univ., College Station.

Poirrier, M. A., 1976, A taxonomic study of the *Spongilla alba, Spongilla cenota, Spongilla wagneri* species group (Porifera: Spongillidae) with ecological observations of *Spongilla alba*. In Harrison, F. W. & R. R. Cowden (eds), Aspects of Sponge Biology. Academic Press, New York: 203–213.

Pozo de la T., C., E. Escobedo C., J. L. Rangel S. & P. Viveros L., 1991. Fauna. In Camarena-Luhrs, T. & S. I. Salazar-Vallejo (eds), Estudios Ecológicos Preliminares de la Zona Sur de Quintana Roo. CIQRO, Chetumal: 49–78.

Reddell, J. A., 1977. A preliminary survey of the caves of the Yucatan Peninsula. Assoc. Mex. Cave Stud. Bull. 6: 215–296.

Reddell, J. A., 1981. A review of the cavernicole fauna of México, Guatemala and Belize. Texas Mem. Mus. Bull. 27: 1–327.

Reid, J. W., 1990. Continental and coastal free-living Copepoda (Crustacea) of Mexico, Central America and the Caribbean region. In Navarro, D. & J. G. Robinson (eds), Diversidad Biológica en la Reserva de la Biosfera de Sian Ka'an, Quintana Roo, México. CIQRO/PSTC/Univ. Florida, México: 175–213.

Sánchez, M., J. Alcocer, A. Lugo, M. R. Sánchez & E. Escobar, 1998. Variación temporal de las densidades bacterianas en cinco cenotes y dos cuevas sumergidas del NE de Quintana Roo, México. In Mancilla D., J. M. & G. Vilaclara F. (eds), Cuadernos de Investigación Interdisciplinaria en Ciencias de la Salud, la Educación y el Ambiente, Vol. 1. UNAM, México: 66–80.

Sánchez, M., J. Alcocer, E. Escobar & A. Lugo, 2002. Phytoplankton of cenotes and anchialine caves along a distance gradient from the northeastern coast of Quintana Roo, Yucatan Peninsula. Hydrobiologia 467 (Dev. Hydrobiol. 163): 79–89.

Sánchez, S. O., E. F. Cabrera C., S. A. Torres P., P. Herrera E., L. Serralta P. & C. Salazar G., 1991. Vegetación. In Camarena-Luhrs, T. & S. I. Salazar-Vallejo (eds), Estudios Ecológicos Preliminares de la Zona Sur de Quintana Roo. CIQRO, Chetumal: 31–48.

Sánchez-Molina, I., 1985. Bacillariophyta. In Zamacona, J. (ed.), Flora Planctónica de los Cenotes de Yucatán. UADY, Mérida: 1–9.

Sarma, S. S. S. & M. Elías-Gutiérrez, 1999. A survey on the rotifer (Rotifera) fauna of Yucatán Peninsula (México). Rev. Biol. trop. 47: 191–200.

Schmitter-Soto, J. J., 1998a. Catálogo de los Peces Continentales de Quintana Roo. ECOSUR, San Cristóbal de las Casas.

Schmitter-Soto, J. J., 1998b. Diagnosis of *Astyanax altior* (Characidae), with a morphometric analysis of *Astyanax* in the Yucatan Peninsula. Ichthyol. Explor. Freshwat. 8: 349–358.

Sharpton, V. L., G. B. Dalrymple, L. E. Marín, G. Ryder, B. C. Schuraytz & J. Urrutia Fucugauchi, 1992, New links between the Chicxulub Impact Structure and the Cretaceous–Tertiary Boundary, Nature 359: 819–821.

Sharpton, V. L., K. Burke, A. Camargo, S. A. Hall, L. E. Marín, G. Suárez, J. M. Quezada, P. D. Spudis & J. Urrutia Fucugauchi, 1993, The gravity expression of the Chicxulub multiring impact basin: size, morphology and basement characteristics. Science 261: 1564–1567.

Steinich, B., 1996. Investigaciones geofísicas e hidrogeológicas en el noroeste de la Península de Yucatán, México. Ph.D. Thesis, Instituto de Geofísica, Universidad Nacional Autónoma de México, México.

Steinich, B. & L.E. Marín, 1996. Hydrogeological investigations in northwestern Yucatán, México, using resistivity surveys. Ground Water 34: 640–646.

Steinich, B., G. Velázquez Olimán, L. E. Marín & E. Perry, 1996. Determination of the ground water divide in the karst aquifer of Yucatan, Mexico, combining geochemical and hydrogeological data. Geofís. Int. 35: 153–159.

228

Stoessel, R. K., W. C. Ward, B. H. Ford & J. D. Schuffert, 1989. Water chemistry and CaCO3 dissolution in the saline part of an open-flow mixing zone, coastal Yucatan Peninsula, Mexico. Bull. Geol. Soc. Am. 10: 159–169.

Stoessel, R. K., Y. H. Moore & J. G. Coke, 1993. The occurrence and effect of sulfide oxidation on coastal limestone dissolution in Yucatan cenotes. Ground Wat. 31: 566–575.

Stringfield, V. T. & H. E. LeGrand, 1974. Karst hydrology of northern Yucatan Peninsula. In Weidie, A. E. (ed.), Field Seminar on Water and Carbonate Rocks of the Yucatan Peninsula, Mexico. New Orleans Geol. Soc., New Orleans: 26–44.

Suárez-Morales, E. & J. W. Reid, 1998. An updated checklist of the free-living copepods (Crustacea) of Mexico. Southw. Nat. 43: 256–265.

Suárez-Morales, E. & E. Rivera-Arriaga, 1998. Hidrología y fauna acuática de los cenotes de la Península de Yucatán. Rev. Soc. Mex. Hist. nat. 48: 37–47.

Suárez-Morales, E., J. W. Reid, T. M. Iliffe & F. Fiers, 1996. Catálogo de los Copépodos (Crustacea) Continentales de la Península de Yucatán, México. CONABIO/ECOSUR, México.

Van der Kamp, G., 1995. The hydrogeology of springs in relation to the biodiversity of spring fauna: a review. J. Kansas Ent. Soc. 68: 4–17.

Velázquez, G., 1995, Estudio geoquímico del anillo de cenotes, M.S. Thesis, Instituto de Geofísica, Universidad Nacional Autónoma de México, México City, México: 77 pp.

Ward, W. C., A. E. Weidie & W. Back, 1985. Geology and hydrogeology of the Yucatan, and Quaternary geology of northeastern Yucatan Peninsula. New Orleans geol. Soc., New Orleans.

Wetzel, R. G. & B. Gopal, 1999. Limnology in developing countries. SIL-International Association for Limnology, Vol. 2: 330 pp.

Wilkens, H., 1982. Regressive evolution and phylogenetic age: the history of colonization of freshwaters of Yucatan by fish and crustacea. Ass. Mex. Cave Stud. Bull. 8: 237–243.

Williams, J. E., J. E. Johnson, D. A. Hendrickson, S. Contreras-Balderas, J. D. Williams, M. Navarro-Mendoza, D. E. McAllister & J. E. Deacon, 1989. Fishes of North America endangered, threatened or of special concern: 1989. Fisheries 14: 2–20.

Wilson, E. M., 1980. Physical geography of the Yucatan Peninsula. In Moseley, E. & E. Terry (eds), Yucatan. A World Apart. Univ. Alabama Press. Alabama: 5–40.

Yager, J., 1987. Speleonectes tulumensis n. sp. (Crustacea, Remipedia) from two anchialine cenotes of the Yucatan Peninsula. Stygologia 3: 160–166.

The manufacturer's authorised representative in the EU is Springer
Nature Customer Service Centre GmbH, Europaplatz 3, 69115 Heidelberg,
Germany. If you have any concerns regarding our products, please
contact ProductSafety@springernature.com

Printed and bound by CPI Group (UK) Ltd, Croydon, CR0 4YY

23/04/2026

02095658-0005